ANNALS OF THE NEW YORK ACADEMY OF SCIENCES

Volume 705

EDITORIAL STAFF
Executive Editor
BILL M. BOLAND
Managing Editor
JUSTINE CULLINAN
Associate Editor
TRUMBULL ROGERS

The New York Academy of Sciences
2 East 63rd Street
New York, New York 10021

THE WORK OF
MARY ELLEN RUDIN

ANNALS OF THE NEW YORK ACADEMY OF SCIENCES
Volume 705

THE WORK OF MARY ELLEN RUDIN

Edited by Franklin D. Tall

The New York Academy of Sciences
New York, New York
1993

Cover: Basins of attraction with the "Wada" property. Courtesy of James A. Yorke, Judy A. Kennedy, and Helena E. Nusse.

Library of Congress Cataloging-in-Publication Data

The Work of Mary Ellen Rudin / editor, Franklin D. Tall.
    p. cm. — (Annals of the New York Academy of Sciences, ISSN 0077-8923; v. 705)
    "Papers presented at the 1991 Summer Conference on General Topology and Applications in honor of Mary Ellen Rudin and her work, held in Madison, Wisconsin, from June 26 to 29, 1991"—Galley.
    Includes bibliographical references.
    ISBN 0-89766-813-8
    1. Topology—Congresses. 2. Set theory—Congresses. I. Tall, Franklin D. II. Rudin, Mary Ellen, 1924– . III. Summer Conference on General Topology and Applications (1991 : Madison, Wisc.) IV. Series.
Q11.N5 vol. 705
[QA611.A1]
500 s—dc20
[514'.322]

94-696
CIP

SP
Printed in the United States of America
ISBN 0-89766-813-8 (cloth)
ISBN 0-89766-814-6 (paper)
ISSN 0077-8923

ANNALS OF THE NEW YORK ACADEMY OF SCIENCES

Volume 705
December 28, 1993

THE WORK OF MARY ELLEN RUDIN[a]

Editor
FRANKLIN D. TALL

Conference Organizers
R. KOPPERMAN, A. MILLER, and F. D. TALL

Advisory Committee
S. ANDIMA, W. W. COMFORT, M. HENRIKSEN, G. ITZKOWITZ, R. KOPPERMAN,
R. LEVY, A. MILLER, P. MISRA, L. NARICI, R. M. SHORTT, A. TODD,
J. E. VAUGHAN, and S. W. WILLIAMS

CONTENTS

[a]The papers in this volume were presented at the 1991 Summer Conference on General Topology and Applications in Honor of Mary Ellen Rudin and Her Work, held in Madison, Wisconsin, from June 26 to 29, 1991.

Financial assistance was received from:
- BARUCH COLLEGE—CUNY
- THE CITY COLLEGE—CUNY
- THE COLLEGE OF STATEN ISLAND—CUNY
- LONG ISLAND UNIVERSITY—C.W. POST CAMPUS
- THE NEW YORK ACADEMY OF SCIENCES (MATHEMATICS SECTION)
- ST. JOHN'S UNIVERSITY
- UNIVERSITY OF WISCONSIN
- WESLEYAN UNIVERSITY

Preface

FRANKLIN D. TALL

Department of Mathematics
University of Toronto
Toronto, ON M5S 1A1, Canada

This volume arose out of the Conference in Honor of Mary Ellen Rudin and her work, held in Madison Wisconsin, June 26–29, 1991. Most of the papers are expanded versions of invited survey talks given at the conference. There are two personal reminiscences, by Burton Jones and Judy Roitman. I would like to thank the organizing committee, especially Ralph Kopperman and Arnie Miller, for the skillful and professional way in which they accomplished the immense amount of work involved in making the conference happen. The great mathematical influence Mary Ellen Rudin has had on the field of set-theoretic topology will be clear to the reader of these surveys. She has also had great influence personally on the practitioners of the field, and we could all tell many anecdotes. But Mary Ellen has only a limited tolerance for that. As she would say, "Enough of such foolishness!" Therefore, on with the mathematics, which is our tribute to her.

MARY ELLEN RUDIN

Conference in Honor
of Mary Ellen Rudin[a]

At the end of June 1991, a conference was held in Mary Ellen Rudin's honor in Madison, Wisconsin, on the occasion of her retirement from teaching (but not from mathematics!). A similar conference had been held earlier in the month to honor her husband, Walter Rudin.

Mary Ellen's conference was mathematically quite rich. Eight survey talks were given on mathematics she has strongly influenced, and there was a large number of nonsurvey talks, including presentations of some strikingly beautiful new results. Most of the speakers had either worked closely at some point with Mary Ellen or were students of people who had. The atmosphere, as one would expect, was encouraging and mathematically vital. Special effort was made to encourage graduate students to come, including graduate student stipends thanks to NSF funding, and there were many graduate students, young post-doctoral mathematicians, and women, including two of the invited speakers. The conference was also graced by the strong and vibrant presence of the Russian mathematician Boris Sapirovski, a remarkable man who died a few months later of cancer and who will be greatly missed. The two major organizers of the conference were Frank Tall and Ralph Kopperman, who did a splendid job. The conference proceedings have been published in two volumes by the New York Academy of Sciences of which this is one.

The Association for Women in Mathematics (AWM) honored Mary Ellen at a party. The citation which follows was delivered on our behalf by Judy Roitman.

I have been asked to express to Mary Ellen Rudin the appreciation, affection, and congratulations of AWM, and it is an honor to do so. Carol Wood, the President of AWM, has sent a message which I would like to read to you:

> AWM sends best wishes to Mary Ellen on this occasion in which her mathematical contributions are being honored. We add our special acknowledgement and appreciation of what she has meant to women mathematicians over the years—both in the ways she has encouraged and supported young women (and men), and also more broadly by her example of a highly productive career, in which her fine enthusiasm for mathematics and her buoyant spirit heartened us all.
>
> On a personal note, I found it a double whammy when I entered graduate school to be a woman *and* a Southerner; that Mary Ellen could carry off both with such aplomb was encouraging, if humbling, to me!

Now I would like to say a few personal words about Mary Ellen. I am lucky to be of those generations of topologists nurtured by Mary Ellen's mathematical vision and personal warmth. Her mathematical vision has been spoken of at length at this conference. So I would like to speak a little about her personal side. I met Mary Ellen over the phone, when I arrived in Madison as a visiting graduate student nearly 20 years ago, along with Bill Fleissner and Aki Kanamori. Mary Ellen's father had just died, and I caught her getting ready to go to his funeral. I think many of us, at a time

[a] Reprinted with permission from the Association for Women in Mathematics Newsletter, Vol. 21, No. 6, November/December 1991.

like that, faced with a call from an unknown graduate student, would understandably get off the phone as quickly as possible; but Mary Ellen made sure that I was all right—did I have an apartment? an office?—she may even have checked to see if I had library privileges!—and apologized profusely for being unable to meet me until she got back. Needless to say, this was not what I was used to at Berkeley.

The next story comes later, at the International Congress of Mathematicians in Vancouver, when I was a freshly minted Ph.D. and extremely active in AWM activities. Mary Ellen did not approve. One morning as we ate breakfast together I tried to convince her of the importance of improving the situation for women in mathematics. *"The best way to help women in mathematics is to do mathematics!"* she roared, pounding her fist on the table so hard that the dishes jumped in the air.

When my first baby died of meningitis, it was Mary Ellen who wrote the letter I came closest to memorizing, about how it was when her son Bobby was born with Down's syndrome, and the terrible things the doctors had told her about his future. "Your life will never quite be the same again," she wrote, and there was more comfort in the acknowledgement of that reality than in all the voices trying to convince me that, indeed, life would go on as before.

Mary Ellen and Walter have, on their kitchen counter, an old AM radio that is probably as old as their marriage, maybe older. I asked Mary Ellen once why she never got a newer, better model, and she responded "because it still works." This radio is, to me, the material symbol of the basic, decent values that Mary Ellen and Walter have exemplified in their lives, without distraction, and from which we all continue to learn.

And so I would like to present to you, Mary Ellen, this material symbol [AWM mug] of appreciation from the Association for Women in Mathematics.

——JUDY ROITMAN
Department of Mathematics
University of Kansas
Lawrence, Kansas 66044

Some Glimpses of the Early Years

When we returned to Austin with our two little girls, Mary Ellen was just starting topology (=point-set theory). She was fond of the girls and saw quickly how she could make a maximal contribution to their happiness and well-being: she loaned them her own elegant antique walnut doll furniture. The younger of those two girls has just had her fiftieth birthday and Mary Ellen has some 80 or so published papers (mostly involving set theory).

Mary Ellen was just as quick with her research and, as her thesis progressed, Moore had her present it to his advanced class (which included Dick Anderson and Ed Moise). I remember Moore remarking that it was hard stuff and he wondered if they could understand it! And I think he was really speaking of himself. (Of course, it didn't help that Mary Ellen had ordered a set in the beginning one way and toward the end inadvertently reversed it!)

Mary Ellen got her Ph.D. degree in 1949 and took a job at Duke. We moved to Chapel Hill the next year. This made communications much easier and more frequent. She really became a member of the family.

Many of us, probably most of us, are absentminded. But Mary Ellen took precautions. She warned the janitor *never* to lock her office door. And she bought a particular car (possibly a Chevy) whose ignition could be turned on and the ignition key removed so that the ignition could then be turned on and off without a key. Once more on absentmindedness. Mary Ellen had worked pretty hard over one weekend. Monday morning she dressed as usual but couldn't find her shoes. She looked "high and low," finally using a spare pair. All through class (was it College Algebra?) she puzzled over what she might have done with them. At last, it came to her what must have happened. She had gone to a movie Friday night where she probably slipped her shoes off and then gone home without them. So as soon as she could she went to the theater fully expecting to get them. But no such luck. They simply were not there. Weeks, possibly months went by: no shoes. Until it rained hard, so hard that Mary Ellen needed to use her galoshes and there were her shoes, neatly tucked inside.

Mary Ellen had come over to have supper with us and to go to a movie. I think this movie was one of a series selected by the student body for "artistic" content. I think the program consisted of two short movies. One of them actually had people in it. The other was made up of what to me seemed to be shots of unrelated things—a steam engine, then a large sunflower, and then a bird walking along the shore. This went on and on: "stream of consciousness" it may be called. I suppose one isn't expected to get much out of it. When it quit, leaving no clue as to what it was about, Mary Ellen almost in the center of the little theater stood up to her full height and in a voice commensurate said, "I demand an explanation!" I don't think one was forthcoming.

Perhaps I should tell you one more—one with some mathematics in it. Because wherever Mary Ellen was there was some mathematics. Our house didn't have a street at this time—only later. You had to drive down a rather small alley and park in the driveway to the "open door" garage. On this particular Saturday or Sunday afternoon I was busy building for those two girls a playhouse out of lumber discarded during the construction of our new home. I looked up to see Mary Ellen tromping her way across what at one time had been a vegetable garden. She was parked in our driveway, and hearing some hammering headed diagonally across our lot behind the garage and toward the far back corner of the lot. I had stopped and met her on a little slope 20 or 30 feet from the partially built playhouse. Almost without any preliminary explanation she was showing me a type of construction she was hopefully using to solve a long-standing problem of Wilder's. I was familiar with the problem, so after a bit of confusion I could see what she was trying to do. (The problem was to construct in the plane a set so thick that although no connected subset contained more than one point, any larger set would—even if you added only one point.) The trouble was that her process didn't quite do it. If you repeated the process once more, that did some good but didn't quite work either. By this time we were squatting down and I was drawing pictures with my hammer on the ground. Well, I thought that if you could tell what you were doing, maybe repeating it once more might just do it. I don't remember for sure but I think Mary Ellen got back in her car and drove off. I didn't hear anything from her all that next week and more. This meant she had solved the problem and indeed she had. Just for fun, you might try it.

She certainly was a member of the family. When Walter came and took her away to Rochester things were never quite the same. But this could be said of Duke or Chapel Hill.

——F. BURTON JONES
3775 Modoc Road, #205
Santa Barbara, California 93105

The Work of Mary Ellen Rudin[a]

FRANKLIN D. TALL

Department of Mathematics
University of Toronto
Toronto, Ontario
Canada M5S 1A1

ABSTRACT: An overview and historical perspective of the work of Mary Ellen Rudin is given.

Set-theoretic topology has been inspired by Mary Ellen Rudin for two decades. Her influence and leadership in North America and worldwide have been central to the development of the field. The purpose of this paper is to introduce this volume by giving an overview and a historical perspective on her career. Since there already is an excellent long biographical article about her [1], we will mainly concentrate on her work.

I have listed all of Mary Ellen's papers at the end of this article, referring to them as, for example, [1971a]. Other works cited are listed separately and referred to as, for example, [26]. I have also appended a record of her doctoral students and their dissertation titles. In picking which results to mention in this admittedly incomplete overview, I have tended to go lightly over areas covered in the other articles in this volume, while still hitting the high points for those who will read only this paper. I have also mentioned a bit more of her earlier work than one might expect since it deserves to be more widely known.

Mary Ellen grew up in a small town in Texas and had no idea she would be a mathematician. When she met R. L. Moore at registration at the University of Texas, he decided she was a good prospect and—without her quite realizing at first what was happening to her—he began reeling her in. In classic Moore fashion, he created a confident, powerful researcher who didn't know any mathematics other than what he fed her. As she said, when she got out she "had a Ph.D. but I hadn't really been to graduate school. I didn't know any algebra, any topology, any analysis" [1]. Her first few papers were written in Moore's antiquated mathematical language, which adds to the difficulty of the mathematics itself. When Mary Ellen went to mathematics meetings and realized no one could understand what she was talking about, she switched to more conventional language. However, as Steve Watson [28] has demonstrated in this volume, those first papers are well worth reading. Mary Ellen's first job was at Duke, arranged by Moore, and her first four papers appeared in the *Duke Journal*. The first one [1950] is a reprint of her thesis, on completeness in Moore spaces. In that paper she gives the first example of a nonseparable Moore space satisfying the countable chain condition.

During the time Mary Ellen was at Duke, Burton Jones was right nearby at the

Mathematics Subject Classification: Primary 54-02, 54-03, 01A75.
[a]The author acknowledges support from Grant A-7354 of the Natural Sciences and Engineering Research Council of Canada.

1

University of North Carolina, in Chapel Hill. He was second only to Moore as a formative mathematical influence on her—in particular, he too spoke "Moore language."

Mary Ellen met Walter Rudin at Duke; they were married in 1953, at which time they moved to Rochester, New York. In 1958 they were at Yale, and in 1959 they moved to Madison, Wisconsin, where they have been ever since, except for various visiting positions. As Mary Ellen details in [1], she was never particularly concerned with questions of career. When she arrived in Rochester with Walter, they found a place for her, and similarly in Madison.

Wisconsin had a nepotism law that prevented Mary Ellen from having a regular position. As a result, she stayed a "lecturer," which had one advantage: she was free of committee work. When the law was finally abolished, she was promoted directly to full professor in 1971. I remember that in 1969 I had to have the department chairman, John Nohel, co-sign my thesis, because, as a lecturer, Mary Ellen was not allowed to be a full-fledged supervisor.

Papers [1950], [1951], and [1952a] in the list of Mary Ellen's papers that occurs at the end of this article form a sequence that is discussed in detail in Steve Watson's paper [28] in this volume. Paper [1950] contains her example of a nonseparable Moore space satisfying the countable chain condition. In [1951] she improved that to obtain a space that in addition is locally connected. In [1952a] she showed that there is a locally connected countable chain condition nonseparable Moore space in which any two points can be *separated by a countable set* if and only if there is a Souslin line. (We say $\{a, b\}$ are *separated by* $S \subseteq X$ if $X - S = U \cup V$, where U and V are disjoint clopen subsets of $X - S$, each containing exactly one of a and b.)

The subject of countable chain condition nonseparable Moore spaces has continued to attract interest, for example [18], [19].

In [1952], Mary Ellen disproved a conjecture of R. L. Wilder by constructing a *primitive dispersion set* in the plane. (If D is a subset of a connected space M, we say D is a *dispersion set* of M if $M - D$ is totally disconnected. If no proper $E \subseteq D$ is a dispersion set of M, D is *primitive*.) Wilder was sufficiently impressed that he applied for and obtained a National Science Foundation (NSF) grant for her.

There is a gap in Mary Ellen's bibliography between 1952 and 1955; contrary to what one might think, that wasn't caused by family responsibilities, but rather was because she was spending her time on the Poincaré conjecture and didn't get very far. It is rare that her efforts are unfruitful—at the end of this article we mention some of her favorite problems that have so far resisted her.

There is another gap between 1958 and 1963. When I asked her about it, she said the existence of the gap had never crossed her mind! She had been busy working: with Kakutani on some difficult problems they didn't get enough on to publish, refereeing a lot of papers sent to R. H. Bing—including Prabir Roy's thesis, solving a problem of Kuratowski that she then found out had already been solved, and constructing a "biconnected subset of the plane having no widely connected subset" that was so complicated she gave up trying to write it up. As she said, "It just happened sort of accidentally that I didn't publish any papers." In these days of excessive publishing, it is refreshing to hear this story. In reflecting on this period, however, Mary Ellen did allow as how three difficult pregnancies, small children, two moves, two long European trips, and two months of high fever with the mumps might be some reasons why those were not her best mathematical years.

The first paper of Mary Ellen's that most set-theoretic topologists have heard of is [1955], in which she used a Souslin tree to construct a *Dowker space,* that is, a normal space X such that $X \times [0, 1]$ is not normal. Such a construction would be quite interesting if done for the first time now; but in 1955, before the consistency of the existence of Souslin trees was known, this was quite extraordinary. Souslin trees have long been favorite combinatorial objects of Mary Ellen, who has used them and offshoots of them in constructing Dowker spaces and S-spaces (see below). Her 1969 *Monthly* article [1969a] on them is a masterful introduction. This was based on a lunch-hour talk she gave in the penthouse lounge of the Wisconsin mathematics department, during which she innocently remarked that Souslin died young and all we remember him for is that problem in the first issue of *Fundamenta.* An agitated Karel Prikry got up and proceeded to reel off a long list of Souslin's other accomplishments, to which the published version of the talk carefully pays homage.

Mary Ellen learned about Souslin trees at Texas. In fact, the one and only time R. L. Moore lectured to her class was when he thought he had solved Souslin's problem. He was wrong.

Whether or not Prikry was responsible, starting with [1969a] there is a notable change in Mary Ellen's papers, in that she refers extensively to the literature. Before that, her papers had one or two references, or even five once, but from 1969 on, one often finds ten or more papers referred to. Mary Ellen had become aware of other mathematics that impinged on her work. Previously she just (!) answered questions that someone asked her or that she thought of. The questioner may have known the connections with the literature, but she didn't. Reading the literature was not what Moore trained his students to do, and Mary Ellen admits that she has never done much reading, although she does not recommend that practice to others.

The students of Moore were familiar with both elementary set-theoretic techniques and geometric ones, especially dealing with connected spaces. The broad river of topologists descending from Moore diverged into these two streams. For many years, the Spring Topology Conferences were in effect divided into these two areas and I got the impression that Mary Ellen, Burton Jones, and a couple of others were the only ones who could understand all the talks. Certainly this geometric knowledge and intuition has proved invaluable in her work in constructing pathological manifolds [1976], [1988], [1989], [1990] that is discussed in Peter Nyikos' article [16] in this volume. Some of her more striking results in this area are:

THEOREM [1976]: $\Diamond$ (with Zenor, CH) implies the existence of a nonmetrizable perfectly normal manifold.

THEOREM [1989]: $\Diamond$ implies there is a Dowker space manifold.

THEOREM [1990]: $\Diamond^+$ implies there is a normal noncollectionwise normal manifold.

Mary Ellen wrote several papers about connectedness early in her career. Some of this work is discussed in Mike Starbird's article [23] in this volume. Here are some sample results. A connected set I is *indecomposable* if $I = A \cup B$, A, B connected, implies $\overline{A} \neq I$ or $\overline{B} \neq I$. To the question of whether there is an indecomposable I

such that $I \cup \{p\}$ is indecomposable for some $p \in \bar{I} - I$, Mary Ellen showed "no" in E^2, but "yes" in E^3 [1957c]. A triangulation $\mathscr{K}$ of a tetrahedron T is *shellable* if the tetrahedra $K_1, \ldots, K_n$ of $\mathscr{K}$ can be ordered so that $\cup_{i \leq n} K_i$ is homeomorphic to T for all i. Mary Ellen constructed a Euclidean triangulation that is not shellable [1958]. In [1963] she answered a different sort of question, proving there does not exist a family of countably many disjoint dense arcwise connected sets with union the plane.

A number of the questions about connectedness Mary Ellen worked on in the 1950s were problems posed to her by P. Erdös, whom she met at various American Mathematical Society meetings. Her paper [1958a] is noteworthy not only because she disproved a conjecture of Erdös but because it was her first published use of the continuum hypothesis:

THEOREM: The continuum hypothesis implies there is a nondegenerate (i.e., more than one point) connected subset M of the plane such that if N is any nondegenerate connected subset of M, then $M - N$ is at most countable.

Mary Ellen doesn't remember how or where she first heard about the continuum hypothesis—possibly it was from Burton Jones. It certainly encouraged the propensity she has had ever since Texas to "well-order everything in sight."

As a good Moore student, Mary Ellen does not have many joint papers. The only ones that are really the result of joint work are those she has done with her students or postdocs; the others are answers to people's questions or joining together of two people's individual work. For example, her first joint paper [1957] was with Victor Klee. She answered a question of his, and he wrote the paper. The result:

THEOREM: Suppose X and Y are separable metric spaces and $C(X, Y)$ is the space of all continuous maps of X into Y, under the topology of pointwise convergence. Then every subspace of $C(X, Y)$ is perfectly normal and Lindelöf.

It's always satisfying to a topologist to prove a theorem about the real line. Here are two such theorems of Mary Ellen. In [1957b] she answered a question of de Groot by proving:

THEOREM: A separable metric space E is homeomorphic to a subset of $\mathbb{R}$ if and only if it satisfies the following conditions:

1. Each component of E is a point or an arc (closed, open, or half-open), and no interior point of an arc-component A is a limit point of $E - A$.
2. Each point of E has arbitrarily small neighborhoods with finite boundaries.

There was also a Dutch connection with the second theorem, for which Mary Ellen received a prize from the Mathematical Society of the Netherlands. In [1965] she answered a question of Korevaar by characterizing those subspaces of the real line such that the interval topology induced by the inherited order coincides with the subspace topology. Call such subspaces *totally orderable.*

THEOREM: Let T be a subspace of $\mathbb{R}$. Taking closures in $\mathbb{R}$, let $Q = \cup \{K: K$ is a nontrivial component of T all of whose endpoints belong to $\overline{\overline{T} - T}\}$.

1. If $T - Q$ is compact and $(T - Q) \cap \overline{Q} = 0$, then either $Q = 0$ or $T - Q = 0$.

2. If I is an open interval of $\mathbb{R}$ and p is an endpoint of I and $\{p\} \cup (I \cap (T - Q))$ is compact and $\{p\} = \overline{I \cap Q} \cap \overline{I \cap (T - Q)}$, then the component of T containing p—if any—is trivial.

In 1955 there was a Summer Institute on Set-theoretic Topology in Madison, which was the progenitor of the Spring Topology Conferences. I was surprised to find that that concept—*set-theoretic topology*—was in use at that time; in fact, though, the usage was totally different than it is now, and meant what we would now call *point-set topology*. Whyburn, in a one-page summary of his talk on set-theoretic topology—present and future [29] mainly writes about contributions of topology to the classical theory of functions of a complex variable. Mary Ellen, in her short paper [1955a], puts in a plug for the study of nonclosed sets, especially in problems involving connectedness. Walter Rudin went to the seminar of Gillman on rings of continuous functions, heard about the question of nonhomogeneity of $\beta N - N$, and proved that from the continuum hypothesis by showing that CH implied the existence of P-points [21]. Mary Ellen came to the subject later, around 1965, and it became one of her continuing interests, especially distinguishing among the various points, or, in other words, distinguishing the different types of ultrafilters on ω. Her first paper on the subject is from the second Wisconsin topology seminar in 1965 [1966]. She is one of the people who independently discovered the Rudin–Keisler and Rudin–Frolik orders on types of ultrafilters; for more on these orders and Mary Ellen's work in this area, see Dow's article [5] in this volume, but let me mention just one of her results: there exist 2^κ Rudin–Keisler types of ultrafilters on κ.

Mary Ellen's fame is largely as a producer of weird and wonderful topological spaces—examples and counterexamples. One often wonders how on earth she was able to construct them. She has an uncanny ability to start off with a space that has some of the properties she wants, and then push it and pull it until she gets exactly what she wants. Mary Ellen cheerfully tells people not to read her papers, but rather the later ones by people who simplify what she has done, but she of course is the one who did it first.

Her most famous example—and the one whose existence is known outside of general topology—is of course her example of a Dowker space, that is, a normal space with nonnormal product with $[0, 1]$. The Borsuk homotopy extension theorem [4] with its hypothesis of $X \times [0, 1]$ being normal had led to Dowker's question [6] of whether X normal implied $X \times [0, 1]$ normal. As mentioned earlier, Mary Ellen obtained a counterexample from a Souslin tree in [1955], and then a "real" (i.e., not relying on additional set-theoretic axioms) Dowker space in [1971c]. Twenty years later, that space is still essentially the only such example. It's an amusing sidelight of mathematical history that Mary Ellen's student Mike Starbird [22] (and, independently, Morita [14]) later proved that the hypothesis of normality of X rather than normality of $X \times [0, 1]$ is sufficient for Borsuk's theorem. However, even if Borsuk had proved the right theorem, the problem of whether X normal implies $X \times [0, 1]$ normal would still eventually have been recognized as a fundamental question. For more on Dowker spaces, see the article of Paul Szeptycki and Bill Weiss [25] in this volume.

The Dowker space led to an invitation to address the International Congress of

Mathematicians in Vancouver in 1974 [1974b]. A recent honor is her election to the American Academy of Arts and Sciences in 1991.

Normality is a pervasive theme in a large proportion of Mary Ellen's papers. I recall her once remarking something to the effect that normality is the boundary where point-set topology changes from analysis to set theory.

The question of Dowker spaces is a particular example of the general question of normality of products, to which Mary Ellen and her students have devoted much effort. In particular, normality of products with a metric or compact factor has been of interest. Here is a typical theorem from Rudin and Starbird [1975], answering a question of Morita:

THEOREM: If X is metric and C is compact and $X \times C \times Y$ is normal, then if Z is the image of Y under a closed map, then $X \times C \times Z$ is normal.

Another question of Morita is answered by:

THEOREM: If Y is T_2 and $X \times Y$ is normal for all normal T_2 X, then Y is discrete.

Beslagic's article [3] in this volume extensively surveys normality of products. What I find most wonderful about Mary Ellen's work in this area is how she proves theorems by constructing examples! The examples she uses are generalizations of Dowker spaces in that they are normal, yet certain open covers cannot be shrunk to closed covers. Let me give you an instance of how a space she and Beslagic [1985b] construct from $V = L$ allows one to prove a conjecture of Morita from that hypothesis. Recall that Z is a (Morita) *P-space* if its product with every metric space is normal.

THEOREM: $V = L$ implies X is metrizable if and only if $X \times Z$ is normal, for all normal P-spaces Z.

The idea of the proof is to suppose there were a nonmetrizable space Y with the property that $Y \times Z$ is normal for all normal P-spaces Z. One then constructs from $\diamondsuit^{++}$ (a combinatorial principle following from $V = L$—see Fleissner's article [8] in this volume) a normal P-space H such that if $Y \times H$ is normal, then there are

$$\{Y_\alpha\}_{\alpha < \omega_1}, \ Y_\alpha \subseteq Y \text{ such that } \cup \{Y_\alpha : \alpha < \omega_1\} \neq 0. \tag{*}$$

Next, one constructs a normal P-space K, such that if Y satisfies (*), then $Y \times K$ is not normal. But that's a contradiction!

Another continuing interest of Mary Ellen (especially in the 1970s) has been box products (see [1971d], [1971e], [1972], [1972c], [1973], [1974]). Her Dowker space [1971] is a subspace of a box product, and, as she writes in her first article on the box topology [1971d],

> I am interested in box products as an example machine. If one is not interested in compactness or first countability and wants to build spaces with properties roughly between normality and paracompactness, box products, and particularly subspaces of box products, may be useful.

Miščenkos's space [13], and her own Dowker space, which is a modification of it, led Mary Ellen to be concerned with the fact that she did not know anything about the normality of box products of even the nicest kinds of spaces. For example, she writes in that first paper: "I have no idea how to prove that the box product of countably many copies of $\omega + 1$ is normal."

Well, a year later she does have an idea, proving under CH that the box product of countably many compact metric spaces is normal [1972c] (announced in [1971e]). This has been generalized by Mary Ellen and others. See Lawrence's article [10] in this volume. (The most exciting mathematical result presented at the Rudin conference was Lawrence's proof that the box product of uncountably many copies of $\omega + 1$ is not normal.)

Another one of Mary Ellen's interests—in fact, of most set-theoretic topologists— is the question of S- and L-spaces.

"Is every regular hereditarily separable space hereditarily Lindelöf?"

"Is every regular hereditarily Lindelöf space hereditarily separable?"

There were many results in the 1970s by members of the Madison seminar concerning these questions. Mary Ellen produced the first *S-space* (a regular hereditarily separable space that is not hereditarily Lindelöf), assuming the existence of a Souslin tree [1972a]. This was a surprising result, since a Souslin line is an *L-space,* that is, a regular hereditarily Lindelöf space that is not hereditarily separable. She later showed how to get Dowker S-spaces from CH [1975b], [1976a]. For more on S- and L-, see Todorčević's article [26] in this volume.

As Mary Ellen has often emphasized, S- and L-questions are not just interesting in themselves; these two patterns keep recurring in other topological problems, notably the metrizability of perfectly normal manifolds. In fact Szentmiklóssy's theorem [24] that Martin's Axiom plus $2^{\aleph_0} > \aleph_1$ implies there are no compact S-spaces is a crucial component of Mary Ellen's proof [1979b] that that hypothesis in fact kills all nonmetrizable perfectly normal manifolds.

Another area that interested Mary Ellen was the normal Moore space conjecture. She herself did not work on it much—there are just the surveys [1975b] and [1975c] and the joint paper with Mike Starbird [1977a] that explored the question of what a counterexample would look like. However, it was the prime concern of two of her students' theses—the author's and Bill Fleissner's—and also received some attention from Starbird.

Mary Ellen wrote a series of three papers [1977b], [1977c], [1977d] on the subject of weakly compact subspaces of Banach spaces, now known as *Eberlein compacts.* The first was joint with Y. Benyamini and her student Mike Wage; the others were joint with Ernie Michael. J. Lindenstrauss [12] in 1972 had asked whether continuous images of Eberlein compacts are Eberlein compact. The first of these papers [1977b] provided a positive answer, using the characterizations due to D. Amir and Lindenstrauss [2] and H. P. Rosenthal [20], respectively: A compact space is Eberlein iff:

(a) It is homeomorphic to a subspace of $c_0(\Gamma) = \{f : \Gamma \to \mathbb{R}$ such that for all $\epsilon > 0$, there are only finitely many $\gamma \in \Gamma$ for which $|f(\gamma)| > \epsilon\}$, where Γ is some arbitrary set.

Or

(b) It has a σ-point-finite separating open cover of open F_σ subsets.

A family $\mathscr{F}$ of subsets of a space X is *separating* if for each $x, y \in X$, there is an $F \in \mathscr{F}$ that contains exactly one of x and y.

The second paper [1977c] provides a purely topological proof that continuous images of Eberlein compacts are Eberlein compact, using only (b), while [1977d] proves that a compact Hausdorff space that is the union of two metrizable subspaces is Eberlein.

For more on Eberlein compacts, see [27].

The 1970s was Mary Ellen's most active decade in terms of number of papers—36 of her 79 publications appeared during that decade—which coincided with the greatest period of activity in the Madison seminar. Her fundamental work on box products, the Dowker space, the work on Eberlein compacts, S-spaces, the undecidability of the existence of perfectly normal nonmetrizable manifolds, the normality of products with compact or metric factors all date from this decade. But the 24 papers since could hardly be considered as resting on one's laurels. The pattern is that Mary Ellen pursues some favorite themes, for example, normality of products and, in particular, Dowker spaces, over many years, but also takes time out to solve particular problems that come her way.

Looking at Mary Ellen's papers in the 1980s and 1990s, the two predominant themes are first, the general area comprising the related topics of normality of products, Dowker spaces, and shrinking open covers, and second, the construction of pathological nonmetrizable manifolds. There are isolated papers on a number of other topics, but the only subject besides the aforementioned themes that crops up more than once is her latest work—on monotone normality. A space is *monotonically normal* if it is T_1 and for every point x and open U containing x, there is an open $\mathscr{N}(x, U)$ such that

1. For $x \notin V$ and $y \notin U$, $\mathscr{N}(x, U) \cap \mathscr{N}(y, V) = 0$.
2. For $U \subseteq V$, $\mathscr{N}(x, U) \subseteq \mathscr{N}(x, V)$.

Monotonically normal spaces generalize linearly ordered spaces; Mary Ellen's student Diana Pike Palenz in her thesis [17] showed that—as was previously known for linearly ordered spaces—monotonically normal spaces are countably paracompact. A very pretty result Mary Ellen obtained jointly with Z. Balogh in [1992] is that the Engelking–Lutzer [7] characterization of paracompactness in linearly ordered spaces, namely the absence of subspaces homeomorphic to stationary sets, also holds for monotonically normal spaces.

Mary Ellen has also recently [1993] constructed a monotonically normal space that is not "acyclically monotonically normal" nor "K_0." We omit the definitions. Suffice it to say this answered a number of questions raised by people working in the area.

The subject mentioned earlier of shrinking open covers is a natural one—$\mathscr{U} = \{U_\alpha\}_{\alpha < \kappa}$ is a shrinking of the open cover $\mathscr{V} = \{V_\alpha\}_{\alpha < \kappa}$ if $\mathscr{U}$ is an open cover and for all α, $\overline{U}_\alpha \subseteq V_\alpha$. In a metacompact normal space, all open covers can be shrunk (we say the space has *the shrinking property*); a normal space is Dowker iff there is a countable open cover that can't be shrunk. Two natural variations of the shrinking property are:

PROPERTY B: Every monotone (i.e., increasing by inclusion) open cover has a monotone shrinking.

WEAK B PROPERTY: Every monotone open cover has a shrinking.

In fact, a normal space is Dowker iff it has a countable monotone open cover with no shrinking, that is, it fails to have what we may call weak $B(\aleph_0)$. Similarly, define weak $B(\kappa)$ to say that every monotone open cover of size κ has a shrinking, and $B(\kappa)$ to say it has a monotone shrinking. A κ-*Dowker* space is then a normal space failing to have weak $B(\kappa)$.

As Mary Ellen notes: "It is strangely difficult to find an example of a normal space without the shrinking property." Just as with Dowker spaces, there is essentially only one class of ZFC examples—generalizations of Mary Ellen's Dowker space. These κ-Dowker spaces are dealt with in [1978] and [1985]. (In [1972b], Mary Ellen noted that a straightforward generalization of [1955] yielded a κ-Dowker space from a κ-Souslin tree.)

In [1983e] Mary Ellen pointed out that the normal, countably paracompact, paralindelöf nonparacompact space constructed by her student Caryn Navy in her thesis [15] is a normal space with property B, which is not paracompact, answering a question of Yasui.

The more interesting questions involving shrinking are related to normality of products—the $\Diamond^{++}$ example of [1985b] referred to earlier is an ω_1-Dowker P-space. A real such space would solve Morita's conjecture. For more on this see Beslagic's article [3] in this volume. For the general topic of shrinking, also see the thesis of Mary Ellen's latest doctoral student—Zorana Lazarevic [11].

Incidentally, there are nontrivial positive results concerning shrinking—for example, in [1983a], Mary Ellen proved that Σ-products of metric spaces have the shrinking property.

Mary Ellen's papers became (and stayed) more set-theoretic in the late 1960s and, especially, the early 1970s. There were a variety of reasons for this in addition to what Marxists might call "historical necessity." The most important was that set theory was being done at Wisconsin. During the late 1960s and early 1970s, Jack Silver, Karel Prikry, and Adrian Mathias were all at Madison for a year or more, while Ken Kunen came there to stay (except for a few years later spent in Texas). David Booth and I were both students in logic who had begun thinking about applications of set theory to topology—in my case, around 1966 when I came across Burton Jones's theorem that $2^{\aleph_0} < 2^{\aleph_1}$ implies that every separable normal Moore space is metrizable. Kunen came to Madison in 1968; David and I frequently went back and forth between Ken and Mary Ellen, and that accelerated the collaboration that made Madison such an exciting place for set-theoretic topology in the 1970s. In 1971 Ken came back from a year at Berkeley with Bill Fleissner and Judy Roitman in tow, and—having attained critical mass—he and Mary Ellen started a seminar in set-theoretic topology for their students and visitors. During the summers, I and others would come and there'd be 5–10 people. Juhász was there one year; Eric van Douwen made his first North American appearance one summer, and we soon learned from his favorite T-shirt that Wisconsin's "state bird" was the mosquito! A lot of exciting mathematics got done during those summers.

The most exciting summer, though, was the one of 1974, when we all went off to the Conference Board of the Mathematical Sciences Regional Conference in Laramie, Wyoming, for Mary Ellen's *Lectures on Set Theoretic Topology,* which was published as [1975c]. We can thank Joe Martin for these lectures. He had been at

Wisconsin for a number of years; there were several sections of the first graduate topology course, each with ten or so students; he would teach one, R. H. Bing would teach one, Mary Ellen another, and I believe there was a fourth. Joe is a master of the Moore method; he converted me into a researcher and a topologist and a practitioner of a modified Moore method. I never did take a course from Mary Ellen, who doesn't believe in the Moore method. Anyway, Joe moved to the University of Wyoming as Chairman and he came up with the idea of Mary Ellen giving a lecture series there for the Conference Board—ten lectures in five days. That was the origin of the book.

The only thing I remember about Laramie itself was that our not very tall dormitory was the tallest building in the state, but I do remember the sense of excitement we felt at Mary Ellen's lectures, which really set the course of the field for the next decade. The chapters on cardinal functions, hereditary separability, and hereditary Lindelöfness, box products, and so forth (see the table of contents we have reproduced at the end of this article), established a framework. And this time she certainly knew the literature—there were 130 references! It was a challenge— and still is—to try to solve a problem on her problem list. If one were really ambitious, one could work on the few problems that were clearly important: the existence of L- and S-spaces, the Normal Moore Space Conjecture, normality of box products, and so forth.

Mary Ellen's little book, along with Juhász' *Cardinal Functions in Topology* [9], proclaimed the existence of a new field, *set-theoretic topology,* which grew out of general topology, but employed new methods, which led to new questions. It was an exciting time.

Mary Ellen has had 16 Ph.D. students. I have listed their names and thesis titles at the end of this article. There is no clear pattern among them. Some worked more or less by themselves, some were true collaborators with Mary Ellen, some were guided by her. Their thesis topics mainly—but not entirely—reflect Mary Ellen's interests. In addition, Bill Fleissner was an unofficial student of Kunen and Mary Ellen, obtaining his degree at Berkeley, but spending a couple of years in Madison. Judy Roitman was another unofficial student. Mary Ellen has also had three postdoctoral students—all well-known topologists:

William Weiss, 1975–1976
Hao-Xuan Zhou, 1981
Stephen Watson, 1982–1983

In addition, of course, Mary Ellen has influenced a whole generation of set-theoretic topologists.

I first met Mary Ellen in the summer of 1967. I was casting around for a supervisor and for problems to work on, since the faculty member I had been talking to had not only left Madison, but had also then solved the problem I'd been researching. Mary Ellen was very approachable, helpful, and encouraging, which was what I needed, and which qualities have supported a long line of students since. At conferences she is always just as happy to talk to beginning graduate students as to well-established researchers.

Some of the things I remember from that period as a graduate student are how her enthusiasm survived what I now realize must have been my pathetic attempt at a

Dowker space, how she arranged funding for me to go to my first topology conference, in Houston in 1968, not canceling it when my proof of the consistency of the Normal Moore Space Conjecture fell through. Her cheerful ignorance or disregard of departmental regulations came in handy for someone who was not very good at jumping through hoops.

Mary Ellen has served the mathematical community, as Vice-president of the American Mathematical Society and on numerous committees. However, one of her really great services has been her activity as purveyor of examples to all sorts of mathematicians. Several times a week she will receive inquiries from mathematicians around the world—often, nontopologists—asking for an example of this or that. Long before other people worried about conservation, she would answer such letters by writing back on the letter itself. I doubt whether any "green" sentiments motivate her; rather, it is a successful attempt to avoid accumulating paper. Similarly, I have often observed her quickly disposing of an unsolicited preprint by unloading it on a graduate student. I am convinced that this avoidance of clutter, especially mentally, is one of the secrets of Mary Ellen's mathematical power. By avoiding filling up her memory, she maximizes processing capability in her brain. I remember her once even forgetting the definition of paracompactness. Another time, I was proving some results relating Souslin lines to Moore spaces, and asked her if she knew if anything had been done along these lines. She said, "no"; later I found she had published my results 20 years earlier!

Mary Ellen has now retired from teaching trigonometry, serving on departmental committees, and other such unrewarding faculty activity. We can therefore antici-pate even more exciting research from her in the future. In particular, there are (only) a few problems that have interested her for many years, and yet have eluded her efforts so far. Here are three she singles out:

1. *The linearly Lindelöf problem:* Is there a normal non-Lindelöf space in which every well-ordered-by-inclusion open cover has a countable subcover?
2. *The σ-disjoint base problem:* Is there a normal space with a σ-disjoint base that is not paracompact?
3. *The $M_1 = M_3$ problem:* Does every space with a σ-closure-preserving closed pseudobase have a σ-closure-preserving open base?

For further information, see, for example, [1990a]. (A *pseudobase* is a collection $\mathscr{P}$ of subsets of a space X such that for each $x \in X$ and each open U containing x, there is a $P \in \mathscr{P}$ such that $x \in \text{int } P \subseteq U$.)

Now that she has more time, one can reasonably expect at least one of these to yield! We are looking forward to it.

THESES SUPERVISED BY MARY ELLEN RUDIN

1969   Franklin David Tall, "Set-theoretic consistency results and topological theo-rems concerning the normal Moore space conjecture and related problems."

1970   Nancy MacMaster Warren, "Extending continuous functions in Stone–Čech compactifications of discrete spaces and in zero-dimensional spaces."

1974   Michael Peter Starbird, "The normality of products with a compact or a metric factor."

1975   Brian Maynard Scott, "The behavior of orthocompactness in products, with emphasis on certain analogies with that of normality under similar circumstances, or orthocompact : metacompact :: normal : paracompact (a.e.)."

1976   Andrew Joseph Berner, "Classes of spaces defined by nested sequences of sets and maps among such spaces."

1976   Lee Parsons, "Normality and countable paracompactness in spaces related to the Stone–Čech compactification of a discrete space."

1976   Michael Lee Wage, "Applications of set theory to analysis and topology."

1979   Roderick Allen Price, "CH (and less) solves a set-theoretic problem of Čech."

1980   Diana Gail Pike Palenz, "Paracompactness in monotonically normal spaces."

1980   George Su-an Lee, "Some combinatorial results on Stone spaces."

1981   Caryn Linda Navy, "Non-paracompactness in para-Lindelöf spaces."

1983   Charles F. Mills, "Supercompact spaces and related structures."

1986   Amer Beslagic, "Products of topological spaces."

1988   Shouli Jiang, "The strict p-space problem and generalized metric spaces as images of metric spaces."

1990   Ning Zhong, "Generalized metric spaces and products."

1992   Zorana Lazarevic, "Shrinking open covers."

CONTENTS OF LECTURES ON SET THEORETIC TOPOLOGY[b]

PUBLICATIONS OF MARY ELLEN RUDIN

[1950]   Concerning abstract spaces. Duke Math. J. **17:** 317–327.

[1951]   Separation in non-separable spaces. Duke Math. J. **18:** 623–629.

[1952]   A primitive dispersion set of the plane. Duke Math. J. **19:** 323–328.

[1952a] Concerning a problem of Souslin's. Duke Math. J. **19:** 629–639.

[1955]   Countable paracompactness and Souslin's problem. Can. J. Math. **7:** 543–547.

[1955a] Connectedness. *In* Summary of Lectures and Seminars, Summer Institute in Set-Theoretic Topology, R. H. Bing, Ed.: 91–92. Madison, Wisconsin, 1955. (Revised 1957.)

[b]Reproduced from *Lectures on Set Theoretic Topology,* by Mary Ellen Rudin, *Conference Board of the Mathematical Sciences Regional Conference Series in Mathematics,* Volume 23, by permission of the American Mathematical Society.

[1956]   A separable normal nonparacompact space. Proc. Am. Math. Soc. **7**: 940–941.

[1957]   A note on certain function spaces (with V. L. Klee). Arch. Math. **7**: 469–470.

[1957a] A subset of the countable ordinals. Am. Math. Mon. **64**: 351.

[1957b] A topological characterization of sets of real numbers. Pac. J. Math. **7**: 1185–1186.

[1957c] A property of indecomposable connected sets. Proc. Am. Math. Soc. **8**: 1152–1157.

[1958]   An unshellable triangulation of a tetrahedron. Bull. Am. Math. Soc. **64**: 90–91.

[1958a] A connected subset of the plane. Fundam. Math. **64**: 15–24.

[1963]   Arcwise connected sets in the plane. Duke Math. J. **30**: 363–368.

[1965]   Interval topology in subsets of totally orderable spaces. Trans. Am. Math. Soc. **112**: 376–389.

[1965a] A technique for constructing examples. Proc. Am. Math. Soc. **16**: 1320–1323.

[1966]   Types of ultrafilters. *In* Topology Seminar, Wisconsin (1965), R. H. Bing and R. J. Bean, Eds.: 147–152. Ann. Math. Studies No. 60. Princeton Univ. Press. Princeton, N.J.

[1969]   A new proof that metric spaces are paracompact. Proc. Am. Math. Soc. **20**: 603.

[1969a] Souslin's conjecture. Am. Math. Mon. **76**: 1113–1119.

[1970]   Composants in βN. *In* Proceedings of the Washington State Univ. Conference on General Topology, March 1970: 117–120.

[1971]   A normal space X for which X × I is not normal. Bull. Am. Math. Soc. **77**: 246.

[1971a] Partial order on the types of βN. Trans. Am. Math. Soc. **155**: 353–362.

[1971b] *Review of* Counterexamples in Topology by L. A. Steen and J. A. Seebach. Am. Math. Mon. **48**: 4140.

[1971c] A normal space X for which X × I is not normal. Fundam. Math. **73**: 179–186.

[1971d] The box topology. *In* Proceedings of the Univ. of Houston Point-set Topology Conference, D. R. Traylor, Ed.: 191–199. Univ. of Houston, Houston, Texas.

[1971e] Box products. *In* General Topology and Its Relations to Modern Analysis and Algebra III, Proceedings of the Third Prague Topological Symposium, J. Novak, Ed.: 385. Academia. Prague.

[1972]   Box products and extremal disconnectedness. *In* Proceedings of the Univ. of Oklahoma Topology Conference, D. Kay, J. Green, L. Rubin, and L. P. Su, Eds.: 274–284. Univ. of Oklahoma. Norman.

[1972a] A normal, hereditarily separable, non-Lindelöf space. Ill. J. Math. **16**: 621–625.

[1972b] Souslin trees and Dowker spaces. Topics in Topology (Proceedings of the Colloquium, Keszthely, 1972), Á. Császár, Ed.: 557–562. Colloquium of the János Bolyai Mathematical Society **8**.

[1972c] The box product of countably many compact metric spaces. Gen. Topol. Appl. **2**: 293–398.

[1973]   A non-normal box product (with P. Erdös). Infinite and Finite Sets (Proceedings of the Colloquium, Keszthely, 1973; Dedicated to P. Erdös on his 60th Birthday), (II), A. Hajnal, R. Rado, and V. T. Sós, Eds.: 629–631. Colloquium of the János Bolyai Mathematical Society 10.

[1973a]   The normality of a product with a compact factor. Bull. Am. Math. Soc. 79: 983–985.

[1973b]   Two problems. In Proceedings of the International Symposium on Topology and its Applications (Beograd 1973), D. R. Kurepa, Ed.: 221.

[1973c]   A separable Dowker space. Symp. Math., Inst. Naz. Alta Math. 16: 125–132.

[1974]   Countable box products of ordinals. Trans. Am. Math. Soc. 190: 1–8.

[1974a]   A non-normal hereditarily separable space. Ill. J. Math. 18: 481–483.

[1974b]   The normality of products. Proceedings of International Congress of Mathematicians (Vancouver, 1974) (2), R. D. James, Ed.: 81–84. Vancouver, British Columbia.

[1975]   Products with a metric factor (with M. Starbird). Gen. Topol. Appl. 5: 235–248.

[1975a]   The normality of products with one compact factor. Gen. Topol. Appl. 5: 45–49.

[1975b]   The metrizability of normal Moore spaces. In Studies in Topology, (Proceedings of the Conference, Univ. of North Carolina, Charlotte, North Carolina, 1974; Dedicated to the Mathematics Section of the Polish Academy of Science), N. M. Stravrakis and K. R. Allen, Eds.: 507–516. Academic Press. New York.

[1975c]   Lectures on Set Theoretic Topology, Conference Board of the Mathematical Sciences Regional Conference Series in Mathematics 23. American Mathematical Society. Providence, R.I.

[1976]   A perfectly normal nonmetrizable manifold (with P. Zenor). Houston J. Math. 2: 129–134.

[1976a]   Two more hereditarily separable non-Lindelöf spaces (with K. Kunen and I. Juhász). Can. J. Math. 28: 998–1005.

[1977]   Martin's Axiom. In Handbook of Mathematical Logic, K. J. Barwise, Ed.: 491–503. North-Holland. Amsterdam, the Netherlands.

[1977a]   Some examples of normal Moore spaces (with M. Starbird). Can. J. Math. 29: 84.

[1977b]   Continuous images of weakly compact subspaces of Banach spaces (with Y. Benyamini and M. Wage). Pac. J. Math. 70: 309–324.

[1977c]   A note on Eberlein compacts (with E. Michael). Pac. J. Math. 72: 487–495.

[1977d]   Another note on Eberlein compacts (with E. Michael). Pac. J. Math. 72: 497–499.

[1977e]   A narrow view of set theoretic topology. In General Topology and Its Relations to Modern Analysis and Algebra IV (Proceedings of the Fourth Prague Topology Symposium, Prague, 1976), Part A, J. Novak, Ed.: 190–195. Lecture Notes Math. 609. Springer-Verlag. Berlin.

[1978]   κ-Dowker spaces. Czech. Math. J. 28: 324–328.

[1978a]   Unordered types of ultrafilters (with S. Shelah). Topol. Proc. 3: 199–204.

[1978b]   Review of Aspects of Topology by Charles O. Christenson and William L.

Voxman *and* Set-theoretic Topology by Gregory L. Naber. Bull. Am. Math. Soc. **84:** 271–272.

[1979a] Hereditary normality and Souslin lines. Gen. Topol. Appl. **10:** 103–106.

[1979b] The undecidability of the existence of a perfectly normal non metrizable manifold. Houston J. Math. **5:** 249–252.

[1980] S & L spaces. *In* Surveys in General Topology, G. M. Reed, Ed.: 431–444. Academic Press. New York.

[1981] Directed sets which converge. *In* General Topology and Modern Analysis, L. F. McAuley and M. M. Rao, Eds.: 295–305. Academic Press. New York.

[1981a] Topology and measure. *In* Measure Theory and its Applications, Proceedings of the 1980 Conference at Northern Illinois University, G. A. Goldin and R. F. Wheeler, Eds.: 121. Northern Illinois Press. DeKalb.

[1983] A normal screenable nonparacompact space. Gen. Topol. Appl. **15:** 313–322.

[1983a] The shrinking property. Can. Math. Bull. **26:** 385–388.

[1983b] Collectionwise normality in screenable spaces. Proc. Am. Math. Soc. **87:** 347–350.

[1983c] Countable products of paracompact scattered spaces (with S. Watson). Proc. Am. Math. Soc. **89:** 551–553.

[1983d] Dowker's set theory question. Q & A Gen. Topol. **1:** 75–76.

[1983e] Yasui's questions. Q & A Gen. Topol. **2:** 122–127.

[1984] Dowker spaces. *In* Handbook of Set-Theoretic Topology, K. Kunen and J. E. Vaughan, Eds.: 761–780. North-Holland. Amsterdam, the Netherlands.

[1984a] Two questions of Dowker. Proc. Am. Math. Soc. **91:** 155–158.

[1985] κ-Dowker spaces. *In* Aspects of Topology, In Memory of Hugh Dowker 1912–1982, I. M. James and E. H. Kronheimer, Eds.: 175–195. London Mathematical Society Lecture Notes **93.** London.

[1985a] A lattice of conditions (with P. Collins, M. Reed and W. Roscoe). Proc. Am. Math. Soc. **94:** 487–496.

[1985b] Set theoretic constructions of non shrinking open covers (with A. Beslagic). Topol. Appl. **20:** 167–177.

[1986] Normality of products and Morita's conjectures (with K. Chiba and T. Przymusiński). Topol. Appl. **22:** 19–32.

[1988] A nonmetrizable manifold from $\Diamond^+$. Topol. Appl. **28:** 105–112.

[1988a] A few topological problems. Comment. Math. Univ. Carolinae. **29:** 743–749.

[1989] Countable point separating open covers for manifolds. Houston J. Math. **15:** 255–266.

[1990] Two nonmetrizable manifolds. Topol. Appl. **35:** 137–152.

[1990a] Some Conjectures. *In* Open Problems in Topology, J. van Mill and G. M. Reed, Eds.: 183–193. North-Holland. Amsterdam, the Netherlands.

[1990b] New classic problems (with Z. Balogh, S. W. Davies, A. Dow, G. Gruenhage, F. D. Tall, S. Watson). Topol. Proc. **15:** 201–220.

[1992] Monotone normality (with Z. Balogh). Topol. Appl. **47:** 115–127.

[1993] A cyclic monotonically normal space which is not K_0. Proc. Am. Math. Soc. In press.

[19??]   The set theory and topology of the real line and plane as seen in the work of G. C. Young and W. H. Young. Submitted for publication.

REFERENCES

1. ALBERS, D. J. & C. REID. 1988. An interview with Mary Ellen Rudin. Coll. Math. J. **2:** 114–135.
2. AMIR, D. & J. LINDENSTRAUSS. 1968. The structure of weakly compact subsets in Banach spaces. Ann. Math. **88:** 35–46.
3. BESLAGIC, A. 1993. Normality in products. This issue.
4. BORSUK, K. 1937. Sur les prolongments des transformations continues. Fundam. Math. **28:** 203.
5. DOW, A. 1993. βN. This issue.
6. DOWKER, C. H. 1951. On countably paracompact spaces. Can. J. Math. **3:** 219–224.
7. ENGELKING, R. & D. J. LUTZER. 1977. Paracompactness in ordered spaces. Fundam. Math. **94:** 49–58.
8. FLEISSNER, W. 1993. Theorems from measure axioms, counterexamples from $\Diamond^{++}$. This issue.
9. JUHÁSZ, I. 1971. Cardinal Functions in Topology. Math. Centrum. Amsterdam, the Netherlands.
10. LAWRENCE, L. B. 1993. Toward a theory of normality and paracompactness in box products. This issue.
11. LAZAREVIC, Z. 1992. Shrinking open covers, Ph.D. Thesis, University of Wisconsin, Madison.
12. LINDENSTRAUSS, J. 1972. Weakly compact sets—Their topological properties and the Banach spaces they generate. Ann. Math. Studies **69:** 235–273.
13. MIŠČENKO, A. 1962. Finally compact spaces. Sov. Math. Dokl. **145:** 1199–1202.
14. MORITA, K. 1975. On generalizations of Borsuk's homotopy extension theorem. Fundam. Math. **88:** 1–6.
15. NAVY, C. L.1981. Non-paracompactness in para-Lindelöf spaces, Ph.D. Thesis, University of Wisconsin, Madison.
16. NYIKOS, P. J. 1993. Mary Ellen Rudin's contributions to the theory of nonmetrizable manifolds. This issue.
17. PALENZ, D. G. P. 1980. Paracompactness in monotonically normal spaces, Ph.D. Thesis, University of Wisconsin, Madison.
18. PIXLEY, C. & P. ROY. 1969. Uncompletable Moore spaces. In Proceedings of the 1969 Auburn Topological Conference, W. R. R. Transue, Ed.: 65–69.
19. PRZYMUSIŃSKI, T. & F. D. TALL. 1974. The undecidability of the existence of a non-separable normal Moore space satisfying the countable chain condition. Fundam. Math. **85:** 291–297.
20. ROSENTHAL, H. P. 1974. The heredity problem for weakly compactly generated Banach spaces. Compos. Math. **28:** 83–111.
21. RUDIN, W. 1956. Homogeneity problems in the theory of Čech compactifications. Duke Math. J. **23:** 409–419.
22. STARBIRD, M. 1975. The Borsuk homotopy extension theorem without the binormality condition. Fundam. Math. **87:** 207–211.
23. ———. 1993. Mary Ellen Rudin as advisor and geometer. This issue.
24. SZENTMIKLÓSSY, Z. 1980. S-spaces and L-spaces under Martin's axiom. In Colloquium János Bolyai Mathematical Society **23** (Budapest, 1978), **II:** 1139–1145. North-Holland. Amsterdam, the Netherlands.
25. SZEPTYCKI, P. J. & W. A. R. WEISS. 1993. Dowker spaces. This issue.
26. TODORČEVIĆ, S. 1993. Some applications of S and L combinatorics. This issue.
27. WAGE, M. L. 1980. Weakly compact subsets of Banach spaces. In Surveys in General Topology, G. M. Reed, Ed.: 479–494. Academic Press. New York.
28. WATSON, S. 1993. Mary Ellen Rudin's early work on Suslin spaces. This issue.
29. WHYBURN, G. 1955. Set theoretic topology; Present and future. Summary of Lectures and Seminars, Summer Institute in Set-Theoretic Topology, R. H. Bing, Ed.:6. Madison, Wis. (Revised 1957).

Normality in Products[a]

AMER BESLAGIC

Department of Mathematics
George Mason University
Fairfax, Virginia 22030

I would like to use this opportunity to thank my teacher and friend Mary Ellen Rudin, for the patience, generosity, and help she gave me over all these years.

0. INTRODUCTION

This is a survey paper on recent results and ones that haven't been discussed in Przymusinski's [47] or Hoshina's [27] surveys. In addition to new results we will be presenting new techniques. Many well-known old results in the first section, in particular, are given new proofs. Many interesting and basic facts are not mentioned and can be found on the papers just cited. Some results that should have been included are omitted due to the incompetence of the author.

The emphasis here is on examples.

We do not discuss three important results by Mary Ellen Rudin. The first two [50, 54] can be found in Weiss's article in this book.

THEOREM 0.1: There is a normal space X such that $X \times [0, 1]$ is not normal.

THEOREM 0.2: Assume that for every normal Y, the product $X \times Y$ is normal. Then the space X is discrete.

Recall that a space Y is a $P(roduct)$-*space* iff $Y \times M$ is normal for every metric space M. (This is *different* from G_δ are open.) The following is done by Rudin, with help from Chiba and Przymusinski [17].

THEOREM 0.3: Assume $V = L$. Assume that for every P-space Y, the product $X \times Y$ is normal. Then the space X is metric.

For the convenience of the reader all papers by Mary Ellen Rudin involving normality in products are listed in the References.

1. RUDIN LEMMA

In this section we illustrate the connection between the normality of products with one compact factor and paracompactness or collectionwise normality of the other factor. The key tool is Theorem 1.2, which Rudin proved in [53].

The most important results are summarized in Theorem 1.18.

Mathematics Subject Classification. Primary 54B10, 54D15, 03E45, 03E50.
Key words and phrases. Normal, product.

DEFINITION 1.1: Let H and K be two subsets of the product $X \times C$. The family $\{U \times O_U : U \in \mathscr{U}\}$ of open subsets of $X \times C$ is called an $\langle H, K \rangle$-*separation* (in $X \times C$) iff

  (i) $\mathscr{U}$ is a locally finite open cover of X;
  (ii) For every $U \in \mathscr{U}$, $H \cap (U \times C) \subset U \times O_U$ and $K \cap U \times O_U = \emptyset$.

Note that if $\{U \times O_U : U \in \mathscr{U}\}$ is an $\langle H, K \rangle$-separation in $X \times C$, then $V = \bigcup \{U \times O_U : U \in \mathscr{U}\}$ is an open subset of $X \times C$ with $H \subset V$ and $K \cap \overline{V} = \emptyset$. Hence if any two closed disjoint subsets of $X \times C$ have a separation, the product $X \times C$ is normal. The converse holds if C is compact.

RUDIN LEMMA 1.2: Let C be compact. The product $X \times C$ is normal iff every two closed disjoint subsets of $X \times C$ have a separation in $X \times C$.

Proof: Assume that the product is normal and let H and K be two disjoint closed subsets of $X \times C$. Let $f : X \times C \to [0, 1]$ be such that $f(H) \subset \{0\}$ and $f(K) \subset \{1\}$. Let $\mathscr{C}(C)$ be the space of all (continuous) maps from C to $\mathbb{R}$ with the supremum norm metric $\rho(g, h) = \max\{|g(y) - h(y)| : y \in C\}$. The function F from X to $\mathscr{C}(C)$ defined by $F(x) = f|(\{x\} \times C)$ is continuous since C is compact.

Let $\mathscr{U}'$ be a locally finite open cover of $\mathscr{C}(C)$ with sets of ρ-diameter $< \frac{1}{3}$. The family $\mathscr{U} = \{F^{-1}(U) : U \in \mathscr{U}'\}$ is a locally finite open cover of X. For each $U \in \mathscr{U}$ we define an open O_U in C such that

$$H \cap (U \times C) \subset U \times O_U \quad \text{and} \quad K \cap (\overline{U \times O_U}) = \emptyset.$$

For $U \in \mathscr{U}$ let $O_U = \{y \in C : \exists x \in \overline{U}(f(x, y) < \frac{1}{3})\}$, so O_U is open in C and $H \cap (U \times C) \subset U \times O_U$ since $f|H = 0$.

To see that $K \cap \overline{U \times O_U} = \emptyset$, let $P = \{y \in C : \exists x \in \overline{U}(f(x, y) > \frac{2}{3})\}$; then P is open in C and $K \cap (\overline{U} \times C) \subset \overline{U} \times P$. We show that $P \cap O_U = \emptyset$, thus showing that $K \cap \overline{U \times O_U} = \emptyset$. If $y \in P \cap O_U$, there are $x, x' \in \overline{U}$ with $f(x, y) < \frac{1}{3}$ and $f(x', y) > \frac{2}{3}$, so $\rho(F(x), F(x')) > \frac{1}{3}$, hence the ρ-diameter of $F(U) \in \mathscr{U}'$ is $> \frac{1}{3}$, a contradiction. $\square$

We now work toward establishing a relationship between paracompactness and normality in products with a compact space.

Recall that a space is κ-*paracompact* iff every open cover of size $\leq \kappa$ has a locally finite open refinement.

Also, an open cover $\{V_\alpha : \alpha < \kappa\}$ of a space X is a *shrinking* of the cover $\{U_\alpha : \alpha < \kappa\}$ iff for every $\alpha < \kappa$, $\overline{V_\alpha} \subset U_\alpha$.

The following is well known (and easy to prove):

LEMMA 1.3: A space is normal and κ-paracompact iff every open cover of size $\leq \kappa$ has a locally finite shrinking.

THEOREM 1.4: Assume that X is normal and κ-paracompact, and C is a compact space of weight $\leq \kappa$. Then $X \times C$ is normal.

Proof: Let $\mathscr{B}$ be a base for C, closed under finite unions, and with $|\mathscr{B}| \leq \kappa$. Let H and K be two disjoint closed subsets of $X \times C$. We find an $\langle H, K \rangle$-separation.

For $B \in \mathscr{B}$ let

$$U_B = \{x \in X : K \cap (\{x\} \times C) \subset B \quad \text{and} \quad H \cap (\{x\} \times \overline{B}) = \varnothing\} =$$

$$X \backslash [\pi(K \cap X \times (C \backslash B)) \cup \pi(H \cap (X \times \overline{B}))],$$

where $\pi : X \times C \to X$ is the projection map. Since C is compact, π is closed, so U_B is open.

Also $\{U_B : B \in \mathscr{B}\}$ covers X since $\mathscr{B}$ is closed under finite unions and C is compact. Let $\{V_B : B \in \mathscr{B}\}$ be a locally finite shrinking of $\{U_B : B \in \mathscr{B}\}$.

Then $\{V_B \times B : B \in \mathscr{B}\}$ is an $\langle H, K\rangle$-separation. $\square$

So if X is normal and sufficiently paracompact and C compact, the product $X \times C$ is normal. Now we show that normality of a product may imply paracompactness of a factor.

LEMMA 1.5: Let X be a space such that every increasing open cover of cardinality $\leq \kappa$ has a locally finite refinement. Then X is κ-paracompact.

Proof: Let κ be the least cardinal for which the lemma fails and let $\{U_\alpha : \alpha < \kappa\}$ be an open cover of X. For $\alpha < \kappa$ let $V_\alpha = \bigcup_{\beta < \alpha} U_\beta$. Then $\{V_\alpha : \alpha < \kappa\}$ is an increasing open cover of size $\leq \kappa$. We first construct a locally finite open shrinking $\{W_\alpha : \alpha < \kappa\}$ of $\{V_\alpha : \alpha < \kappa\}$.

Let $\{O_\alpha : \alpha < \kappa\}$ be a locally finite open refinement of $\{V_\alpha : \alpha < \kappa\}$, with $O_\alpha \subset V_\alpha$ for $\alpha < \kappa$. The family $\{X \backslash \bigcup_{\beta > \alpha} O_\beta : \alpha < \kappa\}$ is an increasing open cover of X.

Let $\{W_\alpha : \alpha < \kappa\}$ be a locally finite open refinement of $\{X \backslash \bigcup_{\beta > \alpha} O_\beta : \alpha < \kappa\}$ with $W_\alpha \subset X \backslash \bigcup_{\beta > \alpha} O_\beta$ for $\alpha < \kappa$. Then

$$\overline{W}_\alpha \subset X \backslash \bigcup_{\beta > \alpha} O_\beta \subset V_\alpha.$$

Each $\overline{W}_\alpha$ is covered by $\{U_\beta : \beta < \alpha\}$, so there is a locally finite open (in $\overline{W}_\alpha$) refinement $\mathscr{W}_\alpha$ of $\{U_\beta : \beta < \alpha\}$ covering $\overline{W}_\alpha$. Then $\mathscr{U}_\alpha = \{W \cap W_\alpha : W \in \mathscr{W}_\alpha\}$ is a locally finite open (in X) cover of W_α so $\mathscr{U} = \bigcup_{\alpha < \kappa} \mathscr{U}_\alpha$ is a locally finite open refinement of $\{U_\alpha : \alpha < \kappa\}$. $\square$

THEOREM 1.6: Let $\kappa + 1$ have the order topology. Then the product $X \times (\kappa + 1)$ is normal iff X is normal and κ-paracompact.

Proof: One direction is Theorem 1.4. For the other direction we use Lemma 1.5. Fix an increasing open cover $\{U_\alpha : \alpha < \kappa\}$ of X. Let $H = X \times \{\kappa\}$ and $K = \bigcup (X \backslash U_\alpha) \times \{\alpha\}$.

We first check that $H \cap K = \varnothing$: if $\langle x, \kappa\rangle \in H$, let $\beta < \kappa$ be such that $x \in U_\beta$. Then $U_\beta \times [(\kappa + 1) \backslash \beta]$ is a neighborhood of $\langle x, \kappa\rangle$ missing $\bigcup_{\alpha < \kappa}(X \backslash U_\alpha) \times \{\alpha\}$; hence, $\langle x, \kappa\rangle \notin K$.

Let $\{V \times O_V : V \in \mathscr{V}\}$ be an $\langle H, K\rangle$-separation. We show that $\mathscr{V}$ refines $\{U_\alpha : \alpha < \kappa\}$. Fix $V \in \mathscr{V}$ and let $\alpha < \kappa$ be such that $(\kappa + 1) \backslash \alpha \subset O_V$, in particular $\alpha \in O_V$. Since $K \cap (V \times O_V) = \varnothing$, we have that $(V \times O_V) \cap [(X \backslash U_\alpha) \times \{\alpha\}] = \varnothing$, so $V \subset U_\alpha$, since $\alpha \in O_V$.

Recall that $\mathscr{V}$ is a locally finite open cover of X, so we are done. $\square$

COROLLARY 1.7: A space X is normal and κ-paracompact iff $X \times 2^\kappa$ is normal iff $X \times I^\kappa$ is normal iff for every compact C with $w(C) \leq \kappa$, $X \times C$ is normal.

THEOREM 1.8: Assume that C is a compactification of X. Then X is paracompact iff $X \times C$ is normal.

Proof: One direction is Theorem 1.4. For the other direction we use Rudin Lemma.

Let $\mathscr{U}$ be an open cover of X, and let $\mathscr{V}$ be a "swelling" of $\mathscr{U}$ to C, that is, let $\mathscr{V} = \{V_U : U \in \mathscr{U}\}$ be a family of open sets in C such that $V_U \cap X = U$ for every $U \in \mathscr{U}$.

Let $H = \{\langle x, y \rangle \in X \times C : x = y\}$ and $K = X \times (C \backslash \cup \mathscr{V})$. Since $X \times C$ is normal, fix an $\langle H, K \rangle$-separation $\{W \times O_W : W \in \mathscr{W}\}$. We show that each $W \in \mathscr{W}$ is covered by finitely many members of $\mathscr{U}$, thus showing that $\mathscr{U}$ has a locally finite open refinement.

Fix $W \in \mathscr{W}$. Since $\{\langle x, x \rangle : x \in W\} \subset H \cap (W \times C) \subset W \times O_W$ we have $W \subset O_W$. Since $K \cap (\overline{W} \times O_W) = \varnothing$ we have $\mathrm{cl}_C W \subset \mathrm{cl}_C O_W \subset \cup \mathscr{V}$. By the compactness of $\mathrm{cl}_C W$ it follows that $\mathrm{cl}_C W$ is covered by finitely many members of $\mathscr{V}$; hence W is covered by finitely many members of $\mathscr{U}$. $\square$

COROLLARY 1.9: A space X is paracompact iff $X \times \beta X$ is normal.

The following is a trivial consequence of the fact that X is a perfect image of the product $X \times C$ when C is compact.

THEOREM 1.10: If X is κ-paracompact and C compact, the product $X \times C$ is κ-paracompact.

In 1.6, 1.7, and 1.8 the compact factor was of a special type. But even if we don't assume anything (special) about the compact factor, the normality of the product gives us some information on the noncompact factor, as the following two theorems show.

THEOREM 1.11: If C is a nondiscrete compact space and $X \times C$ normal, then X is countably paracompact.

Proof: Let $D = \{c_n : n < \omega\}$ be an infinite (relatively) discrete subset of C and let $\{F_n : n < \omega\}$ be a decreasing family of closed sets in X with $\cap_{n < \omega} F_n = \varnothing$.

Let $H = X \times Bd(D)$ and $K = \cup_{n < \omega} (F_n \times \{c_n\})$. Any $\langle H, K \rangle$-separation can be used to get an open family $\{U_n : n < \omega\}$ in X with $\cap_{n < \omega} U_n = \varnothing$ and $F_n \subset U_n$ for each n. $\square$

Another proof of Theorem 1.9 is to map C onto $\omega + 1$ and use Theorem 1.6.

We now look at the connection between normality and collectionwise normality.

THEOREM 1.12: Assume that C is compact and $X \times C$ normal. Then X is $w(C)$-collectionwise normal.

Proof: Let $\kappa = w(C)$, and let $\{F_\alpha : \alpha < \kappa\}$ be a discrete family of closed sets in X. We would like to construct families $\{H_\alpha : \alpha < \kappa\}$ and $\{K_\alpha : \alpha < \kappa\}$ of closed sets in C with $H_\alpha \cap K_\alpha = \varnothing$ for each α, and $(H_\alpha \cap K_\beta \neq \varnothing$ or $H_\beta \cap K_\alpha \neq \varnothing$ for $\alpha \neq \beta$). Then let $H = \cup_\alpha (F_\alpha \times H_\alpha)$ and $K = \cup_\alpha (F_\alpha \times K_\alpha)$, and any $\langle H, K \rangle$-separation will do the trick. These families are easy to get if C is 0-dimensional, if not we have to do this thing twice:

LEMMA 1.13: There are families $\{H_\alpha^i : \alpha < \kappa \wedge i < 2\}$ and $\{K_\alpha^i : \alpha < \kappa \wedge i < 2\}$ of closed subsets of C such that for each i and α, $H_\alpha^i \cap K_\alpha^i = \varnothing$ and if $\alpha \neq \beta$, then either $(H_\alpha^0 \cap K_\beta^0 \neq \varnothing$ or $H_\beta^0 \cap K_\alpha^0 \neq \varnothing)$ or $(H_\alpha^1 \cap K_\beta^1 \neq \varnothing$ or $H_\beta^1 \cap K_\alpha^1 \neq \varnothing)$.

This finishes the proof, since if $H^i = \bigcup_{\alpha < \kappa}(F_\alpha \times H_\alpha^i)$ and $K^i = \bigcup_{\alpha < \kappa}(F_\alpha \times K_\alpha^i)$, and U_α^i the union of the first factors of an $\langle H^i, K^i \rangle$-separation that intersect F_α, we have that $\{U_\alpha^0 \cap U_\alpha^1 : \alpha < \kappa\}$ is locally finite. $\square$

Proof of Lemma 1.13: This holds for any regular space C with $w(C) = \kappa$: construct them by induction on $\alpha < \kappa$. We'll have

$$H_\alpha^0 = \overline{V}_\alpha^0, \qquad K_\alpha^0 = C \backslash V_\alpha^1, \qquad H_\alpha^1 = \overline{V}_\alpha^1, \qquad K_\alpha^1 = C \backslash V_\alpha^2,$$

where $\overline{V}_\alpha^0 \subset V_\alpha^1 \subset \overline{V}_\alpha^1 \subset V_\alpha^2$ and the V_α^i are open and nonempty.

At stage $\alpha < \kappa$ we can find an $x \in C$ and open sets V_α^i for $i < 3$ such that $x \in V_\alpha^0 \subset \overline{V}_\alpha^0 \subset V_\alpha^1 \subset \overline{V}_\alpha^1 \subset V_\alpha^2$ and there is no V_β^i for $\beta < \alpha$ with $x \in V_\beta^i \subset V_\alpha^2$. This will do. $\square$

After seeing the preceding result one is tempted to try to improve Theorem 1.11 to have $X \, w(C)$-paracompact. This cannot be done, however, as Theorem 1.15 shows.

For a cardinal κ let $A(\kappa)$ denote the one-point compactification of the *discrete* topology on κ.

LEMMA 1.14. A normal space X is κ-collectionwise normal and countably paracompact iff for every locally finite family $\{F_\alpha : \alpha < \kappa\}$ of closed subsets of X there is a locally finite family $\{U_\alpha : \alpha < \kappa\}$ of open subsets such that for every $\alpha < \kappa$, $F_\alpha \subset U_\alpha$.

Proof: If there is an $n \in \omega$ such that no point of X is in no more than n sets F_α, then an easy induction on n and the collectionwise normality of X show the existence of the U_α.

For the general case, for $n \in \omega$ let V_n be the set of all points of X that belong to at most n of the F_α. Since $\{V_n : n < \omega\}$ is an open cover of a normal, countably paracompact space X, there is a locally finite shrinking $\{W_n : n < \omega\}$ of it, by Lemma 1.3.

Fix $n \in \omega$. The argument of the first paragraph applied to the family $\{F_\alpha \cap \overline{W}_n : \alpha < \kappa\}$ gives us a locally finite, open family $\{U_{\alpha,n} : \alpha < \kappa\}$. The sets $U_\alpha = \bigcup_{n < \omega} U_{\alpha,n}$ are as required. $\square$

THEOREM 1.15. A space X is κ-collectionwise normal and countably paracompact iff $X \times A(\kappa)$ is normal.

Proof: The direction from right to left follows from Theorems 1.11 and 1.12, and it is alsy easy to prove directly.

Going from left to right, assume that X is κ-collectionwise normal and countably paracompact. Indentify $A(\kappa)$ with $\kappa + 1$ where the point κ is the unique nonisolated point. Observe also that Lemma 1.14 holds. Let $H, K \subset X \times A(\kappa)$ be two closed disjoint sets. For $\alpha \leq \kappa$ let H_α and K_α be such that $H \cap [X \times \{\alpha\}] = H_\alpha \times \{\alpha\}$ and $K \cap [X \times \{\alpha\}] = K_\alpha \times \{\alpha\}$, respectively. Fix an open $U_\kappa \subset X$ with $K_\kappa \subset U_\kappa$ and $U_\kappa \cap \overline{H}_\kappa = \varnothing$. Then the family $\{K_\alpha \backslash U_\alpha : \alpha < \kappa\}$ is locally finite in X. Fix locally finite $\{U_\alpha^0 : \alpha < \kappa\}$ with $K_\alpha \backslash U_\kappa \subset U_\alpha^0$ and $\overline{U}_\alpha^0 \cap H_\alpha = \varnothing$. Let $U_\alpha^1 \subset U_\kappa$ be open with $K_\alpha \backslash U_\alpha^0$

$\subset U^1_\alpha$ and $\overline{U}^1_\alpha \cap H_\alpha = \varnothing$. Let $U_\alpha = U^0_\alpha \cup U^1_\alpha$, and $U = \bigcup \{U_\alpha \times \{\alpha\} : \alpha \leq \kappa\}$. Then $\overline{U} = \overline{\bigcup \{U_\alpha \times \{\alpha\} : \alpha \leq \kappa\}} = \bigcup \{\overline{U}_\alpha \times \{\alpha\} : \alpha \leq \kappa\}$, so $\overline{U} \cap H = \varnothing$ and $K \subset U$. Done. $\square$

The next fact is easy but not as trivial as Theorem 1.10.

THEOREM 1.16: Let X be κ-collectionwise normal and C compact. If the product $X \times C$ is normal, it is κ-collectionwise normal.

Proof: There is nothing to prove if C is discrete. If C is nondiscrete, X is countably paracompact, so the theorem follows from the following easy consequence of Lemma 1.14.

LEMMA 1.17. Assume that a κ-collectionwise normal, countably paracompact space X is a perfect image of a normal space Y. Then Y is κ-collectionwise normal. $\square$

The following summarizes the most important results of this section.

THEOREM 1.18: Let κ be an infinite cardinal.

(1) X is normal and κ-paracompact iff for all compact C with $w(C) = \kappa$, $X \times C$ is normal iff $X \times 2^\kappa$ is normal.
(2) X is countably paracompact and κ-collectionwise normal iff there exists a compact C with $w(C) = \kappa$ such that $X \times C$ is normal iff $X \times A(\kappa)$ is normal.

REMARKS 1.19: The terminology in Definition 1.1 was suggested by Eric van Douwen.

Rudin proved Theorem 1.10 and (a variation) of 1.2 in [53], en route to proving the following conjecture of Morita [41]:

If C is compact, $X \times C$ normal and Y a closed image of X, then $Y \times C$ is normal.

The present proofs of 1.2 and 1.12 are taken from [11], where a simpler proof of Morita's conjecture is given. Almost the same version of Rudin Lemma, with the same proof, appeared in Starbird [64, theorem 3.3] where he also gives another simple proof of Theorem 1.12. The proof of Rudin Lemma is standard (see [22, 5.1.38]).

Lemma 1.5 and Theorem 1.6 are due to Ken Kunen, 1.7 is from [41], and 1.8 is Tamano's theorem [67]. Theorem 1.9 is from [20].

Theorem 1.11 is from [20], and Theorem 1.15 is due to [1]. Lemma 1.14 is due to Dowker [21] and Katetov [32]. Theorem 1.16 is from [65], the present proof is from [8].

Gruenhage, Nogurv, and Purisch [25] give an internal characterization of spaces X with $X \times \omega_1$ normal.

2. PRODUCTS WITH ONE METRIC FACTOR

In this section we assume that *all spaces are nondiscrete*. This, of course, is a natural assumption if one looks at finite products, and it enables us to state the main theorem (2.1) of this section in an elegant way. This result is from Rudin and Starbird [61].

THEOREM 2.1: Assume that X is normal and M metric. The product $X \times M$ is normal iff it is countably paracompact.

Proof: Let $\mathscr{B} = \bigcup_{n < \omega} \mathscr{B}_n$ be a basis for M such that each $\mathscr{B}_n$ is a locally finite family of open sets of diameter $< 1/2^n$ and every discrete member of $\mathscr{B}$ contains exactly one point. By induction we can pick points $p_B \neq q_B \in B$, for each nondiscrete $B \in \mathscr{B}$ so that no point in M is picked twice.

Let $\pi : X \times M \to X$ be the projection map.

Assume first that $X \times M$ is normal and show that it is countably paracompact. Let $\{U_m : m < \omega\}$ be an open cover of $X \times M$. We want to construct an open cover $\{V_{m,n} : m, n < \omega\}$ such that $\overline{V}_{m,n} \subset U_m$ for each m, n. This will show that $X \times M$ is countably paracompact.

For $B \in \mathscr{B}$ and $m < \omega$ let $U_{B,m}$ be the largest open subset of X such that $U_{B,m} \times \overline{B} \subset U_m$, and let $F_B = X \backslash \bigcup_{m < \omega} U_{B,m}$. Observe that each F_B is closed in X and that $F_B = \varnothing$ if B is discrete. Define $H_B = F_B \times \{p_B\}$ and $K_B = F_B \times \{q_B\}$ if B is nondiscrete; otherwise, let $H_B = K_B = \varnothing$. For $n < \omega$, let $H_n = \bigcup\{H_B : B \in \mathscr{B}_n\}$ and $K_n = \bigcup\{K_B : B \in \mathscr{B}_n\}$. Since each H_n and K_n is the union of a locally finite family of closed sets, both of them are closed.

Finally, define $H = \bigcup_{n < \omega} H_n$ and $K = \bigcup_{n < \omega} K_n$ and observe that $H \cap K = \varnothing$ since the points p_B, q_B are distinct. We now show that H and K are closed.

Fix a point $p = \langle x, y \rangle \in X \times M$ and find a $k < \omega$ with $p \notin \bigcup_{n \geq k} H_n$. This shows that H is closed, and then, by symmetry, K is closed. There is an $m < \omega$ with $p = \langle x, y \rangle \in U_m$, so let $A \times B$ be open with $p \in A \times B \subset \overline{A \times B} \subset U_m$. Since the diameter of each member of $\mathscr{B}_n$ is $< 1/2^n$, there is a $k < \omega$ and an open $C \subset M$ containing y such that for every $B' \in \bigcup_{n \geq k} \mathscr{B}_n$ if $B' \cap C \neq \varnothing$, then $B' \subset B$. Then $(A \times C) \cap (\bigcup_{n \geq k} H_n) = \varnothing$.

Since $X \times M$ is normal, let I and J be open subsets of $X \times M$ containing H and K, respectively, with $\overline{I} \cap \overline{J} = \varnothing$. For $B \in \mathscr{B}$ let $U_B = \pi[I \cap (X \times B)] \cap \pi[J \cap (X \times B)]$, so U_B is open and $F_B \subset U_B$. Hence for each $B \in \mathscr{B}$, the family $\{U_{B,m} : m < \omega\} \cup \{U_B\}$ is an open cover of X. Since M is nondiscrete X is countably paracompact, so fix an open cover $\{V_{B,m} : m < \omega\} \cup \{V_B\}$ of X with $\overline{V}_{B,m} \subset U_m$ and $\overline{V}_B \subset U_B$.

For $m, n < \omega$ let $V_{m,n} = \bigcup\{V_{B,m} \times B : B \in \mathscr{B}_n\}$. The local finiteness of $\mathscr{B}_n$ implies $\overline{V}_{m,n} = \bigcup\{\overline{V}_{B,m} \times \overline{B} : B \in \mathscr{B}_n\} \subset U_m$. It remains to show that $\{V_{m,n} : m, n < \omega\}$ covers $X \times M$. Let $\langle x, y \rangle \in X \times M$ be arbitrary. Since $\overline{I} \cap \overline{J} = \varnothing$, we may assume that $\langle x, y \rangle \notin \overline{I}$. Fix an open neighborhood $A \times B$ of $\langle x, y \rangle$ with $B \in \mathscr{B}$ and $(A \times B) \cap \overline{I} = \varnothing$. But then $\pi[I \cap (X \times B)] \cap A = \varnothing$, so $x \notin U_B$ and there is an $m < \omega$ with $x \in V_{B,m}$. Hence $\langle x, y \rangle \in V_{B,m} \times B$, so $\{V_{m,n} : m, n < \omega\}$ covers $X \times M$; thus $X \times M$ is countably paracompact.

Now we assume that $X \times M$ is countably paracompact and show that it is normal. Let H and K by two disjoint closed subsets of $X \times M$.

For $n < \omega$ define

$$F_n = \bigcup\{(\overline{\pi[H \cap (X \times \overline{B})]} \cap \overline{\pi[K \cap (X \times \overline{B})]}) \times \overline{B} : B \in \mathscr{B}_n\}.$$

Note that $\{F_n : n < \omega\}$ is a decreasing closed family with empty intersection. So fix an open sequence $\{O_n : n < \omega\}$ such that each $F_n \subset O_n$ and $\bigcap_{n<\omega} O_n = \varnothing$.

Fix $n < \omega$ and $B \in \mathscr{B}_n$. For $C \in \mathscr{B}$ with $C \subset B$ the sets $A = \pi[H \cap (X \times \overline{C})] \setminus \pi[O_n \cap (X \times \overline{C})]$ and $\pi[K \cap (X \times \overline{C})]$ are closed and disjoint so there is an open $U_{B,C}$ in X containing A and with $\overline{U}_{B,C} \cap \pi[K \cap (X \times \overline{C})] = \varnothing$.

The family $\{U_{B,C} \times C : n, m < \omega \wedge B \in \mathscr{B}_n \wedge C \in \mathscr{B}_m \wedge C \subset B\}$ is a σ-locally finite open family covering H and with the property that $(\overline{U}_{B,C} \times C) \cap K = \varnothing$ for each B, C.

By symmetry there is such a family covering K, so the product is normal. $\square$

COROLLARY 2.2: Let M be metric and C compact. If $X \times M$ and $X \times C$ are normal, so is $X \times M \times C$.

Proof: By Theorem 2.1 $X \times M$ is countably paracompact, then the product $X \times M \times C = (X \times C) \times M$ is countably paracompact, and since $X \times C$ is normal, Theorem 2.1 implies that $(X \times C) \times M$ is normal. $\square$

COROLLARY 2.3: Let X be paracompact (collectionwise normal), M metric, and $X \times M$ normal. Then $X \times M$ is paracompact (collectionwise normal).

Proof: For paracompactness, let $C = \beta(X \times M)$. Corollary 2.2 implies that $X \times M \times C$ is normal, so by Theorem 1.8, $X \times M$ is paracompact.

For collectionwise normality use Theorem 1.15: let $C = A(\kappa)$. $\square$

COROLLARY 2.4: Let X be Lindelöf, M separable metric, and $X \times M$ normal. Then $X \times M$ is Lindelöf.

Proof: Let D be a countable dense subset of M. Then $X \times D$ is a dense Lindelöf subspace of $X \times M$. Since $X \times M$ is paracompact by Corollary 2.3, this implies that it is Lindelöf. $\square$

Note: Corollaries 2.2–2.4 are true even if some of the factors are discrete.

The preceding results raise the following questions:

If the product $X \times Y$ is normal and both factors X and Y satisfy a (paracompactnesslike) property ϕ, does the product satisfy ϕ?

More specifically, can we have a Dowker product $X \times Y$ without either X or Y being Dowker?

These questions are considered in the next section.

Another natural question involves Corollary 2.2; can we relax the conditions on the factors? Maybe we need a restriction on one factor only, or Corollary 2.2 holds if M and C are assumed to both be metric or both compact.

The one-factor restriction is not true even if we assume that the restricted factor is compact metric. The answer in the next section to the Dowker product question just cited gives a counterexample.

We don't know what happens if both factors are compact, but if both factors are metric, Lawrence [37] constructed a counterexample that we now present.

DEFINITION 2.5: For $Y \subset Z \times Z$, the *2-cardinality* of Y, $|Y|_2$, is defined as $|Y|_2 = \min \{|B| : Y \subset (B \times Z) \cup (Z \times B)\}$.

LEMMA 2.6: Assume that $Y \subset \mathbb{R} \times \mathbb{R}$ is closed. Then either $|Y|_2 \leq \omega$ or $|Y|_2 = c$.

Proof: This is like the proof of the Cantor theorem. Assume that $|Y|_2 > \omega$. Without loss of generality we may assume that for every open $U \subset \mathbb{R} \times \mathbb{R}$, either $Y \cap U = \varnothing$ or $|Y \cap U|_2 > \omega$.

Now construct a Cantor tree $T = \{A_s \times B_s : s \in 2^{<\omega}\}$ such that for every $s \in 2^{<\omega}$:

(i) $A_s \times B_s$ is open in $\mathbb{R} \times \mathbb{R}$ with diameter $< 1/2^{|s|}$;
(ii) $Y \cap (A_s \times B_s) \neq \varnothing$;
(iii) If $i = 0, 1, \overline{A_{s^\frown\langle i \rangle}} \times \overline{B_{s^\frown\langle i \rangle}} \subset A_s \times B_s$;
(iv) $\overline{A_{s^\frown\langle 0 \rangle}} \cap \overline{A_{s^\frown\langle 1 \rangle}} = \varnothing$ and $\overline{B_{s^\frown\langle 0 \rangle}} \cap \overline{B_{s^\frown\langle 1 \rangle}} = \varnothing$.

(To get (iv) use that $|Y \cap U|_2 > \omega$ if U is open and $Y \cap U \neq \varnothing$.)

If $f \in 2^\omega$, then $\bigcap_{n<\omega} A_{f|n} \times B_{f|n}$ is a one-point subset of Y, and (iv) shows that $|\bigcup\{\bigcap_{n<\omega} A_{f|n} \times B_{f|n} : f \in 2^\omega\}|_2 = C$. $\square$

If $p = \langle p_i : i < m \rangle \in \mathbb{R}^m$, ran $(p) = \{p_i : i < m\}$.
Let $\pi : \mathbb{R} \times \mathbb{R} \times \mathbb{R} \to \mathbb{R} \times \mathbb{R}$ denote the projection map onto the first two coordinates.

THEOREM 2.7: There is a Lindelöf space X and a separable metric M such that $X \times M$ is Lindelöf, but $X \times M \times M$ is not normal.

This follows from the following.

LEMMA 2.8: There is an uncountable, compact $D \subset \mathbb{R} \times \mathbb{R} \times \mathbb{R}$ and an $A \subset \mathbb{R}$ such that

(i) For every closed $Y \subset \mathbb{R} \times \mathbb{R}$ with $|Y|_2 = c$, $|(A \times A) \cap Y| = c$;
(ii) $(A \times A \times A) \cap D = \varnothing$;
(iii) For every $p \neq q \in D$ (ran $(p) \cap$ ran $(q) = \varnothing$).

Proof of Theorem 2.7: Let A and D be as in Lemma 2.8. Note that A satisfies also

(i′) For every closed $Y \subset \mathbb{R}$ with $|Y| = c$, $|A \cap Y| = c$.

To see this apply (i) to $(Y \times Y) \cap \Delta$, where $\Delta = \{\langle x, y \rangle : x = y\}$.
Let $X = \mathbb{R}$ with points in A having the usual topology and points in $\mathbb{R}\backslash A$ isolated. Let $M = A$.

By induction on $m = 0, 1, 2$, we show that X^m is Lindelöf, hence $X \times M$, being a closed subset of $X \times X$, is.

Since $|X^0| = 1$, X^0 is Lindelöf. Assume $m = 1, 2$, and X^{m-1} Lindelöf, and show that X^m is Lindelöf.

Let $\mathscr{U}$ be an open cover of X^m and $\mathscr{V}$ a countably subfamily of $\mathscr{U}$ with $A^m \subset \bigcup \mathscr{V}$. So $Y = X^m\backslash\bigcup\mathscr{V}$ is closed in $\mathbb{R}^m$ and disjoint from A^m, by (i) or (i′), $|Y|_2 < c$, that is, $|Y|_2 \leq \omega$ (Lemma 2.6) if $m = 2$ or $|Y| < c$, that is, $|Y| \leq \omega$ if $m = 1$. But then Y is Lindelöf since X^{m-1} is.

Since D is closed in $\mathbb{R}^3$, it is closed in X^3. If $d = \langle d_0, d_1, d_2\rangle \in D$, let $d_i \notin A$, so $\{d_i\}$ is open in X, hence if $D_j = X$ for $j \neq i$, $D_i = \{d_i\}$, $D \cap (D_0 \times D_1 \times D_2) = \{d\}$ and $D_0 \times D_1 \times D_2$ is open in X^3, so D is closed discrete in X^3.

Observe that $|\pi(D)|_2 = c$, so $|(A \times A) \cap \pi(D)| = c$, since $\pi(D)$ is closed (compact) in $\mathbb{R} \times \mathbb{R}$.

Let $E = \{d \in D : \pi(d) \in A \times A\}$. Then E is an uncountable, closed discrete subset of $A \times A \times X = M \times M \times X$, hence $X \times M \times M$ is not Lindelöf, and by Corollary 2.4 not normal. ☐

Proof of Lemma 2.8: Let $C \subset (0, 1)$ be a Cantor set and let $D = \{\langle x, x + 1, x + 2 \rangle : x \in C\}$.

List all closed $Y \subset \mathbb{R} \times \mathbb{R}$ with $|Y|_2 = c$ as $\langle Y_\alpha : \alpha < c \rangle$, each Y_α listed c times.

For $\xi < c$ pick $a_\xi \in Y_\xi$ so that if $A_\alpha = \bigcup \{\text{ran } (a_\xi) : \xi < \alpha\}$, then $(A_\alpha \times A_\alpha \times A_\alpha) \cap D = \varnothing$.

At stage α we pick $a_\alpha \in Y_\alpha$ as follows: The set $S = \bigcup \{\text{ran } (x) : x \in \mathbb{R} \times \mathbb{R} \text{ and } (A_\alpha \cup \text{ran } (x))^3 \cap D \neq \varnothing\}$ has size $\leq |A_\alpha| + \omega < c$ (because if $p \in (A_\alpha \cup \text{ran } (x))^3 \cap D$, then ran $(p) \cap A_\alpha \neq \varnothing$), so since $|Y_\alpha|_2 = c$, find $a_\alpha \in Y_\alpha$ with ran $(a_\alpha) \cap S = \varnothing$.

Finally let $A = \bigcup_{\alpha < c} A_\alpha$. ☐

Now we digress a little and present Bill Fleissner's writeup of Alster's theorem [6] that MA (Martin's Axiom) implies that there is Lindelöf X such that $X \times \omega^\omega$ is not normal. Michael [39] used the continuum hypothesis (CH) to get a first such example. No results are known in ZFC. See also Lawrence [36].

For $f, g \in \omega^\omega$ define $f \leq g$ iff $\forall n (f(n) \leq g(n))$, and $f \leq {}^* g$ iff $\exists m \forall n \geq m (f(n) \leq g(n))$.

DEFINITION 2.9: A sequence $\langle f_\alpha : \alpha < \kappa \rangle$ is a *strong scale* in ω^ω iff $\forall \alpha < \beta < \kappa$ $(f_\alpha \leq {}^* f_\beta)$ and $\forall f \in \omega^\omega \exists \alpha < \kappa (f \leq f_\alpha)$.

THEOREM 2.10: Assume MA. Then there is a Lindelöf space X such that $X \times \omega^\omega$ is not Lindelöf (hence not normal).

Proof: Let π denote the product topology on $(\omega + 1)^\omega$, where $\omega + 1$ has the order topology, and let $\langle f_\alpha : \alpha < c \rangle$ be a strong scale on ω^ω.

For $\alpha < c$, let $X_\alpha = \{f \in \omega^\omega : f \leq f_\alpha\}$ and let $X = X_c = (\omega + 1)^\omega$. Observe that each X_α is compact in the product topology, and that for $\alpha < \beta, X_\alpha \cap X_\beta$ is open in X_α.

The topology on X is generated by

$$\{X_\alpha \cap U : \alpha \leq c \wedge U \in \pi\}$$

as a base. (The preceding observation implies that this indeed is a base.)

Note that for each α and $x \in X_\alpha \backslash \bigcup_{\beta < \alpha} X_\beta$, the family

$$\{X_\alpha \cap U : x \in U \wedge U \text{ is } \pi\text{-clopen}\}$$

is a base for x in X. In particular, for $\alpha \leq c$, the π-subspace topology and the X-subspace topology on $X_\alpha \backslash \bigcup_{\beta < \alpha} X_\beta$ coincide.

$1°$ *X is regular*

Since X is Hausdorff, it is enough to show that X is 0-dimensional. Fix an $x \in X$ and let α be the least with $x \in X_\alpha$, then $\{X_\alpha \cap U : x \in U \wedge U \text{ is } \pi\text{-clopen}\}$ is a clopen base for x. ☐

$2°$ *X is Lindelöf*

Let $\mathcal{V}$ be an open cover of X. By induction on $\alpha \leq c$ we show that there is a countable subfamily of $\mathscr{V}$ covering X_α. So let α be the least ordinal for which this fails.

For $x \in X_\alpha \backslash \bigcup_{\beta < \alpha} X_\beta$ fix a π-open U_x and $V_x \in \mathscr{V}$ with $x \in U_x \cap X_\alpha \subset V_x$. Since $X_\alpha \backslash \bigcup_{\beta > \alpha} X_\beta$ is separable metric, there is a countable family $\mathscr{U} \subset \{U_x : x \in X_\alpha \backslash \bigcup_{\beta < \alpha} X_\beta\}$ covering $X_\alpha \backslash \bigcup_{\beta < \alpha} X_\beta$. So $X_\alpha \backslash \bigcup \{V_x : U_x \in \mathscr{U}\} \subset X_\alpha \backslash \bigcup \mathscr{U}$ and $X_\alpha \backslash \bigcup \mathscr{U} \subset \bigcup_{\beta < \alpha} X_\beta$ is π-compact.

We show that $X_\alpha \backslash \bigcup \mathscr{U}$ is covered by countably many X_β, $\beta < \alpha$, thus finishing the proof.

Case 1: $\alpha = c$. Since $X_c \backslash \bigcup \mathscr{U} \subset \bigcup_{\beta < \alpha} X_\beta = \omega^\omega$ is π-compact and $\langle f_\beta : \beta < c \rangle$ is a strong scale, there is a $\beta < c$ with $X_c \backslash \bigcup \mathscr{U} \subset X_\beta$.

Case 2: $\alpha < c$. The following lemma finishes the proof.

LEMMA 2.11: Assume *MA*. Let M be a compact metric space and $\mathscr{K}$ a closed cover of M of cardinality less than c. Then $\mathscr{K}$ has a countable subcover.

Proof: Define

$$F_0 = M$$

$$F_{\xi+1} = F_\xi \backslash \bigcup \{\text{int}_{F_\xi} K : K \in \mathscr{K}\}$$

$$F_\xi = \bigcap_{\eta < \xi} F_\eta \text{ for } \xi \text{ a limit.}$$

Note that since $|\mathscr{K}| < c$, MA implies that if $F_\xi \neq \varnothing$, then $F_{\xi+1} \subsetneq F_\xi$. Hence there is a $\xi < \omega_1$ with $F_\xi = \varnothing$. Then

$$M = \bigcup \{\text{int}_{F_\eta} K : \eta < \xi \wedge K \in \mathscr{K}\}$$

and for each $\eta < \xi$ there is a countably subfamily of $\{\text{int}_{F_\eta} K : K \in \mathscr{K}\}$ covering $\bigcup \{\text{int}_{F_\eta} K : K \in \mathscr{K}\}$ since M is hereditarily Lindelöf.   □

3° $X \times \omega^\omega$ is not Lindelöf

The diagonal $\Delta = \{\langle x, y \rangle \in X \times \omega^\omega : x = y\}$ is a closed subset of $X \times \omega^\omega$, and is homeomorphic to $\bigcup_{\alpha < c} X_\alpha$ (with the topology inherited from X). But $\bigcup_{\alpha < c} X_\alpha$ is not Lindelöf since the cover $\{X_\alpha : \alpha < c\}$ has no countable subcover.   □

It is also easy to show directly that $X \times \omega^\omega$ is not normal. The sets $(X \backslash \omega^\omega) \times \omega^\omega$ and Δ cannot be separated:

Let U be an open set containing Δ, and let D be a countable dense subset of ω^ω (in the product topology). For $p \in D$ let $A_p = \{x \in \omega^\omega : \langle x, p \rangle \in U\}$. Since $\omega^\omega = \bigcup_{p \in D} A_p$, one of the A_p is somewhere dense. Fix a p with such an A_p and let $x \in X \backslash \omega^\omega$ be such that $x \in \text{cl}_X A_p$. Then $\langle x, p \rangle \in \text{cl} U$.

REMARKS 2.12: Rudin and Starbird [61] proved 2.1–2.4. Morita showed [28] one-half of Theorem 2.1, that countable paracompactness implies normality.

In [61], Rudin and Starbird show that the analogue of Morita's conjecture from 1.12 holds if the compact factor C is replaced by a metric factor M.

Corollary 2.4 holds (easily) if M is replaced by a compact C.

Call a space shrinking if every open cover has a shrinking (see the definition between 1.2 and 1.3). The same proof as for 2.1 shows that if X is shrinking, M metric, and $X \times M$ normal, then $X \times M$ is shrinking [8], but it is not known if M can be replaced by a compact C. The strongest result known, due to Lazarevic [38], is:

Let X be a product of cardinals and C compact. If $X \times C$ is normal, then $X \times C$ is shrinking.

A related problem is: Is a normal, perfect preimage of a shrinking space, shrinking?

Lemma 2.6 was first proved by van Douwen [18]. Lawrence [37] proves more than is stated in 2.7: For every $n \leq \omega$ there is a Lindelöf X and a separable metric M such that for every $m < n$, $X \times M^m$ is Lindelöf but $X \times M^n$ is not normal. He also shows that no such M can be complete. The Bernstein set idea is from Michael [40].

Lawrence shows [36] that if $b = \omega_1$, then there is a Lindelöf X such that $X \times \omega^\omega$ is not normal, and Alster and Gruenhage [7] show that if $b > \omega_1$ and X is a Lindelöf non-Archimedean space, then $X \times \omega^\omega$ is Lindelöf.

3. EXAMPLES

Here we present several examples involving products and normalitylike properties. We start with a Kunen line technique, a CH construction, then we do an Eric line, a ZFC construction, and at the end we construct a de Caux-type Dowker space, a $\diamond$ construction. In all these constructions we use an Ostaszewski technique [44].

THEOREM 3.1: Assume CH. There is a non-Lindelöf, 0-dimensional, first countable, locally compact, hereditarily separable, and perfectly normal space K.

Proof: Let $\mathbb{R} = \{x_\alpha : \alpha < \omega_1\}$, and let ρ be the Euclidean topology on $\mathbb{R}$. For $\alpha < \omega_1$, let $\mathbb{R}_\alpha = \{x_\beta : \beta < \alpha\}$ and let ρ_α be the topology on $\mathbb{R}_\alpha$ inherited from ρ. We use an Ostaszewski technique to refine ρ. We want to make each $\mathbb{R}_\alpha$ open, thus making the new topology non-Lindelöf; at the same time we want to change the Euclidean topology on $\mathbb{R} \times \mathbb{R}$ as little as possible. Let τ be the topology on $\mathbb{R}$ such that $K = \langle \mathbb{R}, \tau \rangle$ satisfies the theorem. Ideally, we would like to have:

$$\forall X \subset \mathbb{R} \ \forall x \in \mathbb{R}(x \in \mathrm{cl}_\rho X \to x \in \mathrm{cl}_\tau X).$$

But this, of course, implies that $\rho = \tau$. But we can settle for the next best thing, namely that the preceding formula holds eventually:

$$\forall X \subset \mathbb{R} \ \exists \alpha < \omega_1 \ \forall \beta > \alpha(x_\beta \in \mathrm{cl}_\rho X \to x_\beta \in \mathrm{cl}_\tau X). \tag{*}$$

Now, if τ is a 0-dimensional first countable, locally compact, Hausdorff topology on $\mathbb{R}$ that is finer than the Euclidean topology ρ, and such that each $\mathbb{R}_\alpha$ is open in τ and (*) holds, then τ satisfies the theorem.

To see this we need a couple of lemmas:

LEMMA 3.2: Let ρ be the Euclidean topology on $\mathbb{R}$, and let τ be a finer regular topology. Assume that for every two X, $Y \subset \mathbb{R}$ with $|\mathrm{cl}_\rho X \cap \mathrm{cl}_\rho Y| = c$ we have that $\mathrm{cl}_\tau X \cap \mathrm{cl}_\tau Y \neq \varnothing$.

Then $\langle \mathbb{R}, \tau \rangle$ is normal.

Proof: If $X, Y \subset \mathbb{R}$ are disjoint and τ-closed, then $|\mathrm{cl}_\rho X \cap \mathrm{cl}_\rho Y| < c$, hence is countable.

Let $A = \mathrm{cl}_\rho X \cap \mathrm{cl}_\rho Y$, so A is G_δ in $\langle \mathbb{R}, \rho \rangle$; hence, there are ρ-closed $X_n, n < \omega$, with

$$(\mathrm{cl}_\rho X)\backslash A = \bigcup_{n<\omega} X_n.$$

Since $\langle \mathbb{R}, \rho \rangle$ is normal, for each $n < \omega$, fix a ρ-open $U_n \supset X_n$ with $\mathrm{cl}_\rho U_n \cap \mathrm{cl}_\rho Y = \varnothing$. So each U_n is τ-open and $(\mathrm{cl}_\tau U_n) \cap Y = \varnothing$.

Also, τ is regular and $X \cap A$ countable (Lindelöf is enough) so we can cover $X \cap A$ with τ-open $U'_n, n < \omega$, that satisfy $(\mathrm{cl}_\tau U'_n) \cap Y = \varnothing$. So the family $\{U_n : n < \omega\} \cup \{U'_n : n < \omega\}$ covers X.

By symmetry, there are τ-open $V_n, V'_n, n < \omega$, covering Y with $(\mathrm{cl}_\tau V_n) \cap X = \varnothing = (\mathrm{cl}_\tau V'_n) \cap X$.

Hence X and Y can be separated by open sets, so $\langle \mathbb{R}, \tau \rangle$ is normal. $\square$

Observe that (*) implies the assumptions of Lemma 3.2, so our space $K = \langle \mathbb{R}, \tau \rangle$ is normal.

LEMMA 3.3: Assume that τ satisfies (*). If $X \subset \mathbb{R}$, then $|\mathrm{cl}_\rho X \backslash \mathrm{cl}_\tau X| \leq \omega$.

Proof: Let $X \subset \mathbb{R}$. By (*) there is an $\alpha < \omega_1$ such that for every $\beta > \alpha$, if $x_\beta \in \mathrm{cl}_\rho X$, then $x_\beta \in \mathrm{cl}_\tau X$. So $\mathrm{cl}_\rho X \backslash \mathrm{cl}_\tau X \subset \mathbb{R}_{\alpha+1}$, hence countable. $\square$

The preceding lemma implies that $\langle \mathbb{R}, \tau \rangle$ is hereditarily separable and perfect.

3.4 ASSUME THAT τ SATIFIES (*). THEN $\langle \mathbb{R}, \tau \rangle$ IS HEREDITARILY SEPARABLE: Let $X \subset \mathbb{R}$ and fix a countable $D \subset X$ with $X \subset \mathrm{cl}_\rho D$. Since $\mathrm{cl}_\rho D \backslash \mathrm{cl}_\tau D$ is countable, we have that $X \backslash \mathrm{cl}_\tau D$ is countable; hence, X is a separable subspace of $\langle \mathbb{R}, \tau \rangle$.

3.5 ASSUME THAT τ SATISFIES (*) AND REFINES ρ. THEN $\langle \mathbb{R}, \tau \rangle$ IS PERFECT: Let $X \subset \langle \mathbb{R}, \tau \rangle$ be closed. Then $X = \mathrm{cl}_\tau X \subset \mathrm{cl}_\rho X$, and $\mathrm{cl}_\rho X \backslash X$ is countable. So X is G_δ in $\langle \mathbb{R}, \rho \rangle$; hence, G_δ in $\langle \mathbb{R}, \tau \rangle$.

It remains to construct a 0-dimensional, first countable, locally compact, Hausdorff topology τ on $\mathbb{R}$ that refines ρ and satisfies (*). We do that by induction.

Using CH, list all countable subsets of $\mathbb{R}$ as $\langle A_\alpha : \alpha < \omega_1 \rangle$ so that for each $\alpha < \omega_1$, $A_\alpha \subset \mathbb{R}_\alpha$. (Recall the definitions of $\mathbb{R}_\alpha$ and ρ_α from the beginning of the proof.)

By induction on $\alpha < \omega_1$, we define a topology τ_α on $\mathbb{R}_\alpha$ so that for $\alpha < \omega_1$ the following holds:

1° For every $\beta < \alpha$, $\tau_\beta = \tau_\alpha \cap \mathscr{P}(\mathbb{R}_\beta)$;
2° τ_α is first countable, locally compact, 0-dimensional Hausdorff;
3° $\tau_\alpha \supset \rho_\alpha$;
4° For every $\beta \leq \alpha$ if $x_\alpha \in \mathrm{cl}_\rho A_\beta$, then $x_\alpha \in \mathrm{cl}_{\tau_{\alpha+1}} A_\beta$.

Note that 1° says that $\langle \mathbb{R}_\beta, \tau_\beta \rangle$ is an open subspace of $\langle \mathbb{R}_\alpha, \tau_\alpha \rangle$ for $\beta < \alpha$.
For $0 \leq \alpha < \omega$ let τ_α be discrete.
If $\alpha \leq \omega_1$ is a limit ordinal, let τ_α be the topology generated by $\bigcup_{\beta<\alpha} \tau_\beta$ as a basis. Let $\tau = \tau_{\omega_1}$, and $K = \langle \mathbb{R}, \tau \rangle$.

Note that 4° implies (*): if $X \subset \mathbb{R}$, let $\beta < \omega_1$ be such that $A_\beta \subset X$ is countable and $X \subset \mathrm{cl}_\rho A_\beta$. If $\alpha \geq \beta$ and $x_\alpha \in X$, we have $x_\alpha \in \mathrm{cl}_\rho A_\beta$ so $x_\alpha \in \mathrm{cl}_{\tau_{\alpha+1}} A_\beta$. But by 1°, $\langle \mathbb{R}_{\alpha+1}, \tau_{\alpha+1} \rangle$ is an (open) subspace of $\langle \mathbb{R}, \tau \rangle$, so $x_\alpha \in \mathrm{cl}_\tau A_\beta$. Hence (*) holds.

It remains to do the construction for the case $\alpha + 1$. So we are given τ_β for $\beta \leq \alpha$. We want to make sure that 4° holds, so list $\langle A_\beta : \beta \leq \alpha \wedge x_\alpha \in \mathrm{cl}_\rho A_\beta \rangle$ as $\langle A^n : n < \omega \rangle$, each A_β listed infinitely many times.

For $n < \omega$, pick $x^n \in A^n$ with ρ-distance from x^n to x_α less than $1/2^n$. Then pick a compact, open V_n in $\langle \mathbb{R}_\alpha, \tau_\alpha \rangle$ containing x^n and with ρ-diameter less than $1/2^n$. (This can be done because $\langle \mathbb{R}_\alpha, \tau_\alpha \rangle$ is locally compact, 0-dimensional, and τ_α refines the Euclidean topology ρ_α.)

Note that the sequence $\langle x^n : n < \omega \rangle$ converges to x_α in the Euclidean topology ρ, so the set $\{x^n : n < \omega\}$ is closed discrete in $\langle \mathbb{R}_\alpha, \rho_\alpha \rangle$, hence also in $\langle \mathbb{R}_\alpha, \tau_\alpha \rangle$. Since ρ-diameter of V_n is less than $1/2^n$, for each n the family $\{V_n : n < \omega\}$ is closed discrete in $\langle \mathbb{R}_\alpha, \tau_\alpha \rangle$. (One can also use the normality of $\langle \mathbb{R}_\alpha, \tau_\alpha \rangle$ to get such V_n.)

Let $B_m = \{x_\alpha\} \cup \bigcup_{n \geq m} V_n$, for $m < \omega$. Note that the ρ-diameter of B_m is less than $1/2^{m-1}$.

The topology $\tau_{\alpha+1}$ is generated by $\tau_\alpha \cup \{B_m : m < \omega\}$ as a basis.

Note that 1° holds trivially and that $\tau_{\alpha+1}$ is finer than $\rho_{\alpha+1}$ because of the observation about the diameter of B_m. Hence 3° holds, and therefore $\tau_{\alpha+1}$ is Hausdorff.

Each B_m is compact since any $\tau_{\alpha+1}$-open U containing x_α contains all but finitely many $V_{n,S}$, and each of the remaining $V_{n,S}$ is τ_α; hence, $\tau_{\alpha+1}$, compact.

For 4°, if $x_\alpha \in \mathrm{cl}_\rho A_\beta$ and $m < \omega$, we show $B_m \cap A_\beta \neq \varnothing$. Fix an $n \geq m$ with $A_\beta = A^n$. Then $x^n \in A_\beta$ and $x^n \in V^n \subset B_m$. So $x_\alpha \in \mathrm{cl}_{\tau_{\alpha+1}} A_\beta$.

This finishes the construction and the proof. $\quad\square$

DEFINITION 3.6: Let $\mathbb{R} = \{x_\alpha : \alpha < \omega_1\}$ and $\mathbb{R}_\beta = \{x_\beta : \beta < \alpha\}$. A *Kunen line K* is a space $\langle \mathbb{R}, \tau \rangle$ where τ is a regular topology refining the Euclidean topology ρ on $\mathbb{R}$ with each $\mathbb{R}_\beta$ open in $\langle \mathbb{R}, \tau \rangle$ and satisfying:

$$\forall X \subset \mathbb{R} \exists \alpha < \omega_1 \forall \beta > \alpha (x_\beta \in \mathrm{cl}_\rho X \to x_\beta \in \mathrm{cl}_\tau X).$$

We have seen that any Kunen line is a regular, hereditarily separable non-Lindelöf space (these are called *S*-spaces). Todorcevic [68], [69, theorem 8.9] showed that it is consistent with ZFC that there are no *S*-spaces. Rudin [51, 52] constructed a normal and a nonnormal *S*-space from a Suslin line, see also [69, theorem 5.6]. A regular, hereditarily Lindelöf, nonseparable space is called an *L*-space. It is not known if *L*-spaces exist in ZFC, but a Suslin line is an *L*-space, and CH implies their existence also [35].

Now we show how to use a Kunen line to get interesting product spaces.

THEOREM 3.7: Assume CH. There is a non-Lindelöf, 0-dimensional first countable, locally compact space K such that $K \times K$ is hereditarily separable and perfectly normal.

Proof: As before, let $\mathbb{R} = \{x_\alpha : \alpha < \omega_1\}$, $\mathbb{R}_\alpha = \{x_\beta : \beta < \alpha\}$, ρ the Euclidean topology on $\mathbb{R}$ and ρ_α the topology on $\mathbb{R}_\alpha$ inherited from ρ.

Again use an Ostaszewski technique to refine ρ. Make each $\mathbb{R}_\alpha$ open, thus making the new topology non-Lindelöf. Let τ be the new topology on $\mathbb{R}$ such that $K = \langle \mathbb{R}, \tau \rangle$ satisfies the theorem.

In order to make the product $K \times K$ hereditarily separable and perfectly normal we need an analogue to (*) from Theorem 3.1. What we need is

$$\forall X \subset \mathbb{R} \times \mathbb{R} \; \exists \alpha < \omega, \forall \beta, \gamma > \alpha(\langle x_\beta, x_\gamma \rangle \in \text{cl}_{\rho \times \rho} X \to \langle x_\beta, x_\gamma \rangle \in \text{cl}_{\tau \times \tau} X). \quad (**)$$

Where, of course, $\rho \times \rho$ and $\tau \times \tau$ denote the product topologies on $\langle \mathbb{R}, \rho \rangle \times \langle \mathbb{R}, \rho \rangle$ and $\langle \mathbb{R}, \tau \rangle \times \langle \mathbb{R}, \tau \rangle$, respectively.

So we'll aim for (**).

List all countable subsets of $\mathbb{R} \times \mathbb{R}$ as $\langle A_\alpha : \alpha < \omega_1 \rangle$ so that $\forall \alpha < \omega_1, A_\alpha \subset \mathbb{R}_\alpha \times \mathbb{R}_\alpha$.

By induction on $\alpha < \omega_1$, we define a topology τ_α on $\mathbb{R}_\alpha$ so that for $\alpha < \omega_1$ the following holds:

1° $\forall \beta < \alpha \, (\tau_\beta = \tau_\alpha \cap \mathscr{P}(\mathbb{R}_\beta))$;
2° τ_α is first countable, locally compact, 0-dimensional Hausdorff;
3° $\tau_\alpha \supset \rho_\alpha$; and
4° $\forall \beta \leq \alpha \, \forall \gamma \leq \beta \, ([\langle x_\alpha, x_\beta \rangle \in \text{cl}_{\rho \times \rho} A_\gamma \to \langle x_\alpha, x_\beta \rangle \in \text{cl}_{\tau_{\alpha+1} \times \tau_{\alpha+1}} A_\gamma]$ and $[\langle x_\beta, x_\alpha \rangle \in \text{cl}_{\rho \times \rho} A_\gamma \to \langle x_\beta, x_\alpha \rangle \in \text{cl}_{\tau_{\alpha+1} \times \tau_{\alpha+1}} A_\gamma])$.

In order to be able to get 4° we need one more induction hypothesis, but we postpone stating it until the end of this proof where we perform the construction for the case $\alpha + 1$.

For $0 \leq \alpha < \omega$ let τ_α be discrete.

For $\alpha \leq \omega_1$, a limit, let τ_α be the topology generated by $\bigcup_{\beta < \alpha} \tau_\beta$ as a basis, let $\tau = \tau_{\omega_1}$ and $K = \langle \mathbb{R}, \tau \rangle$.

Observe that we have (*) from 3.1, that is:

$$\forall X \subset \mathbb{R} \; \exists \alpha < \omega_1 \forall \beta > \alpha (x_\beta \in \text{cl}_\rho X \to x_\beta \in \text{cl}_\tau X).$$

To see this, let $B \subset X$ be countable with $X \subset \text{cl}_\rho B$. Then the set $A = \{\langle b, b \rangle \in \mathbb{R} \times \mathbb{R} : b \in B\}$ is listed as some A_γ. So if $\beta \geq \gamma$, 4° shows that $(x_\beta \in \text{cl}_\rho B = \text{cl}_\rho X) \to (x_\beta \in \text{cl}_\tau B \subset \text{cl}_\tau X)$.

Hence we have:

3.8. K IS HEREDITARILY SEPARABLE AND PERFECTLY NORMAL.

Recall Definition 2.5. The following plays the role of Lemma 3.2.

LEMMA 3.9: If $X \subset \mathbb{R} \times \mathbb{R}$, then

$$|\text{cl}_{\rho \times \rho} X \backslash \text{cl}_{\tau \times \tau} X|_2 \leq \omega.$$

Proof: Let A_γ be a countable $\rho \times \rho$-dense subset of X. If $\alpha, \beta \geq \gamma$, then 4° tells us that $\langle x_\alpha, x_\beta \rangle \in \text{cl}_{\rho \times \rho} A_\gamma \to \langle x_\alpha, x_\beta \rangle \in \text{cl}_{\tau \times \tau} A_\gamma$; hence, if $\langle x_\alpha, x_\beta \rangle \in \text{cl}_{\rho \times \rho} X$, then $\langle x_\alpha, x_\beta \rangle \in \text{cl}_{\tau \times \tau} X$.

So $\text{cl}_{\rho \times \rho} X \backslash \text{cl}_{\tau \times \tau} X \subset (\mathbb{R}_\gamma \times \mathbb{R}) \cup (\mathbb{R} \times \mathbb{R}_\gamma)$, that is,

$$|\text{cl}_{\rho \times \rho} X \backslash \text{cl}_{\tau \times \tau} X|_2 \leq \omega. \quad \square$$

The preceding lemma gives us (**).

3.10. K × K IS HEREDITARILY SEPARABLE: Let $Y \subset K \times K$ and X be a countable $\rho \times \rho$-dense subset of Y. By Lemma 3.9, $|Y \backslash \text{cl}_{\tau \times \tau} X|_2 \leq \omega$, so there is an $\alpha < \omega_1$ with

$$Y \backslash \text{cl}_{\tau \times \tau} X \subset (\mathbb{R}_\alpha \times K) \cup (K \times \mathbb{R}_\alpha).$$

But $\mathbb{R}_\alpha$ is countable and K is hereditarily separable, so $(\mathbb{R}_\alpha \times K) \cup (K \times \mathbb{R}_\alpha)$ is hereditarily separable.

Let Z be a countable $\tau \times \tau$-dense subset of $Y \backslash \mathrm{cl}_{\tau \times \tau} X$. Then $X \cup Z$ is a countable $\tau \times \tau$-dense subset of Y. $\square$

Observe now that Theorem 1.4 implies that perfectly normal spaces have normal products with compact metric spaces. Also countable spaces are perfect. So since each $\mathbb{R}_\alpha$ is a countable union of countable, clopen, compact metric spaces, the subspace $(K \times \mathbb{R}_\alpha) \cup (\mathbb{R}_\alpha \times K)$ of $K \times K$ is a pairwise disjoint countable union of clopen perfectly normal subspaces, in particular is perfectly normal.

3.11. $K \times K$ IS PERFECT: Let $X \subset K \times K$ be $\tau \times \tau$-closed. By Lemma 3.9 there is an $\alpha < \omega_1$ with

$$(\mathrm{cl}_{\rho \times \rho} X) \backslash X \subseteq (\mathbb{R}_\alpha \times K) \cup (K \times \mathbb{R}_\alpha).$$

We show that $(K \times K) \backslash X$ is F_σ. Let $\{F_n : n < \omega\}$ be a $\rho \times \rho$-closed family with $\bigcup_{n < \omega} F_n = (K \times K) \backslash (\mathrm{cl}_{\rho \times \rho} X)$.

It remains to cover

$$(\mathrm{cl}_{\rho \times \rho} X) \backslash X \subset [(\mathbb{R}_\alpha \times K) \cup (K \times \mathbb{R}_\alpha)] \backslash X$$

with countably many closed subsets of $K \times K$.

By the preceding observation $[(\mathbb{R}_\alpha \times K) \cup (K \times \mathbb{R}_\alpha)] \backslash X$ is F_σ in $K \times K$. Done.

LEMMA 3.12: If X and Y are two closed disjoint subsets of $K \times K$, then $|\mathrm{cl}_{\rho \times \rho} X \cap \mathrm{cl}_{\rho \times \rho} Y|_2 \leq \omega$.

Proof: By Lemma 3.9 fix an $\alpha < \omega_1$ such that

$$(\mathrm{cl}_{\rho \times \rho} X \backslash X) \cup (\mathrm{cl}_{\rho \times \rho} Y \backslash Y) \subset (\mathbb{R}_\alpha \times \mathbb{R}) \cup (\mathbb{R} \times \mathbb{R}_\alpha).$$

Then

$$\mathrm{cl}_{\rho \times \rho} X \cap \mathrm{cl}_{\rho \times \rho} Y \subset (\mathbb{R}_\alpha \times \mathbb{R}) \cup (\mathbb{R} \times \mathbb{R}_\alpha). \square$$

3.13 $K \times K$ IS NORMAL: Let X and Y be two closed disjoint subsets of $K \times K$. By Lemma 3.12 fix an $\alpha < \omega_1$ with

$$A = \mathrm{cl}_{\rho \times \rho} X \cap \mathrm{cl}_{\rho \times \rho} Y \subset (\mathbb{R}_\alpha \times \mathbb{R}) \cup (\mathbb{R} \times \mathbb{R}_\alpha).$$

Let $\{F_n : n < \omega\}$ be a $\rho \times \rho$-closed family with $(K \times K) \backslash A = \bigcup_{n < \omega} F_n$. For $n < \omega$ let V_n be a $\rho \times \rho$-open set with $(\mathrm{cl}_{\rho \times \rho} X) \cap F_n \subset V_n$ and $(\mathrm{cl}_{\rho \times \rho} V_n) \cap (\mathrm{cl}_{\rho \times \rho} Y) = \emptyset$.

Since $(\mathbb{R}_\alpha \times K) \cup (K \times \mathbb{R}_\alpha) = \bigcup_{n < \omega} C_n$, where each C_n is a perfectly normal clopen subspace of $K \times K$ (by the observation before 3.11), for each $n < \omega$ let V'_n be an open subset of C_n with $X \cap C_n \subset V'_n$ and $(\mathrm{cl}_{\rho \times \rho} V'_n) \cap (Y \cap C_n) = \emptyset$.

So $\{V_n : n < \omega\} \cup \{V'_n : n < \omega\}$ is an open cover of X such that $(\mathrm{cl}_{\tau \times \tau} V'_n) \cap Y = \emptyset = (\mathrm{cl}_{\tau \times \tau} V'_n) \cap Y$ for each $n < \omega$.

By symmetry we can find a similar open cover of Y, so X and Y can be separated. $\square$

Finally, we show how to construct $\tau_{\alpha+1}$ given τ_α. Recall our induction conditions $1°$–$4°$ from the beginning of the proof. We need: For $\beta \le \alpha$ define

(i) $D^0_{\beta,\alpha}$, a countable dense subset of $(\mathrm{cl}_{\rho \times \rho} A_\beta) \cap (\{x_\alpha\} \times \mathbb{R})$;
(ii) $D^1_{\beta,\alpha}$, a countable dense subset of $(\mathrm{cl}_{\rho \times \rho} A_\beta) \cap (\mathbb{R} \times \{x_\alpha\})$;
(iii) $\{B_{n,\alpha} : n < \omega\}$, a decreasing base for x_α (in $\tau_{\alpha+1}$), so that

$5° \; \forall \beta \le \alpha \forall d \in D^0_{\beta,\alpha} \forall n < \omega \exists s \in B_{n,\alpha} \exists t (\langle s, t \rangle \in A_\beta \wedge \rho \times \rho$-distance between $\langle s, t \rangle$ and d is $< 1/2^n)$, and $\forall \beta \le \alpha \; \forall d \in D^1_{\beta,\alpha} \forall n < \omega \exists s \in B_{n,\alpha} \exists t (\langle t, s \rangle \in A_\beta \wedge \rho \times \rho$-distance between $\langle t, s \rangle$ and d is $< 1/2^n)$.

Observe that $d \in D^0_{\beta,\alpha} \to d = \langle x_\alpha, a \rangle$ and $d \in D^1_{\beta,\alpha} \to d = \langle a, x_\alpha \rangle$.
We construct points $\{p_n : n < \omega\}$ by induction on $n < \omega$, so that $\{p_n : n < \omega\} \subseteq \mathbb{R}_\alpha$ and ρ-distance from p_n to x_α is $< 1/2^n$. Then we fix a ρ_α-open-discrete family $\{U_n : n < \omega\}$ with each $p_n \in U_n$. Let $\{V_n : n < \omega\}$ be a family of τ_α-compact-open sets with $p_n \in V_n \subset U_n$ for $n < \omega$.
For $n < \omega$, let $B_{n,\alpha} = \{x_\alpha\} \cup \bigcup_{m \ge n} V_m$.
List $\{\langle d, \beta \rangle : \beta \le \alpha \wedge d \in D^0_{\beta,\alpha} \cup D^1_{\beta,\alpha}\} \cup \{\langle \gamma, \alpha, \beta \rangle : \gamma \le \beta \le \alpha\} \cup \{\langle \gamma, \beta, \alpha \rangle : \gamma \le \beta \le \alpha\}$ as $\langle r_n : n < \omega \rangle$, each point listed ω-many times.
At stage n,

(a) $r_n = \langle \gamma, \alpha, \beta \rangle$. Want to pick a p_n with

$$\exists s \in B_{n,\beta} (\langle p_n, s \rangle \in A_\gamma \wedge \rho\text{-distance from } p_n \text{ to } x_\alpha < 1/2^n),$$

if $\langle x_\alpha, x_\beta \rangle \in \mathrm{cl}_{\rho \times \rho} A_\gamma$. Otherwise, pick any p_n close to x_α.

Assume that $\langle x_\alpha, x_\beta \rangle \in \mathrm{cl}_{\rho \times \rho} A_\gamma$. By the second case of $5°$ applied to x_β and A_γ we fix a $d \in D^1_{\gamma,\beta}$ with $\rho \times \rho$-distance from d to $\langle x_\alpha, x_\beta \rangle < 1/2^{n+1}$ and an $s \in B_{n+1,\beta} (\subset B_{n,\beta})$ so that

$$\exists t (\langle t, s \rangle \in A_\gamma \wedge \rho \times \rho\text{-distance from } \langle t, s \rangle \text{ to } d < 1/2^{n+1}).$$

Set $p_n = t$ (t from above).
Observe that $\langle p_n, s \rangle \in (B_{n,\alpha} \times B_{n,\beta}) \cap A_\gamma$, so since $\langle \gamma, \alpha, \beta \rangle$ is listed ω-many times, we have that

$$\langle x_\alpha, x_\beta \rangle \in \mathrm{cl}_{\rho \times \rho} A_\gamma \to \langle x_\alpha, x_\beta \rangle \in \mathrm{cl}_{\tau_{\alpha+1} \times \tau_{\alpha+1}} A_\gamma,$$

that is, $4°$ holds.
(b) $r_n = \langle \gamma, \beta, \alpha \rangle$. Want to pick a p_n with

$$\exists a \in B_{n,\beta} (\langle a, p_n \rangle \in A_\gamma \wedge \rho\text{-distance from } p_n \text{ to } x_\alpha < 1/2^n),$$

if $\langle x_\beta, x_\alpha \rangle \in \mathrm{cl}_{\rho \times \rho} A_\gamma$. Same as (a).
(c) $r_n = \langle d, \beta \rangle$. Let $d \in D^0_{\beta,\alpha}$; the case $d \in D^1_{\beta,\alpha}$ is the same. We want p_n with

$$\exists a (\langle p_n, a \rangle \in A_\beta \text{ and } \rho \times \rho\text{- distance from } \langle p_n, a \rangle \text{ to } d < 1/2^n).$$

Easy since $D^0_{\beta,\alpha} \subset \mathrm{cl}_{\rho \times \rho} A_\beta$.
This takes care of $5°$ since each $\langle d, \beta \rangle$ is listed ω-many times, and finishes the proof of the theorem. $\square$

A space X is a *strong S(L)-space* iff $\forall n < \omega$ (X^n is an $S(L)$-space). It is easy to see that X^ω is hereditarily separable (Lindelöf) iff $\forall n < \omega$ [X^n is hereditarily separable (Lindelöf)]. Zener [74] showed that there is a strong S-space iff there is a strong L-space, see [69, theorem 9.2]. The construction in Theorem 3.7 can be modified (easily) to get a strong S-space. Kunen [34] showed that MA $+ \neg$CH implies that X^ω is hereditarily separable iff X^ω is hereditarily Lindelöf, that is, that there are no strong S (or L)-spaces. For a simple proof see [69, corollary 7.11].

Kunen's construction can be modified a little so that it works in ZFC. Doing that, one gets an Eric line, a space with fewer interesting properties than a Kunen line, but still interesting and useful in constructing examples involving normality and products.

Recall that a space $X = \{x_\alpha : \alpha < c\}$ is right separated of type c iff for each $\alpha < c$, $\{x_\beta : \beta < \alpha\}$ is open in X.

THEOREM 3.14: There is a topology τ on $\mathbb{R}$, finer than the Euclidean topology and such that $\langle \mathbb{R}, \tau \rangle$ is normal, first countable, locally compact, locally countable 0-dimensional, separable, and right separated of type c.

Proof: Lemma 3.2 is the key for getting normality. So we want the assumptions of Lemma 3.2 to hold for τ. Letting ρ be the Euclidean topology on $\mathbb{R}$, we need

$$\forall X, Y \subset \mathbb{R} \ (|\mathrm{cl}_\rho X \cap \mathrm{cl}_\rho Y| = c \to \mathrm{cl}_\tau X \cap \mathrm{cl}_\tau Y \neq \varnothing).$$

Since $\langle \mathbb{R}, \rho \rangle \rangle$ is hereditarily separable, in order to get the preceding condition it is enough to have

For every two *countable* $X, Y \subset \mathbb{R}$, $(|\mathrm{cl}_\rho X \cap \mathrm{cl}_\rho Y| = c \to \mathrm{cl}_\tau X \cap \mathrm{cl}_\tau Y \neq \varnothing)$.   (†)

List $\mathbb{R}$ and $\mathscr{A} = \{\langle A, B \rangle : A, B$ are countable subsets of $\mathbb{R}$ and $|\mathrm{cl}_\rho A \cap \mathrm{cl}_\rho B| = c\}$ as $\mathbb{R} = \{x_\alpha : \alpha < c\}$ and $\mathscr{A} = \{\langle A_\alpha, B_\alpha \rangle : \omega \leq \alpha < c\}$ so that:

(a) $\mathbb{Q} = \{x_n : n < \omega\}$;
(b) $\forall \alpha \geq \omega (A_\alpha \cup B_\alpha \subset \{x_\beta : \beta < \alpha\})$;
(c) $\forall \alpha \geq \omega (x_\alpha \in \mathrm{cl}_\rho A_\alpha \cap \mathrm{cl}_\rho B_\alpha)$.

For $\alpha < c$, let $\mathbb{R}_\alpha = \{x_\beta : \beta < \alpha\}$ and let ρ_α be the topology on $\mathbb{R}_\alpha$ inherited from α. By induction on α we construct topologies τ_α on $\mathbb{R}_\alpha$ so that for each $\alpha < c$:

1° $\forall \beta < \alpha [\tau_\beta = \tau_\alpha \cap \mathscr{P}(\mathbb{R}_\beta)]$;
2° τ_α is first countable, locally compact, locally countable, 0-dimensional, Hausdorff;
3° $\tau_\alpha \supset \rho_\alpha$.

Furthermore, if $\alpha \geq \omega$, then:

4° $\mathbb{Q} \ (= \mathbb{R}_\omega)$ is dense in $\langle \mathbb{R}_\alpha, \tau_\alpha \rangle$;
5° $x_\alpha \in \mathrm{cl}_{\tau_{\alpha+1}} A_\alpha \cap \mathrm{cl}_{\tau_{\alpha+1}} B_\alpha$.

CONSTRUCTION: For $\alpha < \omega$, let τ_α be discrete.
Let $\alpha \geq \omega$. If α is a limit ordinal, let τ_α be the topology on $\mathbb{R}_\alpha$ generated by $\bigcup_{\beta < \alpha} \tau_\beta$ as a basis.
Assume that we have τ_α. We show how to construct $\tau_{\alpha+1}$.

We want 4° and 5° to hold for $\tau_{\alpha+1}$. So let $\{p_n : n < \omega\} \subset \mathbb{Q} \cup A_\alpha \cup B_\alpha$ be such that $|x_\alpha - p_n| < 1/2^n$ for $n < \omega$, and there are infinitely many p_n in $\mathbb{Q}$, and in A_α and in B_α.

For $n < \omega$, pick a compact open V_n in $\langle \mathbb{R}_\alpha, \tau_\alpha \rangle$ containing p_n and with ρ-diameter $< 1/2^n$.

Define, for $m < \omega$,

$$B_m = \{x_\alpha\} \cup \bigcup_{n \geq m} V_n,$$

and let $\tau_{\alpha+1}$ be the topology generated by $\tau_\alpha \cup \bigcup \{B_m : m < \omega\}$ as a basis.

It is easy to check that 1°–5° hold.

Let τ be the topology on $\mathbb{R}$ generated by $\bigcup_{\alpha < c} \tau_\alpha$ as a basis. This finishes the construction.

Our condition 4° implies that $\mathbb{Q}$ remains dense in $\langle \mathbb{R}, \tau \rangle$, and 5° implies (†); hence, $E = \langle \mathbb{R}, \tau \rangle$ is normal. □

DEFINITION 3.15: An Eric line E is a space $\langle \mathbb{R}, \tau \rangle$ where τ is a regular topology refining the Euclidean topology ρ and that satisfies:

$$\forall X, Y \subset \mathbb{R}\,(\,|\mathrm{cl}_\rho X \cap \mathrm{cl}_\rho Y| = c \to \mathrm{cl}_\tau X \cap \mathrm{cl}_\tau Y \neq \varnothing).$$

It is easy to see that each Eric line is countably paracompact.

THEOREM 3.16: There is a Lindelöf space E such that the product $E \times E$ is normal but not paracompact.

Proof: We use an Eric line technique. We find two Eric lines E_0 and E_1 that are separable Lindelöf, so that the diagonal $\Delta = \{\langle x, y \rangle \in E_0 \times E_1 : x = y\}$ is right separated of type c, hence, non-Lindelöf. That shows that the product $E_0 \times E_1$ is not paracompact (since it is separable and non-Lindelöf).

We make sure that the products $E_0 \times E_0$, $E_0 \times E_1$, and $E_1 \times E_1$ are normal. Finally the space $E = E_0 \oplus E_1$ satisfies the theorem.

To get each E_i Lindelöf, use the same idea as in Theorem 2.7, a Bernstein subset of $\mathbb{R}$.

To get the normality of $E_i \times E_j$, Lemma 3.12 should hold for the product, that is, if ρ is the Euclidean topology on $\mathbb{R}$, then:

$$\forall X, Y \subset E_i \times E_j (\,|\mathrm{cl}_{\rho \times \rho} X \cap \mathrm{cl}_{\rho \times \rho} Y|_2 = c \to \mathrm{cl}_{E_i \times E_j} X \cap \mathrm{cl}_{E_i \times E_j} Y \neq \varnothing). \qquad (\dagger\dagger)$$

Since $\rho \times \rho$ is hereditarily separable, we need the preceding to hold only for *countable* X and Y.

First, as in Lemma 2.8, construct pairwise disjoint $A_0, A_1, A_2 \subset \mathbb{R}\backslash\mathbb{Q}$ so that $\mathbb{R} = \mathbb{Q} \cup A_0 \cup A_1 \cup A_2$ and for every $\rho \times \rho$-closed $Y \subset \mathbb{R} \times \mathbb{R}$ with $|Y|_2 = c$ we have that $|(A_1 \times A_1) \cap Y|_2 = c$ for each $i = 0, 1, 2$.

Recall that this automatically gives that for every ρ-closed $Y \subset \mathbb{R}$ with $|Y| = c$, we have that $|Y \cap A_i| = c$ for each i.

If α is an ordinal, we say that $\alpha \equiv i \pmod 3$ if $\alpha = \beta + n$, where $\beta = 0$ or a limit ordinal, $n < \omega$, and $n \equiv i \pmod 3$, where $i = 0, 1, 2$.

Observe that we can list $\mathscr{A} = \{\langle B, C \rangle : B, C$ are countable subsets of $\mathbb{R} \times \mathbb{R}$ and $|\mathrm{cl}_{\rho \times \rho} B \cap \mathrm{cl}_{\rho \times \rho} C|_2 = c\}$ and $\mathbb{R}$ as $\mathscr{A} = \langle \langle B_\alpha, C_\alpha \rangle : \omega \leq \alpha < c \wedge \alpha \equiv 0 \pmod 3 \rangle$ and

$\mathbb{R} = \langle x_\alpha : \alpha < c \rangle$, with $\mathbb{R}_\alpha = \{x_\beta : \beta < \alpha\}$ so that:

(a) $\mathbb{Q} = \{x_n : n < \omega\}$;
(b) Every $B_\alpha \cup C_\alpha \subset \mathbb{R}_\alpha \times \mathbb{R}_\alpha$;
(c) For every $\langle B_\alpha, C_\alpha \rangle$, the point $\langle x_\alpha, x_{\alpha+1} \rangle \in A_2 \times A_2$; and $\langle x_\alpha, x_{\alpha+1} \rangle \in \mathrm{cl}_{\rho \times \rho} B_\alpha \times \mathrm{cl}_{\rho \times \rho} C_\alpha$;
(d) If $\alpha < \beta$ and $x_\alpha = x_\beta$, then $\alpha \equiv 0 \pmod 3$ and $\beta = \alpha + 1$.

Condition (d) tells us that the listing $\langle x_\alpha : \alpha < c \rangle$ of $\mathbb{R}$ is "almost" $1 - 1$, we would like to have it $1 - 1$, but (c) sometimes requires that $x_\alpha = x_{\alpha+1}$ (say $B_\alpha = C_\alpha = \{\langle q, q \rangle : q \in \mathbb{Q}\}$).

We first do the construction to make $E_0 \times E_1$ normal and non-Lindelöf, and at the end we point out which (minor) changes are needed in order to have both $E_0 \times E_0$ and $E_1 \times E_1$ normal.

We construct two topologies τ^0 and τ^1 on $\mathbb{R}$, refining ρ by specifying an open basis for $x_\alpha, x_{\alpha+1}$, and $x_{\alpha+2}$ in τ^0 and τ^1. We do this by induction on $\alpha \geq \omega$, $\alpha \equiv 0 \pmod 3$.

To start, all points x_n, $n < \omega$ are isolated in both τ^0 and τ^1.

At stage α we deal with $x_\alpha, x_{\alpha+1}, x_{\alpha+2}$.

First $x_{\alpha+2}$: Let $i = 0, 1$. If $x_{\alpha+2} \in A_i$, it will have a Euclidean neighborhood base in τ^1, that is, let $B^i_m(x_{\alpha+2}) = \{x \in \mathbb{R} : |x - x_{\alpha+2}| < 1/2^m\}$ for $m < \omega$. Then $\{B^i_m(x_{\alpha+2}) : m < \omega\}$ is a basis for $x_{\alpha+2}$ in τ^i.

If $x_{\alpha+2} \notin A_i$, let $\{q_n : n < \omega\} \subset \mathbb{Q}$ coverage (in ρ) to $x_{\alpha+2}$. Let $B^i_m(x_{\alpha+2}) = \{x_{\alpha+2}\} \cup \{q_n : n \geq m\}$. Then $\{B^i_m(x_{\alpha+2}) : m < \omega\}$ is a basis for $x_{\alpha+2}$ in τ^i.

We do x_α and $x_{\alpha+1}$ simultaneously. We want

$$\langle x_\alpha, x_{\alpha+1} \rangle \in \mathrm{cl}_{\tau^0 \times \tau^1} B_\alpha \cap \mathrm{cl}_{\tau^0 \times \tau^1} C_\alpha.$$

Since $\{x_\alpha, x_{\alpha+1}\} \cap \mathbb{R}_\alpha = \varnothing$ and $B_\alpha \cup C_\alpha \subset \mathbb{R}_\alpha \times \mathbb{R}_\alpha$, we can pick a sequence $\{\langle p_n, q_n \rangle : n < \omega\} \subset B_\alpha \cup C_\alpha$ converging (in $\rho \times \rho$) to $\langle x_\alpha, x_{\alpha+1} \rangle$ such that $m \neq n$ implies $p_m \neq p_n$, $q_m \neq q_n$, and $p_m \neq q_n$. (We cannot exclude $p_m = q_m$; see the comment after the list (a)–(d).) We also require that infinitely many $\langle p_n, q_n \rangle$ are in B_α and in C_α.

Now pick a τ^0 open U_n containing p_n with $U_n \subset B^0_0(p_n)$ and a τ^1 open V_n containing q_n with $V_n \subset B^1_0(q_n)$ so that ρ-diameter of each U_n and V_n is $< 1/2^n$ and if $m \neq n$, $U_m \cap U_n = \varnothing$, $V_m \cap V_n = \varnothing$, $U_m \cap V_n = \varnothing$, and if $p_m \neq q_m$, $U_m \cap V_m = \varnothing$.

Finally, define $B^i_m(x_\alpha)$ and $B^i_m(x_{\alpha+1})$.

Case $1°$: $x_\alpha = x_{\alpha+1}$. Let $B^0_m(x_\alpha) = \{x_\alpha\} \cup \bigcup_{n \geq m} U_n$, and $B^1_m(x_\alpha) = \{x_\alpha\} \cup \bigcup_{n \geq m} V_n$.

Case $2°$: $x_\alpha \neq x_{\alpha+1}$. Let $B^0_m(x_\alpha) = \{x_\alpha\} \cup \bigcup_{n \geq m} U_n$, and $B^1_m(x_{\alpha+1}) = \{x_{\alpha+1}\} \cup \bigcup_{n \geq m} V_n$.

Let $\{r_n : n < \omega\} \subset \mathbb{Q}$ converge (in ρ) to x_α and $\{s_n : n < \omega\} \subset \mathbb{Q}$ converge (in ρ) to $x_{\alpha+1}$.

Let $B^1_m(x_\alpha) = \{x_\alpha\} \cup \{r_n : n \geq m\}$ and $B^0_m(x_{\alpha+1}) = \{x_{\alpha+1}\} \cup \{s_n : n \geq m\}$.

Let τ^i be the topology generated by $\{B^i_m(x_\alpha) : \alpha < c \wedge m < \omega\}$ as a basis. Let $E_0 = \langle \mathbb{R}, \tau^0 \rangle$ and $E_1 = \langle \mathbb{R}, \tau^1 \rangle$. Note that $\mathbb{Q}$ is dense in both E_0 and E_1, so they are separable.

Observe that (††) holds for $i = 0$ and $j = 1$: if B_α and C_α are countable $\rho \times \rho$-dense subsets of X and Y, respectively, then

$$\langle x_\alpha, x_{\alpha+1} \rangle \in \mathrm{cl}_{E_0 \times E_1} B_\alpha \cap \mathrm{cl}_{E_0 \times E_1} C_\alpha \subset \mathrm{cl}_{E_0 \times E_1} X \cap \mathrm{cl}_{E_0 \times E_1} Y.$$

Since the points of A_i have the Euclidean topology in E_i, the same proof as in Theorem 2.7 shows that *each E_i is Lindelöf.*

To show that $E_0 \times E_1$ is non-Lindelöf we show that the diagonal $\Delta = \{\langle x, y \rangle \in E_0 \times E_1 : x = y\}$ is non-Lindelöf (and Δ is closed in $E_0 \times E_1$ since it is closed in $\rho \times \rho$). We show that each $(\mathbb{R}_\alpha \times \mathbb{R}_\alpha) \cap \Delta$ is open in Δ.

It is enough to show that for each α

$$[B_0^0(x_\alpha) \times B_0^1(x_\alpha)] \cap \Delta \subset \mathbb{R}_{\alpha+1} \times \mathbb{R}_{\alpha+1}.$$

So let α be the least where this fails. Then:

(i) $\alpha < \omega$, hence $x_\alpha \in \mathbb{Q}$, $B_0^0(x_\alpha) = B_0^1(x_\alpha) = \{x_\alpha\} \subset \mathbb{R}_{\alpha+1}$;

(ii) $\alpha \geq \omega$ and $\alpha \equiv 2 \pmod 3$, then since $A_0 \cap A_1 = \varnothing$, there is a $i = 0, 1$ with $x_\alpha \notin A_i$ so $B_0^i(x_\alpha) \subset \{x_\alpha\} \cup \mathbb{Q} \subset \mathbb{R}_{\alpha+1}$;

(iii) $\alpha \geq \omega$ and $\alpha \not\equiv 2 \pmod 3$, so either Case 1° or Case 2° held in our construction. If Case 2° held, then for an $i = 0, 1$, $B_0^i(x_\alpha) \subset \{x_\alpha\} \cup \mathbb{Q} \subset \mathbb{R}_{\alpha+1}$. So assume that Case 1° held.

Then $B_0^0(x_\alpha) = \{x_\alpha\} \cup \bigcup_{n < \omega} U_n$ and $B_0^1(x_\alpha) = \{x_\alpha\} \cup \bigcup_{n < \omega} V_n$. Since U_n is disjoint form V_m, unless $n = m$ and $p_n = q_n$, we have that

$$[B_0^0(x_\alpha) \times B_0^1(x_\alpha)] \cap \Delta = ([\bigcup\{U_n \times V_n : p_n = q_n\}] \cap \Delta) \cup \{\langle x_\alpha, x_\alpha \rangle\}$$

$$\subseteq ([\bigcup\{B_0^0(p_n) \times B_0^1(p_n) : n < \omega\}] \cap \Delta) \cup \{\langle x_\alpha, x_\alpha \rangle\} \subset \mathbb{R}_{\alpha+1} \times \mathbb{R}_{\alpha+1}.$$

Where the first inclusion holds since $U_n \subset B_0^0(p_n)$ and $V_n \subset B_0^1(q_n)$, and the second holds by induction, since $\{p_n : n < \omega\} \subset \mathbb{R}_\alpha$.

So *$E_0 \times E_1$ is not Lindelöf.*

To see that $E_0 \times E_1$ is normal, proceed as in Theorem 3.7. Our condition (††) is the analogue of Lemma 3.12.

Now we proceed almost verbatim as in 3.2 (or 3.13):

Let X, Y be two disjoint closed subsets of $E_0 \times E_1$. By (††) and Lemma 2.6 there is a countable $B \subset \mathbb{R}$ with

$$A = \mathrm{cl}_{\rho \times \rho} X \cap \mathrm{cl}_{\rho \times \rho} Y \subset (B \times E_1) \cup (E_0 \times B).$$

But A is Lindelöf, being a closed subset of a countable union of Lindelöf spaces, so one can proceed as in 3.2, since $X \cap A$ is Lindelöf in $E_0 \times E_1$ and, as pointed out in 3.2, that was enough.

So *$E_0 \times E_1$ is normal.*

To get $E_0 \times E_0$ normal, when constructing bases for x_α and $x_{\alpha+1}$, we want to make sure that

$$\langle x_\alpha, x_{\alpha+1} \rangle \in \mathrm{cl}_{E_0 \times E_0} B_\alpha \cap \mathrm{cl}_{E_0 \times E_0} C_\alpha.$$

So we need to pick, for infinitely many $\langle p_n, q_n \rangle$ in B_α and in C_α, V_n a τ^0 open set containing q_n with $V_n \subset B_0^0(q_n)$ and satisfying the rest of the properties of V_n from our construction.

Do a similar thing for $E_1 \times E_1$ normal.

This finishes the proof. $\square$

We next show how to use $\diamondsuit$ to get countably paracompact spaces X and Y so that $X \times Y$ is normal but not countably paracompact.

Recall that a space X is *Dowker* iff X is normal but not countably paracompact.

THEOREM 3.17. Assume $\diamondsuit$. There is a Dowker space.

Proof. We use an Ostaszewski technique. The underlying set will be $\omega_1 \times \omega$. To make the space noncountably paracompact, the sets $U_n = \omega_1 \times n, n < \omega$, will be open. We want that the cover $\{U_n : n < \omega\}$ cannot be shrunk, that is, that there are no closed $F_n \subset U_n$ with $\bigcup_{n<\omega} F_n = \omega_1 \times \omega$, showing that the space is not countably paracompact.

By induction on $\alpha < \omega_1$, we topologize $\alpha \times \omega$; the space will be first countable, locally compact, 0-dimensional, so the initial segments $\alpha \times \omega$ will be metrizable (being countable, regular, first countable).

We make the segments $(\beta + 1) \times \omega$ clopen.

To make the space normal, assume that $X, Y \subset \omega_1$ are closed and disjoint. If one of them, X say, is countable, then there is a $\beta < \omega_1$ with $X \subset (\beta + 1) \times \omega = A$. Since A is metrizable, X and $Y \cap A$ can be separated in A, and since $X \backslash A = \emptyset, X \backslash A$ and $Y \backslash A$ can be separated in $(\omega_1 \times \omega) \backslash A$. But A is clopen, so X and Y can be separated in $\omega_1 \times \omega$.

We don't know what to do if both X and Y are uncountable, so we use $\diamondsuit$ to make sure that this doesn't happen, that is, we want:

$$\text{If } X, Y \subset \omega_1 \times \omega \text{ are uncountable, then } \text{cl}X \cap \text{cl}Y \neq \emptyset. \tag{$\ddagger$}$$

Incidentally, ($\ddagger$) implies that the space is not countably paracompact, since if $F_n \subset U_n = \omega_1 \times n, n < \omega$, are closed and $\bigcup_{n<\omega} F_n = \omega_1 \times \omega$, one of them, F_m say, is uncountable. But then ($\ddagger$) shows that

$$\emptyset \neq \text{cl}F_m \cap \text{cl}[(\omega_1 \times \omega) \backslash U_m] = F_m \cap [(\omega_1 \times \omega) \backslash U_m],$$

so $F_m \not\subset U_m$.

So all we need is a first countable, Hausdorf, 0-dimensional topology on which $\omega_1 \times \omega$ are open, each $(\alpha + 1) \times \omega$ is clopen, and ($\ddagger$) holds.

Recall that $\text{dom}(A) = \{a : \exists b(\langle a, b \rangle \in A]\}, \text{ran}(A) = \{b : \exists a(\langle a, b \rangle \in A)\}$.

Using $\diamondsuit$ fix sequences $\langle A_\alpha^0 : a < \omega_1 \wedge \alpha \text{ limit} \rangle$ and $\langle A_\alpha^1 : a < \omega_1 \wedge \alpha \text{ limit} \rangle$ such that for each α:

(a) $A_\alpha^0 \cup A_\alpha^1 \subset \alpha \times \omega$;
(b) $\text{ran}(A_\alpha^0 \cup A_\alpha^1) \subset \omega$ is finite;
(c) $\text{dom}(A_\alpha^0)$ and $\text{dom}(A_\alpha^1)$ are sequences of type ω converging to α;
(d) For every two uncountable $X, Y \subset \omega_1 \times \omega$, there is an α with

$$A_\alpha^0 \subset X \quad \text{and} \quad A_\alpha^1 \subset Y.$$

By induction on $\alpha < \omega_1$, for $\langle \alpha, n \rangle \in \omega_1 \times \omega$ we construct families $\{U_{\langle \alpha, n \rangle}^k : k < \omega\}$ of subsets of $(\alpha + 1) \times (n + 1)$ such that if τ_α is the topology on $\alpha \times \omega$ having $\{U_p^k : p \in \alpha \times \omega \wedge k \in \omega\}$ as a (sub)basis then we automatically have that for $\beta < \alpha$, $\beta \times \omega$ is open in $\langle \alpha \times \omega, \tau_\alpha \rangle$, and we require:

1° τ_α is a Hausdorff topology on $\alpha \times \omega$;

2° $\forall p \in \alpha \times \omega$ ($\{U_p^k : k \in \omega\}$ is a decreasing clopen basis for p consisting of compact sets (in τ_α));

3° $\forall \beta < \alpha$ [$(\beta + 1) \times \omega$ is clopen in $\langle \alpha \times \omega, \tau \rangle$];

4° For every limit α there is $n > \max[\mathrm{ran}(A_\alpha^0 \cup A_\alpha^1)]$ such that

$$\langle \alpha, n \rangle \in \mathrm{cl}_{\tau_{\alpha+1}} A_\alpha^0 \cap \mathrm{cl}_{\tau_{\alpha+1}} A_\alpha^1.$$

At the end, let τ be the topology on $\omega_1 \times \omega$ generated by $\{U_p^k : p \in \omega_1 \times \omega \wedge k \in \omega\}$ as a (sub)basis (it will be a basis, actually). The space $\langle \omega_1 \times \omega, \tau \rangle$ will satisfy all that we need. To see (‡), observe that it is implied by (d) and 4°.

If α is 0 or a successor, all points $\langle \alpha, n \rangle$ are isolated, that is, $U_{\langle \alpha, n \rangle}^k = \{\langle \alpha, n \rangle\}$ for each $k < \omega$.

For α a limit ordinal, fix an increasing sequence $\langle \beta_n : n < \omega \rangle$ converging to α. Pick an $n_\alpha > \max[\mathrm{ran}(A_\alpha^0 \cup A_\alpha^1)]$.

If $n \neq n_\alpha$, let $\langle \alpha, n \rangle$ be isolated, that is, $U_{\langle \alpha, n \rangle}^k = \{\langle \alpha, n \rangle\}$ for each $k < \omega$.

By (c) and the fact that each $\beta \times \omega$ is open in $\langle \alpha \times \omega, \tau_\alpha \rangle$ the set $A_\alpha^0 \cup A_\alpha^1$ is closed discrete in $\langle \alpha \times \omega, \tau_\alpha \rangle$. List $A_\alpha^0 \cup A_\alpha^1$ as $\{p_n : n < \omega\}$. Since $\langle \alpha \times \omega, \tau_\alpha \rangle$ is countable and regular, it is normal so let $\{U_n : n < \omega\}$ be a τ_α-discrete family of compact open sets (in τ_α) such that: if $p_n = \langle \gamma_n, m_n \rangle$, then $p_n \in U_n \subset (\gamma_n + 1) \times (m_n + 1)$ and if $\beta_k < \gamma_n$, then $(\beta_k \times \omega) \cap U_n = \varnothing$ (this can be done by 3°).

Let $U_{\langle \alpha, n_\alpha \rangle}^k = \{\langle \alpha, n_\alpha \rangle\} \cup \bigcup_{n \geq k} U_n$.

Observe that for each β_n there is a $k < \omega$ with $U_{\langle \alpha, n_\alpha \rangle}^k \cap (\beta_n \times \omega) = \varnothing$. This implies 3° and 1°, since $\langle \beta_n : n < \omega \rangle$ is increasing and converges to α. Condition 2° is easy to check, and $\langle \alpha, n_\alpha \rangle \in \mathrm{cl}_{\tau_{\alpha+1}} A_\alpha^0 \cap \mathrm{cl}_{\tau_{\alpha+1}} A_\alpha^1$, so 4° holds.

Finally, each $\omega_1 \times n$ is open in τ since for each $\langle \alpha, m \rangle$, $U_{\langle \alpha, m \rangle}^0 \subset (\alpha + 1) \times (m + 1) \subset \omega_1 \times (m + 1)$.

This finishes the proof. $\square$

Now we use the space from Theorem 3.17 to show the following.

THEOREM 3.18: Assume $\diamondsuit$. There are countably paracompact spaces X and Y such that $X \times Y$ is Dowker.

Proof: We construct two topologies τ^0 and τ^1 on $\omega_1 \times \omega$, so that if $X = \langle \omega_1 \times \omega, \tau^0 \rangle$ and $Y = \langle \omega_1 \times \omega, \tau^1 \rangle$, the diagonal $\Delta = \{\langle x, y \rangle \in X \times Y : x = y\}$ is closed in $X \times Y$ and not countably paracompact. Since Δ is homeomorphic to $\omega_1 \times \omega$ with the topology generated by $\tau^0 \cup \tau^1$ as a subbasis, we want this space to look like the space from 3.17.

The topologies will be constructed by an Ostaszewski technique, so the initial segments $\alpha \times \omega$ will be open, and we make $(\alpha + 1) \times \omega$ clopen. Also the spaces will be Hausdorff, first countable, locally compact, and 0-dimensional.

We used (‡) in 3.17 to make that space normal. If we strenghten (‡) a little, we get both X and Y normal and countably paracompact:

For every family $\{F_n : n < \omega\}$ of uncountable subsets of $\omega_1 \times \omega$,

$$\bigcap_{n < \omega} \mathrm{cl}_{\tau^i} F_n \neq \varnothing, \quad \text{for } i = 0, 1. \quad (‡‡)$$

Since (‡‡) clearly implies (‡), we get that both X and Y are normal.

For countable paracompactness, say $\{F_n : n < \omega\}$ is a decreasing family of closed subsets of X with $\bigcap_{n < \omega} F_n = \varnothing$. We want to find open sets $U_n \supset F_n$ with $\bigcap_{n < \omega} U_n = \varnothing$, thus showing that X is countably paracompact.

By (‡‡) one of the F_n, say F_m, is countable, so fix an α with $F_m \subset \alpha \times \omega$. Since $\alpha \times \omega$ is Lindelöf, it is countably paracompact so there are open (in $\alpha \times \omega$, hence in X) sets $U_n \supset F_n$ for $n \geq m$ with $\bigcap_{n \geq m} U_n = \varnothing$. For $n < m$ let $U_n = X$. So $\bigcap_{n < \omega} U_n = \varnothing$.

So (‡‡) implies that X and Y are normal and countably paracompact.

To get $X \times Y$ normal we need an analogue of (‡) for products, that is, we need two-cardinality, see Definition 2.5. We want:

For every two $H, K \subset X \times Y$ if $|H|_2 = |K|_2 = \omega_1$, then

$$\mathrm{cl}_{X \times Y} H \cap \mathrm{cl}_{X \times Y} K \neq \varnothing. \quad (\text{‡‡‡})$$

Now (‡‡‡) implies that $X \times Y$ is normal. Since if H and K are two disjoint closed subsets of $X \times Y$, one of them, H say, is such that $|H|_2 = \omega$, so fix an $\alpha < \omega_1$ such that $H \subset [(\alpha + 1) \times \omega) \times Y \cup X \times [(\alpha + 1) \times \omega] = A$. Then A is clopen, and A is normal since each $(\alpha + 1) \times \omega$ is a countable union of clopen compact metric subspaces, so both $[(\alpha + 1) \times \omega] \times Y$ and $X \times [(\alpha + 1) \times \omega]$ are unions of clopen normal subspaces since by Theorem 1.4 compact metric cross normal countably paracompact (and both X and Y are) is normal.

But then, as in 3.17, H and $K \cap A$ can be separated in A and $H\backslash A = \varnothing$ and $K\backslash A$ can be separated in $(X \times Y)\backslash A$, so since A is clopen, H and K can be separated in $X \times Y$.

To make sure that Δ is closed in $X \times Y$, we construct a Hausdorff topology τ^2 on $\omega_1 \times \omega$ coarser than each τ^0 and τ^1.

Using $\diamondsuit$ fix sequences $\langle A_\alpha^n : n < \omega \wedge \alpha < \omega_1 \wedge \alpha \text{ limit}\rangle$, $\langle B_\alpha^n : n < \omega \wedge \alpha < \omega_1 \wedge \alpha \text{ limit}\rangle$, $\langle C_\alpha : \alpha < \omega_1 \wedge \alpha \text{ limit}\rangle$ and $\langle D_\alpha : \alpha < \omega_1 \wedge \alpha \text{ limit}\rangle$, such that for each α:

(a) $(\bigcup_{n < \omega} A_\alpha^n) \cup (\bigcup_{n < \omega} B_\alpha^n) \subset \alpha \times \omega$;
(b) $C_\alpha \cup D_\alpha \subset (\alpha \times \omega) \times (\alpha \times \omega)$;

furthermore, if $C_\alpha^0 = \mathrm{dom}\,(C_\alpha) \subset \alpha \times \omega$, $C_\alpha^1 = \mathrm{ran}\,(C_\alpha) \subset \alpha \times \omega$, $D_\alpha^0 = \mathrm{dom}\,(D_\alpha) \subset \alpha \times \omega$, and $D_\alpha^1 = \mathrm{ran}\,(D_\alpha) \subset \alpha \times \omega$, then

(c) $\{\mathrm{dom}\,(A_\alpha^n) : n < \omega\} \cup \{\mathrm{dom}\,(B_\alpha^n) : n < \omega\} \cup \{\mathrm{dom}\,(C_\alpha^0 \cup C_\alpha^1), \mathrm{dom}\,(D_\alpha^0 \cup D_\alpha^1)\}$ is a pairwise disjoint family of sets;
(d) Each $\mathrm{dom}\,(A_\alpha^n)$, $\mathrm{dom}\,(B_\alpha^n)$, $\mathrm{dom}\,(C_\alpha^i)$, $\mathrm{dom}\,(D_\alpha^i)$ is a sequence of type ω converging to α;
(e) Each $\mathrm{ran}\,(A_\alpha^n)$, $\mathrm{ran}\,(B_\alpha^n)$, $\mathrm{ran}\,(C_\alpha^i)$, $\mathrm{ran}\,(D_\alpha^i)$ is bounded (i.e., finite).

Finally:

(f) For every family $\{F_n : n < \omega\}$ of uncountable subsets of $\omega_1 \times \omega$ there is an α with $A_\alpha^n \cup B_\alpha^n \subset F_n$ for each n;
(g) For every two $H, K \subset (\omega_1 \times \omega) \times (\omega_1 \times \omega)$ with $|H|_2 = \omega_1$ and $|K|_2 = \omega_1$ there is an α with $C_\alpha \subset H$ and $D_\alpha \subset K$.

By induction on $\alpha < \omega_1$, for $\langle \alpha, n\rangle \in \omega_1 \times \omega$ and $i = 0, 1, 2$, we construct families $\{U_{\langle \alpha, n\rangle}^{i,k} : k < \omega\}$ of subsets of $(\alpha + 1) \times \omega$ such that if τ_α^i is the topology on $\alpha \times \omega$ having $\{U_p^{i,k} : p \in \alpha \times \omega \wedge k \in \omega\}$ as a (sub)basis, then we automatically have that for $\beta < \alpha$, $\beta \times \omega$ is open in $\langle \alpha \times \omega, \tau_\alpha^i\rangle$, and we require:

1° τ_α^i is a Hausdorff topology on $\alpha \times \omega$;
2° $\forall p \in \alpha \times \omega$ ($\{U_p^{i,k} : k \in \omega\}$ is a decreasing clopen basis for p consisting of compat sets (in τ_α^i));

$3°$ $\forall \beta < \alpha \, [(\beta + 1) \times \omega$ is clopen in $\langle \alpha \times \omega, \tau^i_\alpha \rangle]$;

$4°$ $\forall p \in \alpha \times \omega \; \forall k \in \omega (U^{0,k}_p \cup U^{1,k}_p \subset U^{2,k}_p)$;

$5°$ $\forall \langle \beta, n \rangle \in \alpha \times \omega \, (U^{0,0}_{\langle \beta, n \rangle} \cap U^{1,0}_{\langle \beta, n \rangle} \subset (\beta + 1) \times (n + 1))$;

$6°$ For every limit α, $\langle \alpha, 0 \rangle \in \bigcap_{n < \omega} [\mathrm{cl}_{\tau^0_{\alpha+1}} (A^n_\alpha) \cap \mathrm{cl}_{\tau^1_{\alpha+1}} (B^n_\alpha)]$;

$7°$ For every limit α, there is $n > \max [\mathrm{ran} \, (C^0_\alpha \cap C^1_\alpha \cup D^0_\alpha \cup D^1_\alpha)]$ such that

$$\langle \langle \alpha, n \rangle, \langle \alpha, n \rangle \rangle \in (\mathrm{cl}_{\tau^0_{\alpha+1} \times \tau^1_{\alpha+1}} C_\alpha) \cap (\mathrm{cl}_{\tau^0_{\alpha+1} \times \tau^1_{\alpha+1}} D_\alpha).$$

Let τ^i be the topology on $\omega_1 \times \omega$ generated by $\{U^{i,k}_p : p \in \omega_1 \times \omega\}$ as a (sub)basis, and $X = \langle \omega_1 \times \omega, \tau^0 \rangle$, $Y = \langle \omega_1 \times \omega, \tau^1 \rangle$. Observe that $4°$ implies that $\tau^2 \subset \tau^0 \cap \tau^1$, so the diagonal Δ is closed in $X \times Y$.

Identify Δ with $\omega_1 \times \omega$ with the topology generated by $\tau^0 \cup \tau^1$ as a subbasis. Then $5°$ implies that the sets $\omega_1 \times n$ are open in Δ for each n, and since $7°$ implies (‡‡‡) and any uncountable $H \subset \Delta$ satisfies $|H|_2 = \omega_1$, we have that (‡) holds for Δ, so Δ is Dowker.

CONSTRUCTION: If α is 0 or a successor each $\langle \alpha, n \rangle$ is isolated, let $U^{i,k}_{\langle \alpha, n \rangle} = \{\langle \alpha, n \rangle\}$ for each i, k.

If α is a limit, pick $n_\alpha > \max [\mathrm{ran} \, (C^0_\alpha \cup C^1_\alpha \cup D^0_\alpha \cup D^1_\alpha)]$.

If $n \neq 0$ and $n \neq n_\alpha$, $\langle \alpha, n \rangle$ is isolated, let $U^{i,k}_{\langle \alpha, n \rangle} = \{\langle \alpha, n \rangle\}$ for each i, k.

Fix an increasing sequence $\langle \beta_n : n < \omega \rangle$ converging to α. The set $E = \bigcup_{n < \omega}(A^n_\alpha \cup B^n_\alpha) \cup C^0_\alpha \cup C^1_\alpha \cup D^0_\alpha \cup D^1_\alpha$ is closed discrete in $\langle \alpha \times \omega, \tau^2_\alpha \rangle$ by (c), (d), (e), and the fact that each $\beta \times \omega$ is τ^2_α open.

List E as $\{p_n : n < \omega\}$ and pick τ^2_α compact open U_n so that if $p_n = \langle \gamma_n, m_n \rangle$, then $p_n \in U_n \subset (\gamma_n + 1) \times \omega$ and for any k, if $\beta_k < \gamma_n$, then $(\beta_k \times \omega) \cap U_n = \varnothing$.

Also pick U^i_n for $i = 0, 1$ with

$$U^0_n = U^{0,k}_{p_n} \quad \text{and} \quad U^1_n = U^{1,k}_{p_n} \quad \text{for some } k \text{ so that } U^0_n \cup U^1_n \subset U_n.$$

Let

$$U^{2,k}_{\langle \alpha, 0 \rangle} = \{\langle \alpha, 0 \rangle\} \cup \bigcup_{n \geq k} \{U_n : p_n \in \bigcup_{m < \omega} (A^m_\alpha \cup B^m_\alpha)\},$$

$$U^{2,k}_{\langle \alpha, n_\alpha \rangle} = \{\langle \alpha, n_\alpha \rangle\} \cup \bigcup_{n \geq k} \{U_n : p_n \in C^0_\alpha \cup C^1_\alpha \cup D^0_\alpha \cup D^1_\alpha\},$$

$$U^{0,k}_{\langle \alpha, 0 \rangle} = \{\langle \alpha, 0 \rangle\} \cup \bigcup_{n \geq k} \{U^0_n : p_n \in \bigcup_{m < \omega} A^m_\alpha\},$$

$$U^{0,k}_{\langle \alpha, n_\alpha \rangle} = \{\langle \alpha, n_\alpha \rangle\} \cup \bigcup_{n \geq k} \{U^0_n : p_n \in C^0_\alpha \cup D^0_\alpha\},$$

$$U^{1,k}_{\langle \alpha, 0 \rangle} = \{\langle \alpha, 0 \rangle\} \cup \bigcup_{n \geq k} \{U^1_n : p_n \in \bigcup_{m < \omega} B^m_\alpha\},$$

$$U^{1,k}_{\langle \alpha, n_\alpha \rangle} = \{\langle \alpha, n_\alpha \rangle\} \cup \bigcup_{n \geq k} \{U^1_n : p_n \in C^1_\alpha \cup D^1_\alpha\}.$$

We only check that $5°$ holds: First $U^{0,0}_{\langle \alpha, 0 \rangle} \cap U^{1,0}_{\langle \alpha, 0 \rangle} = \{\langle \alpha, 0 \rangle\}$ since

$$\bigcup_{n < \omega} \{U^0_n : p_n \in \bigcup_{m < \omega} A^m_\alpha\} \cap \bigcup_{n < \omega} \{U^1_n : p_n \in \bigcup_{m < \omega} B^m_\alpha\} = \varnothing$$

because

$$(\bigcup_{m < \omega} A^m_\alpha) \cap (\bigcup_{m < \omega} B^m_\alpha) = \varnothing.$$

For $\langle \alpha, n_\alpha \rangle$:

$$U^{0,0}_{\langle \alpha, n_\alpha \rangle} \cap U^{1,0}_{\langle \alpha, n_\alpha \rangle} = \{\langle \alpha, n_\alpha \rangle\} \cup [\bigcup_{n < \omega} \{U^0_n : p_n \in C^0_\alpha \cup D^0_\alpha\} \cap \bigcup_{n < \omega}$$

$$\{U^1_n : p_n \in C^1_\alpha \cup D^1_\alpha\}] = \{\langle \alpha, n_\alpha \rangle\} \cup \bigcup_{n < \omega} \{U^0_n \cap U^1_n : p_n \in (C^0_\alpha \cup D^0_\alpha) \cap (C^1_\alpha \cup D^1_\alpha)\}.$$

If $p_n = \langle \gamma_n, m_n \rangle$, since $U^i_n = U^{i,k}_{p_n} \subset U^{i,0}_{\langle \gamma_n, m_n \rangle}$ we have

$$U^0_n \cap U^1_n \subset U^{0,0}_{\langle \gamma_n, m_n \rangle} \cap U^{1,0}_{\langle \gamma_n, m_n \rangle} \subset (\gamma_n + 1) \times (m_n + 1).$$

by 5°, hence $U^0_n \cap U^1_n \subset \alpha \times n_\alpha$ since $n_\alpha > m_n$.
This finishes the proof.   □

Theorem 3.18 answers, consistently, some of the questions after 2.4.

REMARKS 3.19: Theorem 3.1 is Kunen's construction form [30], and 3.7 is an unpublished result of Kunen [33], where the case K^n is considered, also.

Theorem 3.14 is from van Douwen [18], and 3.16 from Przymusiński [46], where he gives construction for any n. Przymusiński [47] is also relevent. See also Michael [40].

The space from 3.17 is a variation of a de Caux [14] Dowker space and is a part of folklore. Of course, there is Dowker space in ZFC; this was done by Rudin [50].

Theorem 3.18 is from [9]; one can easily get one space D so that $D \times D$ is Dowker but D isn't: let $D = X \oplus Y$ and make sure that $X \times X$ and $Y \times Y$ are also normal. The contunuum hypothesis (CH) suffices for 3.18, [10], and $\diamondsuit$ also shows that X and Y can be perfectly normal [13]. Wage [70] announced that CH implies 3.18.

The idea of making the diagonal Δ of the product satisfy what we need (say by Dowker) is from [70].

4. HEREDITARILY NORMAL PRODUCTS

It was tempting to make this into a long section by giving proofs of the facts listed below. Fortunately, I came to my senses after I remembered what Rolfsen wrote in his book [48]: "Finally I want to thank Mary Ellen Rudin for her advice, which I should have followed sooner: 'Don't try to get everything in that book'."

THEOREM 4.1: If $X \times Y$ is hereditarily normal, then either X is perfectly normal or every countable subset of Y is closed discrete.

COROLLARY 4.2: If X and Y are countably compact and $X \times Y$ hereditarily normal, then both X and Y are perfectly normal.

THEOREM 4.3: If X is compact and $X \times X$ perfectly normal, then X is metrizable.

THEOREM 4.4: If X is countably compact and $X \times X$ perfectly normal, then X is compact, hence metrizable.

THEOREM 4.5: If X is countably compact and $X \times X \times X$ is hereditarily normal, then X is compact metrizable.

Proof: By Corollary 4.2 $X \times X$ is perfectly normal. The conclusion follows from Theorem 4.4   □

In light of Theorem 4.5 it is natural to ask if perfect normality in 4.3 and 4.4 can be replaced by hereditary normality.

We don't know the answer to 4.3 in ZFC, but if MA holds, we have the following.

THEOREM 4.6: Assume MA. Then there is a compact nonmetrizable X with $X \times X$ hereditarily normal.

This if from Gruenhage and Nyikos [26] and it consists of two different examples. The first example, under MA $+ \neg$CH, was announced by Nyikos [43], and it uses the double arrow construction, which, in hind sight, can be viewed as a special case of the Wage [71] method for constructing compact spaces. The second example, under CH, is due to Gruenhage; the construction also uses the Wage method.

Wage's idea is an extension of Kunen and Eric lines. List the Cantor set as $\{x_\alpha : \alpha < c\}$. Double each point x_α to x_α^0 and x_α^1, the neighborhoods of each x_α^0 and x_α^1 are of the form $\{x^i : i = 0, 1 \text{ and } |x_\alpha - x| < 1/2^n\}$ for $n < \omega$, that is, just doubled Euclidean neighborhoods. This space is not Hausdorff, but every open cover has a finite subcover. Now refine this topology by induction, so that at stage α we construct disjoint neighborhoods of x_α^0 and x_α^1 (to get the new topology Hausdorff). To keep the new topology compact we make sure that the union of any new neighborhoods of x_α^0 and x_α^1, respectively, contains an old neighborhood (i.e., a doubled Euclidean neighborhood). The new space will not be second countable (we need c open sets to separate all pairs x_α^0, x_α^1), hence not metrizable. The initial segments $\{x_\beta^i : i = 0, 1 \wedge \beta < \alpha\}$ *will not* be open. For details see the previously mentioned papers [26, 71]. They are well written and worth studying.

The countable compactness, Theorem 4.4, has a complete answer.

THEOREM 4.7: The following statement is independent from ZFC: Every countably compact X with $X \times X$ hereditarily normal is compact.

One direction follows from Corollary 4.2 and Weiss [72] theorem that MA $+ \neg$CH implies that any perfectly normal countably compact X is compact.

Using $\diamondsuit$, one constructs a noncompact, countably compact X with $X \times X$ hereditarily normal [12].

Finally, we have:

THEOREM 4.8: Assume $\diamondsuit$. There exists compact nonmetrizable spaces $X_n, n < \omega$, such that the product $\Pi_{n < \omega} X_n$ is perfectly normal.

This is due to Ivanov [29]; compare it to Theorem 4.3.

MORE REMARKS 4.9: Wage construction [71] is actually a special case of Fedorčuk's [23] inverse limit construction. See also Kunen [35] where he constructs a compact L-space from CH using an inverse limit construction.

Katetov proved 4.1, 4.2, and 4.5 in [31]. Theorem 4.3 is from [63] and Chaber generalized it to 4.4 in [15].

Also if X is compact and $X \times X$ hereditarily paracompact, X is metric [24].

5. INFINITE PRODUCTS

Infinite products are often nonnormal, as shown by Stone [66] (ω has the order, in this case discrete, topology):

THEOREM 5.1: The product ω^{ω_1} is nonnormal.

But if an infinite product is normal, it will be quite paracompact.

THEOREM 5.2: Assume that each X_α for $\alpha < \kappa$, has at least two points. If the product $\prod_{\alpha < \kappa} X_\alpha$ is normal it is κ-paracompact.

The hard part of the proof is the case $\kappa = \omega$, due to Zenor [73] and Nagami [42]. The case $\kappa > \omega$ follows easily [8]. This theorem is fairly useful; for example, it implies the following.

THEOREM 5.3: Assume that the product $\prod_{\alpha < \kappa} X_\alpha$ is normal and that all finite subproducts are paracompact (collectionwise normal) then $\prod_{\alpha < \kappa} X_\alpha$ is paracompact (collectionwise normal).

The same proof gives [16] the following.

THEOREM 5.4: Assume that the product $\prod_{\alpha < \kappa} X_\alpha$ is κ-paracompact and all finite subproducts are normal, then the product $\prod_{\alpha < \kappa} X_\alpha$ is normal.

Finally, the following is a complete characterization of normal products of cardinals from Lazarevic [38]. Here a cardinal has the order topology.

THEOREM 5.5: Let λ be a cardinal and $\langle \kappa_\alpha : \alpha < \lambda \rangle$ a sequence of cardinals. The product $\prod_{\alpha < \kappa} \kappa_\alpha$ is normal iff one of the following holds:

1° λ is countable and for each $\alpha \in \lambda$, $cf(\kappa_\alpha) = \omega$.
2° λ is countable and there is a $\beta \in \lambda$ such that for every $\alpha \in \lambda$ with $\alpha \neq \beta$, $cf(\kappa_\alpha) = \omega$ and $\kappa_\beta > cf(\kappa_\beta) > \sup\{\kappa_\alpha : \alpha \in \lambda \wedge \alpha \neq \lambda\} + \omega$.
3° There is a countable $S \subset \lambda$ such that for each $\alpha \in S$, $cf(\kappa_\alpha) = \omega$ and there is a regular $\kappa > \lambda$ such that for each $\alpha \in \lambda \backslash S$, $\kappa_\alpha = \kappa$ and $\kappa > \sup\{\kappa_\alpha : \alpha \in S\} + \omega$.

REMARKS 5.6: In their paper van Douwen and Vaughan investigate infinite products of cardinals.

We haven't discussed the interesting work of Alster involving Lindelöf products, see [2–5].

REFERENCES

1. ALAS, O. T. 1971. On a characterization of collectionwise normality. Can. Math. Bull. **14**: 13–15.
2. ALSTER, K. 1984. On Michael's problem concerning Lindelof property in the Cartesian products. Fundam. Math. **121**: 149–167.
3. ———. 1987. On the product of a perfect paracompact space and a countable product of scattered paracompact spaces. Fundam. Math. **127**: 241–246.
4. ———. 1987. On spaces whose product with every Lindelöf space is Lindelöf. Colloq. Math. **54**: 171–178.
5. ———. 1988. On the class of all spaces of weight not greater than ω_1 whose Cartesian product with every Lindelöf space is Lindelöf. Fundam. Math. **129**: 133–140.
6. ———. 1990. The product of a Lindelöf space with the space of irrationals under Martin's Axiom. Proc. Am. Math. Soc. **110**: 543–545.
7. ALSTER, K. & G. GRUENHAGE. Products of Lindelöf spaces with the irrationals. In press.
8. BESLAGIC, A. Normality in products. Topol. Appl. **22**: 71–82.
9. ———. 1985. A Dowker product. Trans. Am. Math. Soc. **292**: 519–530.

10. ———. 1990. Another Dowker product. Topol. Appl. **36:** 253–264.
11. ———. 1993. The normality of products with one compact factor, revisited. Topol. Appl. **52:** 121–126.
12. ———. A hereditarily normal square. Topol. Appl. In press.
13. ———. Yet another Dowker product. Topol. Appl. In press.
14. DE CAUX, P. 1976. A collectionwise normal, weakly θ-refinable Dowker space which is neither irreducible nor realcompact. Topol. Proc. **1:** 66–77.
15. CHABER, J. 1976. Conditions which imply compactness in countably compact spaces. Bull. Acad. Pol. Sci. **24:** 993–998.
16. CHIBA, K. 1989. A remark on the normality of infinite products. Proc. Am. Math. Soc. **105:** 310–312.
17. CHIBA, K., T. PRZYMUSIŃSKI & M. E. RUDIN. 1986. Normality of product spaces and Morita's conjectures. Topol. Appl. **22:** 19–32.
18. VAN DOUWEN, E. K. A technique for constructing honest, locally compact, submetrizable examples. Topol. Appl. In press.
19. VAN DOUWEN, E. K. & J. E. VAUGHAN. 1989. Some subspaces of ordinals with normal products. Ann. N.Y. Acad. Sci. **552:** 169–172.
20. DOWKER, C. H. 1951. On countably paracompact spaces. Can. J. Math. **3:** 219–224.
21. ———. 1956. Homotopy extension theorems. Proc. London Math. Soc. **6:** 100–116.
22. ENGELKING, R. 1989. General Topology. Haldermann Verlag. Berlin.
23. FEDORČUK, V. V. 1976. Fully closed mappings and the consistency of some theorems general topology with the axioms of set theory. Math. USSR **28:** 1–26.
24. GRUENHAGE, G. 1984. Covering properties on X^2/Δ, W-sets, and compact subsets of Σ-products. Topol. Appl. **17:** 287–304.
25. GRUENHAGE, G., T. NOGURA & S. PURISCH. 1991. Normality of $X \times \omega_1$. Topol. Appl. **39:** 263–275.
26. GRUENHAGE, G. & P. J. NYIKOS. Normality in X^2 for compact X. Trans. Am. Math. Soc. In press.
27. HOSHINA, T. 1989. Normality of product spaces II. *In* Topics in General Topology, K. Morita and J. Nagata, Eds.: 121–160. North-Holland. Amsterdam, the Netherlands.
28. ISHII, T. 1966. On product spaces and product mappings. J. Math. Soc. Japan **18:** 166–181.
29. IVANOV, A. V. 1978. On bicompacta all finite powers of which are hereditarily separable. Sov. Math. Dokl. **19.**
30. JUHASZ, I., K. KUNEN & M. E. RUDIN. 1976. Two more hereditarily separable non-Lindelof spaces. Can. J. Math. **28:** 998–1005.
31. KATETOV, M. 1948. Complete normality of Cartesian products. Fundam. Math. **35:** 271–274.
32. ———. 1958. Extension of locally finite coverings. Coll. Math. **6:** 145–151.
33. KUNEN, K. Products of S-spaces. Handwritten notes.
34. ———. 1977. Strong S and L spaces under MA. *In* Set Theoretic Topology, G. M. Reed, Ed.: 265–268. Academic Press. New York.
35. ———. 1981. A compact L-space. Topol. Appl. **12:** 283–287.
36. LAWRENCE, L. B. 1990. The influence of a small cardinal on the product of a Lindelöf space and the irrationals. Proc. Am. Math. Soc. **110:** 535–542.
37. ———. 1992. Lindelöf spaces concentrated on Bernstein subsets of real line. Proc. Am. Math. Soc. **114:** 211–215.
38. LAZAREVIC, Z. 1991. Some shrinking spaces. Ph.D. Thesis, Univ. of Wisconsin, Madison.
39. MICHAEL, E. 1963. The product of a normal space and a metric space need not be normal. Bull. Am. Math. Soc. **69:** 357–376.
40. ———. 1971. Paracompactness and the Lindelöf property in finite and countable Cartesian products. Compos. Math. **23:** 199–214.
41. MORITA, K. 1961. Paracompactness and product spaces. Fundam. Math. **53:** 223–236.
42. NAGAMI, K. 1972. Countable paracompactness of inverse limits and products. Fundam. Math. **73:** 261–270.
43. NYIKOS, P. J. 1977. A compact, non-metrizable space P such that P^2 is completely normal. Topol. Proc. **2:** 359–364.

44. OSTASZEWSKI, A. 1976. On countably compact, perfectly normal spaces. J. London Math. Soc. **14**: 505–516.

45. PRZYMUSIŃSKI, T. 1980. Normality and paracompactness in finite and countable Cartesian products. Fundam. Math. **105**: 87–104.

46. ———. 1980. Products of perfectly normal spaces. Fundam. Math. **108**: 129–136.

47. ———. 1984. Products of normal spaces. *In* Handbook of Set-Theoretic Topology, K. Kunen and J. Vaughan, Eds.: 781–826. North-Holland. Amsterdam, the Netherlands.

48. ROLFSEN, D. 1976. Knots and Links. Publish or Perish. Wilmington, Del.

49. RUDIN, M. E. 1955. Countable paracompactness and Souslin's problem. Can. J. Math. **7**: 543–547.

50. ———. 1971. A normal space X for which X × I is not normal. Fundam. Math. **73**: 179–186.

51. ———. 1972. A normal, hereditarily separable, non-Lindelöf space. Ill. J. Math. **16**: 621–625.

52. ———. 1974. A non-normal hereditarily separable space. Ill. J. Math. **18**: 481–483.

53. ———. 1975. The normality of products with one compact factor. Gen. Topol. Appl. **5**: 45–59.

54. ———. 1978. κ-Dowker spaces. Czech. Math. J. **28**: 324–328.

55. ———. 1979. Hereditary normality and Suslin lines. Gen. Topol. Appl. **10**: 103–106.

56. ———. 1983. The shrinking property. Can. Math. Bull **26**: 385–388.

57. ———. 1985. κ-Dowker spaces. *In* Aspects of Topology—In Memory of Hugh Dowker 1912–1982, I. M. James and E. H. Kronheimer, Eds. Cambridge Univ. Press. London/New York.

58. RUDIN, M. E. & A. BESLAGIC. 1985. Set-theoretic constructions of nonshrinking open covers. Topol. Appl. **20**: 167–177.

59. RUDIN, M. E., K. CHIBA & T. PRZYMUSIŃSKI. 1986. Normality of product spaces and Morita's conjectures. Topol. Appl. **22**: 19–32.

60. RUDIN, M. E., I. JUHASZ & K. KUNEN. 1976. Two more hereditarily separable non-Lindelöf spaces. Can. J. Math. **28**: 998–1005.

61. RUDIN, M. E. & M. STARBIRD. 1975. Products with a metric factor. Gen. Topol. Appl. **5**: 235–248.

62. RUDIN, M. E. & WATSON. 1983. Countable product of paracompact scattered spaces. Proc. Am. Math. Soc. **89**: 551–553.

63. SNEIDER, V. 1945. Continuous images of Suslin and Borel sets; Metrization theorems. Dokl. Akad. Nauk SSSR **50**: 77–79.

64. STARBIRD, M. 1976. Products with a compact factor. Gen. Topol. Appl. **6**: 297–303.

65. ———. 1974. The normality of products with a compact or a metric factor. Thesis, Univ. of Wisconsin, Madison.

66. STONE, A. 1948. Paracompactness and product spaces. Bull. Am. Math.Soc. **54**: 997–982.

67. TAMANO, H. 1962. On compactifications. J. Math. Kyoto Univ. **1**: 162–193.

68. TODORCEVIC, S. 1983. Forcing positive partition relations. Trans. Am. Math.Soc. **280**: 703–720.

69. ———. 1989. Partition Problems in Topology. American Mathematical Society. Providence, R.I.

70. WAGE, M. L. 1978. The dimension of product spaces. Proc. Natl. Acad. Sci. U.S.A. **75**: 4671–4672.

71. ———. 1980. Products of Radon spaces. Russ. Math. Surv. **35**: 185–187.

72. WEISS, W. 1978. Countably compact spaces and Martin's Axiom. Can. J. Math. **30**: 243–249.

73. ZENOR, P. 1971. Countable paracompactness in product spaces. Proc. Am. Math. Soc. **30**: 199–201.

74. ———. 1980. Hereditary m-separability and hereditary m-Lindelöf property in product spaces and function spaces. Fundam. Math. **106**: 175–180.

βN^a

ALAN DOW

Department of Mathematics
York University
North York, ONT M3W 1P3
Canada

ABSTRACT: A brief overview is given of those aspects of the study of βN, the Stone–Čech compactification of the integers, which are most closely connected with Mary Ellen Rudin's considerable contributions in this area. Orderings of the ultrafilter types are discussed, a topological dynamics-based presentation of the construction of incomparable types is given, as is an application involving Blass's principle NCF to the Scarborough–Stone problem.

1. INTRODUCTION

Mary Ellen Rudin is, of course, widely recognized as a major contributor to the knowledge of βN, chiefly through her early involvement in the study of the partial orders of the types of ultrafilters. Her influential and well-known paper [28] gave wide publicity to the study of these orders and developed many of the basic properties (many of which had appeared in her earlier but poorly circulated paper [27]). The monograph [29] was the standard reference for set-theoretic topology for many years and, together with her Martin's Axiom survey article [30], helped break the floodgates of independence results about βN.

In this paper we will first go back to the paper [28], learn a few tricks about constructing filters, and present the basic properties of the orders discussed, including the Rudin–Keisler and Rudin–Frolík orderings. Our discussion of the Rudin–Frolík ordering leads us to Kunen's construction of weak P-points and modifications of it by van Mill. We put a different perspective on the Rudin–Shelah construction of 2^c pairwise incomparable Rudin–Keisler ultrafilters, and finish with a brief discussion of the combinatorial principle NCF (Near Coherence of Filters), which is related to Mary Ellen's work on composants of the Stone–Čech remainder of the real-line.

2. PARTIAL ORDERINGS ON βN

βN is the space of ultrafilters of N, where the fixed ultrafilters are identified with the points in N, and the topology is the usual Stone space topology on the Boolean algebra $\mathscr{P}(N)$. That is, the closure of $A \subset N$, $\overline{A}$, consists of A together with the set of free ultrafilters including A, denoted A^*. Therefore, the topology can be character-

Mathematics Subject Classification. Primary 54G15, 54D35.
Key words and phrases. Stone–Čech compactification, orderings of ultrafilters.
[a]Research supported by the National Sciences and Engineering Research Council of Canada.

ized by the fact that a subset of N and its complement have complementary compact open closures.

Since βN is the Stone–Čech compactification of N, it follows that any continuous function f from N into a compact space X, extends uniquely to a continuous function βf from βN into X. Clearly $\beta f(p)$ will be a certain kind of limit point in X of the sequence $\{f(n) : n \in N\}$. We discuss this in more detail below by introducing Bernstein's concept of a p-limit [2].

Two ultrafilters are said to be of the same *type* (in βN) if there is a permutation f of N such that βf takes one to the other. Equivalently, p is of the same type as q if there is an autohomeomorphism of βN sending p to q. Since there are only $\mathfrak{c}$ such autohomeomorphisms of βN but $2^{\mathfrak{c}}$ many points in βN, βN is very much nonhomogeneous in the sense that there are $2^{\mathfrak{c}}$ distinct types. Although there may be $2^{\mathfrak{c}}$ autohomeomorphisms of N^* (under continuum hypothesis (CH), for example), N^* was proved not to be homogeneous by studying partial orderings of the types of βN.

DEFINITION 2.1: $p \leq_{RK} q$ if there is a function $f : N \to N$ such that $f^{-1}(I) \in q$ for each $I \in p$ — (equivalently, $\beta f(q) = p$).

$p \leq_{RF} q$ if there is a function $f : N \to \beta N$ such that $\beta f(p) = q$ *and* βf is an embedding.

To obtain a clearer picture let us recall some of the basic facts found in [28].

PROPOSITION 2.1: (1) An injective function, $f : N \to \beta N$, extends to an embedding precisely when its range is discrete.

(2) If $p \leq_{RF} q$, then $p \leq_{RK} q$.
(3) If $p < q$ and $q < r$, then $p < r$ (where $<$ is either $\leq_{RF}$ or $\leq_{RK}$).
(4) If $f : N \to N$ and $\beta f(p) = p$, then $\{n : f(n) = n\} \in p$.
(5) If $p \leq_{RK} q$ and $q \leq_{RK} p$, then p and q have the same type.

Proof: Let f be an injective function from N into βN with discrete range. Therefore, for each n, we may choose an element A_n of the ultrafilter $f(n)$ such that $\overline{A_n} \cap \{f(m) : m \in N - \{n\}\} = \emptyset$. Since $A_n - \bigcup_{m < n} A_m$ will still be a member of $f(n)$, we may assume that the A_n are pairwise disjoint. Now if I and J are disjoint subsets of N, then $\{f(n) : n \in I\}$ and $\{f(n) : n \in J\}$ have disjoint closures in βN since they are contained in the closures of $\bigcup_{n \in I} A_n$ and $\bigcup_{n \in J} A_n$, respectively, and these sets, being disjoint subsets of N, have disjoint closures. It follows easily that βf is injective.

Now suppose that $\beta f(p) = q$—that is, $p \leq_{RF} q$. Define a function g from N to N as follows: $g(m) = n$ for all n and $m \in A_n$ and $g(m) = 0$ for all $m \in N - \bigcup_n A_n$. Now, since $\beta f(p) = q$, $q \in \overline{\{f(n) : n \in I\}} \subset g^{-1}(I)$ for each $I \in p$. Hence it follows that $\beta g(q) = p$ and $p \leq_{RK} q$.

It is immediate that $\leq_{RF}$ and $\leq_{RK}$ are transitive. Suppose that $\beta f(p) = p$ for some $f : N \to N$. Let I_0 be the set of fixed points of f. It is a classical result, that we can find a partition of N into I_1, I_2, I_3 such that $f[I_i - I_0] \cap I_i$ is empty for $i = 1, 2, 3$. Perhaps the simplest is to notice that one can apply Zorn's Lemma to the (nonempty) family of pairwise disjoint triples that have the aforementioned property and whose union is closed under f. A maximal triple is easily seen to be a partition. Now $I_i \in p$ for some i and, then, so is $f(I_i) \cap I_i \subset I_0$.

The final item follows from the previous as follows. Fix functions f and g such that $\beta f(p) = q$ and $\beta g(q) = p$. Note that $\beta(g \circ f)(p) = p$. Therefore there is an $I \in p$ such that $f \upharpoonright I$ is injective. We may assume that $N - (I \cup f[I])$ is infinite, hence $f \upharpoonright I$ can be extended to a permutation whose extension takes p to q. □

For some of us it is difficult to remember Definition 2.1 because the direction of the maps is reversed. There is a useful alternate view of these orders using the Bernstein notion of p-limit. Call a sequence of sets (or points) *discrete* if each member of the sequence has a neighborhood whose closure is disjoint from all the other members of the sequence.

DEFINITION 2.2: Let $p \in N^*$ and let $\{A_n : n \in N\}$ be a collection of subsets of a space X. We call a point $x \in X$ a p-limit of the sequence $\{A_n : n \in N\}$ if x is a limit point of $\{A_n : n \in I\}$ for each $I \in p$. Equivalently, x is a p-limit point of $\{A_n : n \in N\}$ if, for each neighborhood U of x, the set $\{n \in N : U \cap A_n \neq \varnothing\}$ is in p. We let p-lim $\{A_n : n \in N\}$ denote the set of all p-limits of $\{A_n : n \in N\}$, *hence*

$$p\text{-lim } \{A_n : n \in N\} = \bigcap_{I \in p} \text{cl}_X \left[\bigcup_{n \in I} A_n \right].$$

We also say that x is a p-limit of $\{x_n : n \in N\}$ if $x \in p$-lim $\{\{x_n\} : n \in N\}$.

PROPOSITION 2.2: If p, q are in βN, then $p \leq_{RK} q$ iff q is a p-limit of some discrete sequence of subsets of βN. Also, $p \leq_{RK} q$ iff q is a p-limit of a discrete sequence of points in βN.

Proof: If $p \leq_{RK} q$, then there is a function f from N to N such that $f^{-1}(I) \in q$ for each I in p. This is easily seen to be a restatement of the fact that q is the p-limit of the sequence $\{f^{-1}(n) : n \in N\}$. Conversely, suppose that q is the p-limit of a discrete sequence $\{K_n : n \in N\}$. Since this sequence is discrete, there is a neighborhood U_0 of K_0 whose closure misses $\bigcup_{n>0} K_n$. Let $A_0 = U_0 \cap N$ and notice that $\overline{A_0} \subset \overline{U_0}$. Clearly, the sequence $\{\overline{A_0}\} \cup \{K_n : n > 0\}$ is again a discrete sequence. Hence, we can inductively choose pairwise disjoint $A_n \subset N$ so that $K_n \subset \overline{A_n}$. Let $f \in {}^N N$ be any function such that $f[A_n] = \{n\}$ for each $n \in N$. To show f witnesses that $p \leq_{RK} q$, it suffices to show that $f(J)$ is a member of p for each $J \in q$. Since q is the p-limit of $\{K_n : n \in N\}$, it follows that $f(J) = \{n : J \cap A_n \neq \varnothing\} \supset \{n : \overline{J} \cap K_n \neq \varnothing\}$ is a member of p. □

One of the fundamental and remarkable properties of $\leq_{RF}$ is that it is treelike, that is, for each $q \in \beta N$, the set $\{p \in \beta N : p \leq_{RF} q\}$ is linearly ordered.

This is basically a consequence of the following.

PROPOSITION 2.3: If A, B are σ-compact subsets of βN and q is a common limit point, then there is an $I \in q$ such that either $(A \cap \overline{I}) \subset \overline{B}$ or $(B \cap \overline{I}) \subset \overline{A}$.

A pair of sets is said to be *separated* if each set is disjoint from the closure of the other; that is, separated σ-compact subsets of βN have disjoint closures. Now to prove that $\{p \in \beta N : p \leq_{RF} q\}$ is linearly ordered by $\leq_{RF}$, assume that q is a p-limit of a discrete set A and an r-limit of a discrete set B and (wlog) that $B \subset \overline{A}$. Now the canonical homeomorphism from βN to $\overline{A}$ will take p to the p-limit of A, namely q. Since q is an r-limit of a discrete subset of $\overline{A}$, it follows that p is an r-limit of a discrete subset of βN—hence $r \leq_{RF} p$.

By reversing our perspective in the preceding paragraph we come to another extremely important result of Frolík's: (in Frolík's notation) a type cannot *produce* itself. Let us choose a discrete B contained in $\overline{A} - A$, for some discrete $A \subset \beta N$ and fix q, r, p as earlier. The result is that p is not equal to r. That is, p is not the p-limit of any discrete subset of $\beta N - N$. We deduce this from the following result of Rudin.

PROPOSITION 2.4: q is $\leq_{RK}$-minimal in $\{q : p \leq_{RK} q$ and $r \leq_{RK} q\}$ if and only if whenever there are pairwise disjoint $\{A_n : n \in N\}$ and pairwise disjoint $\{B_n : n \in N\}$ such that q is a p-limit of $\{A_n : n \in N\}$ and an r-limit of $\{B_n : n \in N\}$, there is an I in q such that $I \cap A_n \cap B_m$ is at most a singleton for each pair n and m.

We can deduce Frolík's result from the preceding as follows. Suppose that p is a p-limit of a discrete set $\{x_n : n \in N\} \subset N^*$. Let $A_n \in x_n$ be chosen pairwise disjoint as in Proposition 2.1. Now let $B_n = A_n$ and apply the preceding proposition with $q = r = p$ to find $I \in p$ such that $I \cap A_n$ is at most a singleton for each n. We now have a contradiction, since it follows that $\overline{I}$ is a neighborhood of p that does not contain any of the x_n.

The $\leq_{RK}$ ordering is most definitely not treelike. In fact, it is as upwards directed as it can get. It follows immediately from the definition that no point can have more than c predecessors. Rudin showed that the $\leq_{RK}$ order is directed, Blass [3] showed that it was ω_1-directed, and Comfort and Negrepontis (and independently Kunen, see [11, 10.10]) showed that any set of at most c types has an upper bound (this is proved in Proposition 3.6 below). Kunen [17] showed that the $\leq_{RK}$ ordering is not a chain (hence neither is the $\leq_{RF}$ ordering). The fact that the $\leq_{RF}$ ordering is treelike but is not a chain has many applications, especially in the class of those spaces that share the property of Proposition 2.3.

PROPOSITION 2.5: No infinite homogenous compact space has the property that all pairs of countable separated subsets have disjoint closures.

Proof: Assume that X is an infinite compact space, has the property, and choose an infinite discrete set $A \subset X$. Fix $\leq_{RF}$-incomparable p, q in N^*. Let x be a (unique) p-limit of A and let y be a q-limit. Assume that f is a homeomorphism of X into itself which takes x to y. Let D be the dense set of isolated points in $A \cup f[A]$ and identify $\overline{D}$ with βN. In this copy of βN, the point y is both a p-limit (of $f[A]$) and a q-limit of A. It then follows that p and q are $\leq_{RF}$-comparable. Since there are $\leq_{RF}$-incomparable points, X is not homogeneous. □

Kunen has found substantial strengthenings of the preceding result using *weak P-points* in [20] (also see Section 3). Kunen defined weak P-points in [19] as a point of N^* that is not the limit point of any countable subset of N^* and showed that there are 2^c such points (see Proposition 3.5). We can use this to answer a recent question of Pelant and Weiss (answered independently by van Mill).

PROPOSITION 2.6: If $\mathscr{D}$ is any family of at most c separable subsets of N^*, then there is an infinite closed set contained by $N^* - \bigcup \mathscr{D}$.

Proof: Fix a set $D \subset \bigcup \mathscr{D}$ of cardinality c such that D contains a dense subset of each member of $\mathscr{D}$. We may assume that D has the property that for each $d \in D$, if there is a countable set of weak P-points with d as a limit, then there is such a subset of D. Since there are 2^c weak P-points, fix a countable set $E \subset N^* - D$ of weak P-points. We check that $\overline{E}$ is disjoint from the closure of every countable subset of D—hence $\overline{E} \subset N^* - \bigcup \mathscr{D}$. Assume that $A \subset D$ is countable and that $\overline{E} \cap \overline{A}$ is not empty. Then, by Proposition 2.3 and the fact that the members of E are weak P-points, it follows that $A \cap \overline{E}$ is not empty. Therefore each member of $A \cap \overline{E}$ is a limit of a countable set of weak P-points; hence, there is a countable set of weak P-points in D, say B, such that $\overline{B} \cap \overline{E}$ is not empty. However, this contradicts Proposition 2.3, because every point of $B \cup E$ is a weak P-point.   □

Much of [28] is devoted to showing that we cannot expect (unfortunately) that the structure of the partial orders on βN will lead to a characterization of the types. This is done primarily by constructing several different ultrafilters (assuming CH). We focus on the constructions and the information gained about the orders and not discuss the possibility of characterizing the types. Rudin actually also defines a third order that we denote by . The definition of is the condition 3 below.

PROPOSITION 2.7: If $p, q \in βN$ are not the same type, then the following are equivalent.

(1) p q;
(2) q is a p-limit of a pairwise disjoint sequence of clopen subsets of N^*;
(3) There is a function $f : N \to N$ such that $βf(q) = p$ and, for each $I \in q$, the set $\{n : I \cap f^{-1}(n)$ is infinite$\}$ is a member of p;
(4) There is a sequence $\{A_n : n \in N\}$ of disjoint subsets of N, such that for each $I \in q$, the set $\{n : I \cap A_n$ is infinite$\}$ is a member of p;
(5) q is a p-limit of a sequence $\{A_n : n \in N\}$ of disjoint subsets of N, such that for each sequence $A'_n \in [A_n]^{<\omega}$, q is not a p-limit of $\{A'_n : n \in N\}$;
(6) q is a p-limit of a discrete sequence $\{K_n : n \in N\}$ of subsets of N^*.

It is not immediately clear that the three orders are in fact distinct. It is clear that $\leq_{RF} \subseteq\ \subseteq\ \leq_{RK}$. Indeed, if p and q are not the same type then, using p-limits:

(1) $p \leq_{RF} q$ if q is a p-limit of a discrete sequence of points in N^*.
(2) p q if q is a p-limit of a discrete sequence of subsets of N^*.
(3) $p \leq_{RK} q$ if q is a p-limit of a discrete sequence of subsets of $βN$.

One can establish that $\leq_{RK}\ \not\subseteq\ \leq_{RF}$ is several distinct ways.

PROPOSITION 2.8:

(1) If q is a weak P-point that is not a P-point, then there is a $p \leq_{RK} q$ but q is $\leq_{RF}$-minimal.
(2) (CH) For each $p \in βN$, there are $\leq_{RK}$-incomparable q, r such that $p \leq_{RK} q$, $p \leq_{RK} r$, and q is $\leq_{RF}$ minimal. In fact q can be chosen so that there is no r that is strictly $\leq_{RK}$ between p and q.
(3) If p and r are $\leq_{RF}$-incomparable, then any q that is $\leq_{RK}$ above them both is $\leq_{RF}$ above at most one.

Rudin established (CH) that all three orderings are distinct. The relation can be substituted in for $\leq_{RK}$ in item (1). Interestingly, it seems to have been open if it can be shown in ZFC that and $\leq_{RK}$ are distinct. This fact was brought to my attention by S. Garcia-Ferreira. Even the statement that they are distinct can be reformulated to sound like a much more interesting question.

Recall that a point q is a P-point if and only if whenever q is a limit point of a pairwise disjoint sequence $\{A_n : n \in N\}$ of subsets of N, there is an $I \in q$ such that $I \cap A_n$ is finite for each n but one.

Call q a p-P-point if whenever q is a p-limit point of a pairwise disjoint sequence $\{A_n : n \in N\}$ of subsets of N, there is an $I \in q$ such that $I \cap A_n$ is finite for each n. Since it may happen that q is not a p-limit of any sequence of sets (i.e., in case $p \not\leq_{RK} q$) and we already know that p is a p-P-point, we will say that q is a strict p-P-point if q is strictly above p in the $\leq_{RK}$ ordering.

Therefore the question of whether the two orders, and $\leq_{RK}$, are equal can be restated as: "Is there a strict p-P-point for some $p \in N^*$?" Andreas Blass [7] observed that there are such points (in fact, our earlier construction from a P-point only required a weak P-point). Recall that Kunen showed that there are many $\leq_{RK}$-incomparable weak P-points.

PROPOSITION 2.9: Suppose that p is a weak P-point that is not the minimum in the $\leq_{RK}$ order. Then there is a strict p-P-point (i.e., $\leq_{RK}$ is not contained in).

Proof: Let $\{A_n : n \in N\}$ be any sequence of pairwise disjoint infinite subsets of N. For each n, let p_n be the unique p-limit of the sequence obtained by enumerating A_n in increasing order. Let r be any point that is not $\leq_{RK}$ above p. Let q be the r-limit of the sequence $\{p_n : n \in N\}$ (q is equal to $r \otimes p$ according to the terminology of Proposition 4.8). To see that $p \leq_{RK} q$, note that q is the p-limit of the sequence $\{I_m : m \in N\}$, where I_m is the set that selects the mth element from each A_n.

Now suppose that q is the p-limit of a disjoint sequence $\{B_n : n \in N\} \subset \mathscr{P}(N)$. Without loss of generality $\bigcup_n B_n = N$. Define $g \in {}^N N$ by $g(B_n) = n$ for each n and let βg denote the continuous extension from βN onto βN. It follows that $\beta g(q) = p$ and that p is the r-limit of the sequence $\{\beta g(p_n) : n \in N\}$. Clearly $\{\beta g(p_n) : n \in N\} \cap N$ is a discrete set, hence p cannot be the r-limit of it (since $p \not\leq_{RK} r$). However, neither is p a limit of any countable subset of $N^* - \{p\}$. Therefore we have that $J = \{n : \beta g(p_n) = p\}$ is a member of r.

Now fix an n in J and let f_n denote the unique order-preserving map from N onto A_n, hence $p_n = \beta f_n(p)$. Since $\beta(g \circ f_n)(p) = p$, it follows from Proposition 2.1 that $\bar{A}_n = \{m : g \circ f_n(m) = m\}$ is a member of p. Therefore $I = \bigcup_{n \in J} f_n[\bar{A}_n - n]$ is a member of q. Finally, we finish by observing that $|I \cap B_k| \leq k$ for each k; indeed by the definition of g, we have that, for each $n \in J, f_n[\bar{A}_n] \cap B_k \subseteq \{f_n(k)\}$. $\square$

It appears very unlikely but it is not known if it is consistent that the $\leq_{RK}$ order has a minimum. A related question raised by Shelah is if there is some ultrafilter p that is $\leq_{RK}$ comparable with all others. It has been shown that there is no such p if $\mathfrak{c}$ is regular (Hindman [15]; see also Butkovičová [10]). Keisler proved that it follows from CH that there are minimals in $\leq_{RK}$. An ultrafilter p is $\leq_{RK}$-minimal if and only if it is *selective*. The definition "follows" from Proposition 2.4: whenever p is a limit of pairwise disjoint $\{A_n : n \in N\}$, there is an $A \in p$, such that $|A \cap A_n| \leq 1$ for all n. Kunen [18] proved that it is consistent that there are no $\leq_{RK}$-minimal elements. The next result is probably the main result of Rudin's paper [28]. The construction of the point that is a weak P-point but not a P-point is quite ingenious.

THEOREM 2.10 (CH): There are p, q, r such that $p \leq_{RK}$-minimal and, for each of q and r, p is the only type that is below. But, in addition and by contrast, $p \leq_{RF} r$ while q is a weak P-point and p q (hence not a P-point).

Proof: Rather than repeating the proof from [28] of this theorem, we give a sketch by isolating the different aspects of the proof that are designed to accomplish the separate goals of the construction.

It is actually rather straightforward to construct r. Suppose we have found p—an $\leq_{RK}$-minimal type. Take r to be the p-limit of any countable discrete set of points, $\{p_n : n \in N\}$, each of the same type as p—clearly $p \leq_{RF} r$. Now we suppose that r is an s-limit of a sequence $\{B_m\}_m$, and define $g[B_m] = m$ as in Proposition 2.9. Hence $\beta g(r) = s$, and consider the set $\{\beta g(p_n) : n \in N\}$. By the minimality of p we may as well assume that $\{n : \beta g(p_n) = s\}$ is empty. Therefore s is the p-limit of the sequence $\{\beta g(p_n) : n \in N\}$, which we may assume by the minimality of p, are all distinct. Furthermore, each $\beta g(p_n)$ is either in N or of the same type as p. If "p-many" are in N, the s is equal to p. Otherwise, we may assume that each $\beta g(p_n)$ is the same type as p (hence a P-point), and it follows that s is the same type as r.

Now we analyze (topologically) what we need to construct q. Of course, q will be constructed in an induction of length ω_1 such that at each stage of the induction the approximation to q will be countably generated. The trick is to find ω_1 "doable" tasks that will successfully construct q.

Since p is to be below q, we must fix infinite A_n and guarantee that q will be a p-limit of $\{A_n^* : n \in N\}$. Let $X_n = A_n^*$ and let $X = \bigcup_n X_n$. Since $\overline{X}$ is homeomorphic to βX, let us, at the risk of confusion, use X^* to denote $\overline{X} - X$.

It is a routine part of the induction to ensure that the filter p will be $\leq_{RK}$-minimal.

It is equally routine to ensure that q will be a P-point of the boundary X^*—that is—q will not be a limit point of σ-compact subset of X^*. This goes a long way toward ensuring both that q will be a weak P-point and that p will be the only predecessor to q. Furthermore, for these two properties, we really only have to worry about subsets of $\overline{X}$ since q will be a limit of a σ-compact set C if and only if q is a limit of $C \cap \overline{X}$. More is needed to guarantee that p is the only predecessor to q. It is easily checked that the following suffices, and it is clearly "doable." For every sequence $\{B_m : m \in N\}$ of pairwise disjoint subsets of N, there is a member I of q such that, for each n there is at most one m such that $I^* \cap X_n \cap B_m^* \neq \emptyset$.

Now all that remains is to make q a weak P-point. Kunen had a different CH construction of weak P-points at the time and later, of course, a ZFC construction. This step is the least obvious because, as Kunen remarks, one needs a base-type property because one cannot simply list all the countable sets and avoid them one at a time.

Rudin uses an extremely novel idea. As we carry out the induction, we pick P-points p_α in X (say $p_{\beta+n} \in X_n$ for each limit β) together with enumerations $\{A_{\alpha,\beta} : \alpha \leq \beta < \omega_1\}$ of a descending (mod finite) base for p_α such that

$$\bigcup_n A_{\beta+n,\beta+n} \in q \text{ for each limit } \beta < \omega_1.$$

We may obtain the following two properties for each limit β: for each $n \in N$, there is an $\alpha_n < \beta$ such that $A_{\beta+n,\beta+n} \subset A_{\alpha_n,\beta}$ and $\{A_{\alpha,\beta} : \alpha < \beta\}$ are pairwise disjoint.

Amazingly, this guarantees that q is not the limit point of any countable subset of X. Indeed, suppose that D is a countable subset of X. Since each p_α is a P-point, there is an $f(\alpha) \in \omega_1$ such that $A_{\alpha,f(\alpha)}^*$ is disjoint from $D - \{p_\alpha\}$. Since D is countable, there is a $\gamma < \omega_1$ such that $D \cap \{p_\alpha : \gamma \leq \alpha < \omega_1\}$ is empty. Let β be a limit greater than

$f(\alpha)$ for all $\alpha < \gamma$. Then $\bigcup_n A_{\beta+n,\beta+\omega}$ is a member of q that misses D since

$$A^*_{\beta+n,\beta+\omega} \cap D \subset [A^*_{\alpha_n^\beta,\beta} - A^*_{\alpha_n^\beta,\beta+\omega}] \cap D \subset [A^*_{f(\alpha_n^\beta)} - \{p_{\alpha_n^\beta}\}] \cap D = \varnothing. \quad \square$$

The fact that the preceding proof could be broken into separate parts reflects a more general situation in how the constructions of special filters is often carried out. Given a collection $\mathscr{C}$ of subsets of a space X, a collection $\mathscr{F}$ (usually a filter) is $\mathscr{C}$-remote if for each $Q \in \mathscr{C}$, there is a member of $\mathscr{F}$ whose closure in βX is disjoint from the closure of Q. Therefore a weak P-point is $\mathscr{C}$-remote, where $\mathscr{C}$ is the collection of countable sets. Other interesting classes for $\mathscr{C}$ are "countable and nowhere dense," "countable and discrete," and for the original remote points, simply, "nowhere dense."

Let X be a space that can be partitioned into a countable locally finite collection of closed sets X_n. Call a filter $\mathscr{F}$ of subsets of X nice, if each member of $\mathscr{F}$ is disjoint from at most finitely many of the X_n. As usual, let us denote $\beta X - X$ by X^*. J. van Mill has adapted Kunen's construction of weak P-points and shown how to extend any nice filter $\mathscr{F}$ to a filter $\mathscr{F}^*$ such that the unique point in the intersection of the βX closure of members of $\mathscr{F}^*$ is not the limit of any countable subset of X^* (the adaptation of Kunen's technique for those X^* with weight $\mathfrak{c}$ is relatively straightforward, but van Mill has a very clever innovation for larger spaces).

Therefore, the current situation in the problems of constructing points in βX that are $\mathscr{C}$-remote, where $\mathscr{C}$ is some collection of countable sets, is that it is only necessary to construct nice ($\mathscr{C} \cap \mathscr{P}(X)$)-remote filters on X. The major remaining open problem is to determine if every such βX will have a point that is not the limit of any countable discrete set.

3. OTHER CONSTRUCTIONS

Some obvious questions we can ask about a partial order: Are there minimal/maximal elements? If so, what are their properties? Can an element have an immediate successor? Does every element have an immediate predecessor?

We already discussed the problem of $\leq_{RK}$-minimals. Their existence is not provable in ZFC and if they exist they are selective. One of the basic properties of all the orders is that they do not have maximal elements.

The minimals in the $\leq_{RF}$ order are points that are not the limit of any countable discrete set. They exist in ZFC. There are even $\leq_{RF}$-minimal points that are not weak P-points.

It is not known to me if, in ZFC, there is an element that has no immediate $\leq_{RK}$-predecessor. However Butkovačová [9] showed that there is such a point for the $\leq_{RF}$ order. Consider the space $^{<\omega}N$ with the following topology. Let p be a weak P-point ultrafilter and take the largest topology such that each $t \in {}^{<\omega}N$ is the p-limit of $\{t^\frown n : n \in N\}$. It turns out (see [13]) that this space is regular, Hausdorff, and extremally disconnected (i.e., the closure of every open set is open). Therefore it can be embedded into N^* as $\{x_t : t \in {}^{<\omega}N\}$. Consider the $\leq_{RF}$-predecessors of $x_\varnothing$. Certainly $p \leq_{RF} x_\varnothing$ and so is p^2, where $p^{k+1} = p^k\text{-lim}\{p_n : n \in N\}$ and $\{p_n : n \in N\}$ is a discrete set of points each of type p. Indeed, $x_\varnothing$ is the p^2-limit of the discrete set $\{x_t : t \in {}^2N\}$. More generally, if $x_\varnothing$ is a q-limit of a discrete subset D of $\{x_t : t \in {}^{<\omega}N\}$, then each $x \in D$ is a q-limit of a discrete subset of $\{x_t : t \in {}^{<\omega}N\}$. It follows easily then that $x_\varnothing$ is a q^2-limit of a discrete subset of $\{x_t : t \in {}^{<\omega}N\}$. Therefore, if we are able to ensure that $x_\varnothing$ is not a limit of any discrete set that is disjoint from $\{x_t : t \in {}^{<\omega}N\}$, then no q is an immediate predecessor of $x_\varnothing$, since $q \leq_{RK} x_\varnothing$ implies $q \leq_{RK} q^2 \leq_{RK} x_\varnothing$.

It is shown in [9] and in [13] that no point of ${}^{<\omega}N$ is a limit point (in this topology) of any countable subset of $\beta({}^{<\omega}N) - {}^{<\omega}N$. Hence if we embed ${}^{<\omega}N$ (with the topology as earlier) into N^* so that its closure is a weak P-set, then $x_\varnothing$ (indeed, x_t for each $t \in {}^{<\omega}N$) will have no immediate $\leq_{RF}$-predecessor.

This is what Butkovačová did, but we will see that we can do this by Theorem 3.5—a special case of a result of van Mill [24]. In the previous proof we saw that p^2 is an immediate $\leq_{RF}$-successor of p.

COROLLARY 3.1: There are points p, q such that q is an immediate $\leq_{RF}$-successor of p.

To prove the theorem of van Mill (Theorem 3.5) we first obtain a simple generalization of Kunen's result that there is a c by c independent matrix.

LEMMA 3.2: If N^* maps onto a space X, then N^* also maps onto $X \times (c + 1)^c$.

Proof: We prove this algebraically using Simon's proof that N^* maps onto $(c + 1)^c$ (which was originally proved by Kunen). We may assume that X is zero-dimensional and fix a subalgebra $\mathscr{X} \subset \mathscr{P}(N)$ such that $\mathscr{X}/fin$ is isomorphic to $CO(X)$. We will produce an appropriate subalgebra of the power set of

$$\bigcup_{n \in N} {}^{\mathscr{P}(n)}\mathscr{P}(n).$$

For each pair $S, T \subset N$, let

$$A_{S,T} = \bigcup_{n \in N} \{s \in {}^{\mathscr{P}(n)}\mathscr{P}(n) : s(S \cap n) = T \cap n\},$$

and for each $X \in \mathscr{P}(N)$, let

$$A_x = \bigcup_{n \in X} {}^{\mathscr{P}(n)}\mathscr{P}(n).$$

It should be clear that $\{A_x : X \in \mathscr{X}\}$ is isomorphic to $\mathscr{X}$. We can repeat the proof from [19] that for each $S \subset N$, the family

$$\{A_{S,T} : T \subset N\}$$

is pairwise almost disjoint and if $\mathscr{S}$ is any finite subset of $\mathscr{P}(N)$ and, for each $S \in \mathscr{S}$, a set $T(S) \subset N$ is chosen, then for each n such that $\{S \cap n : S \in \mathscr{S}\}$ are pairwise distinct (i.e., all but finitely many n), there is an $s_n \in {}^{\mathscr{P}(n)}\mathscr{P}(n)$ such that $s_n \in A_{S,T(S)}$ for each $S \in \mathscr{S}$. Therefore, it follows that for each infinite $X \in \mathscr{X}, A_x \cap \bigcap_{S \in \mathscr{S}} A_{S,T(S)}$ is infinite. $\square$

Next it is useful to isolate a couple more of Kunen's wonderful ideas for constructing weak P-points.

DEFINITION 3.1: For a cardinal κ and a space X, call a set $K \subset X$, a κ-OK set if for every increasing sequence $\{C_n : n \in \omega\}$ of regular-closed sets disjoint from K, there is a set $\mathscr{U}$, of neighborhoods of K, such that $|\mathscr{U}| = \kappa$ and, for each $n \in N$, the intersection of any n-many members of $\mathscr{U}$ is disjoint from C_n. When there is such a family $\mathscr{U}$, let us say that K is κ-OK with respect to $\{C_n : n \in N\}$.

LEMMA 3.3: If K is κ-OK for some uncountable κ, then no point of K is a limit point of any countable subset of $X - \overline{K}$.

Proof: Let $\{x_n : n \in N\}$ be any countable subset of $X - \overline{K}$ and fix, for each $n \in N$, a regular-open neighborhood, W_n, of x_n whose closure misses $\overline{K}$. Let $C_n = \bigcup_{k \le n} \overline{W}_k$ and let $\mathscr{U}$ witness that K is κ-OK with respect to $\langle C_n : n \in N \rangle$. Then some $U \in \mathscr{U}$ must be disjoint from $\{x_n : n \in N\}$ since otherwise a single x_n will be a member of infinitely many (in fact, uncountably many) members of $\mathscr{U}$. Take any n of these to obtain an intersection that is not disjoint from C_n (i.e., it contains x_n). $\square$

The next lemma is the main idea behind the construction.

LEMMA 3.4: Suppose that φ is a mapping from N^* onto $(\kappa + 1)^I$ (with I infinite) and that $K \subset N^*$ is a closed subspace such that φ also maps K onto $(\kappa + 1)^I$. Finally, suppose that $\{C_n : n \in N\}$ is an increasing sequence of closed sets that are disjoint from K. Then there is a countable subset I' of I and a closed set $K' \subset K$ such that φ composed with the projection map, maps K' onto $(\kappa + 1)^{I-I'}$ and such that K' is κ-OK with respect to $\{C_n : n \in N\}$.

Proof: Fix $I' = \{i_n : n \in N\} \subset I$ arbitrarily. We may suppose that each C_n is clopen. For each $\xi \in \kappa$, define $Y(\xi, n) = \varphi^{-1}([(i_n, \xi)]) - C_{n+1}$. Now, $\bigcup_n Y(\xi, n)$ is a cozero subset of N^* that meets each C_n in a clopen set. It follows that there is a clopen set Y_ξ such that, for each n,

$$Y_\xi \cap C_n = C_n \cap \bigcup_n Y(\xi, n).$$

Now K' is simply $K \cap \bigcap_{\xi \in \kappa} Y_\xi$ and the Y_ξ witness that K' is κ-OK with respect to $\{C_n : n \in N\}$. To check this, fix any $\xi_0 < \xi_1 < \cdots < \xi_{n-1}$. Now $C_n \cap \bigcap_{i<n} Y_{\xi_i}$ is not empty if and only if there is a sequence j_i ($i < n$) of integers less than $n - 1$ such that $C_n \cap \bigcap_{i<n} Y(\xi_i, j_i)$ is not empty. But clearly, there must be $i < i' < n$ such that $j = j_i = j_{i'}$, which means that $Y(\xi_i, j) \subset \varphi^{-1}([(j, \xi_i)])$ and $Y(\xi_{i'}, j) \subset \varphi^{-1}([(j, \xi_{i'})])$ are disjoint.

It remains to show that K' maps onto $(\kappa + 1)^{I-I'}$. It suffices to show that for any basic clopen set $B \subset (\kappa + 1)^{I-I'}$ and any increasing finite sequence $(\xi_k : k < n)$,

$$\varphi^{-1}(B) \cap \bigcap_{k<n} Y(\xi_k) \cap K$$

is not empty. This follows from the fact that

$$\varphi^{-1}(B) \cap \bigcap_{k<n} Y(\xi_k) \supset \varphi^{-1}(B \cap \bigcap_{k<n} [(i_k, \xi_k)]) - C_{n+1},$$

which meets K since the preimage of every nonempty basic clopen subset of $(\kappa + 1)^I$ meets K while C_{n+1} is disjoint from K. $\square$

Before proving the next lemma we recall the notion of an irreducible map. A map f from X to Y is *irreducible* if it is surjective and no proper closed subset of X maps onto Y. A simple Zorn's Lemma argument can be used to establish that for every continuous function f with compact domain X there is a closed subspace K of X such that $f \upharpoonright K$ is irreducible and $f[K] = f[X]$.

THEOREM 3.5: If N^* maps onto an extremally disconnected space X, then X can be embedded in N^* as a weak P-set.

Proof: Let φ be a mapping onto $X \times (\mathfrak{c} + 1)^\mathfrak{c}$ as in Lemma 3.2. For a subset I of $\mathfrak{c}$ let π_I denote the projection from $X \times (\mathfrak{c} + 1)^\mathfrak{c}$ onto $X \times (\mathfrak{c} + 1)^I$. Now fix an enumeration $\{\langle C^*\alpha, n \rangle : n \in N\rangle : \alpha < \mathfrak{c}\}$ of all increasing sequences of subsets N with each listed $\mathfrak{c}$ times.

We inductively construct a sequence K_α for $\alpha < \mathfrak{c}$ of closed subsets of N^* together with sets $I_\alpha \subset \mathfrak{c}$ such that $\pi_{I_\alpha} \circ \varphi$ maps K onto $X \times (\mathfrak{c} + 1)^{I_\alpha}$ (this is just dual to Kunen's notion of an independent matrix modulo a filter). In order to be sure the induction will go $\mathfrak{c}$ steps, we ensure that $\mathfrak{c} - I_\alpha$ has cardinality at most $|\alpha| \cdot \aleph_0$. Since Gleason showed that compact extremally disconnected spaces are projective, it follows that any irreducible map onto a compact extremally disconnected space is a homeomorphism. Therefore, to ensure that $K_\mathfrak{c}$ will be homeomorphic to X, we guarantee that $\pi_\varnothing \circ \varphi$ is an irreducible mapping onto X. To do this, at each stage we will have that either $C(\alpha, 0)^*$ contains $K_{\alpha+1}$, or that $\pi_\varnothing \circ \varphi$ does not map $K_{\alpha+1} \cap C(\alpha, 0)^*$ onto X. That is, we can take $K_{\alpha+1}$ to be $K_\alpha \cap C(\alpha, 0)^*$ if $\pi_{I_\alpha} \circ \varphi$ maps it onto $X \times (\mathfrak{c} + 1)^{I_\alpha}$. If it does not then choose nonempty clopen subsets $B \subset (\mathfrak{c} + 1)^{I_\alpha}$ and $A \subset X$ such that $\varphi^{-1}(A \times B) \cap K_\alpha$ is disjoint from $C(\alpha, 0)^*$. Let $K_{\alpha+1} = \varphi^{-1}(B) \cap K$ and let $I_{\alpha+1}$ be I_α minus the support of B. It is straightforward to check that the induction hypotheses hold.

Finally, to ensure that $K_\mathfrak{c}$ is a weak P-set, we just apply the construction of Lemma 3.4 if each $C(\alpha, n)^*$ is disjoint from K_α—hence ensuring that $K_{\alpha+1}$ is $\mathfrak{c}$-OK with respect to $\langle C(\alpha, n) : n \in N \rangle$. It is enough to only worry about clopen sets since N^* is compact and zero-dimensional. It should be clear that in Lemma 3.4, we can add that K' will map onto $X \times (\mathfrak{c} + 1)^{I'}$. $\square$

One of the difficulties with the $\leq_{RK}$ ordering in controlling the set of predecessors is that it is not treelike. Laflamme [21] has shown that the predecessors of a type can form a decreasing chain of any countable well-ordered type. On the other hand, there does not seem to be an easily stated property that would guarantee that a (nonminimal) point does not have an immediate predecessor. For example, a point that is simultaneously a q-point and a P-point is minimal, but there are points with any other combination of these two properties that have immediate predecessors. There is a fairly routine CH construction of a point with no immediate $\leq_{RK}$ predecessors. Are either of the following consistent for the $\leq_{RK}$ order? No point has an immediate predecessor. Every point has an immediate predecessor.

We finish this section with the proof that the $\leq_{RK}$ order is upward directed.

PROPOSITION 3.6: Any set of at most $\mathfrak{c}$ types has an upper bound in the $\leq_{RK}$ order.

Proof: Fix any function φ from N onto a dense subset of the separable space $N^\mathfrak{c}$. Extend φ to a continuous function from βN onto $(\beta N)^\mathfrak{c}$ (still call it φ). Let $\{p_\alpha : \alpha < \mathfrak{c}\}$ be a set of free ultrafilters on N and let $\vec{p} \in (\beta N)^\mathfrak{c}$ denote the point $\langle p_\alpha : \alpha \in \mathfrak{c} \rangle$. Fix a closed subspace $K \subset N^*$ such that φ maps K irreducibly onto $(\beta N)^\mathfrak{c}$. We claim that each point p of $K \cap \varphi^{-1}(\vec{p})$ is $\leq_{RK}$ above each p_α. Indeed, $\vec{p}$ is a

p_α-limit of the sequence $\{[(\alpha, n)] : n \in N\}$; hence, by the irreducibility of $\varphi \restriction K, p$ is a p_α-limit of $\varphi^{-1}([(\alpha, n)])$. $\square$

4. INDEPENDENT FAMILIES AND MANY TYPES

Kunen constructed two $\leq_{RK}$-incomparable ultrafilters by using an independent family to *guide* an induction through c steps. Rudin and Shelah [32] refined this technique to construct 2^c pairwise incomparable types. Simon [33] has a nice construction that combines the Rudin–Shelah construction with Kunen's weak P-point construction to prove a result that has frequent applications: there are 2^c pairwise $\leq_{RK}$-incomparable weak P-points.

DEFINITION 4.1: A family $\mathscr{I} \subset \mathscr{P}(N)$ is *independent* if for each disjoint pair of finite subfamilies $\mathscr{D}$ and $\mathscr{E}$, the set $\cap \mathscr{D} - \cup \mathscr{E} = \cap \mathscr{D} \cap \cap_{E \in \mathscr{E}} N - E$ is not empty. Given a filter $\mathscr{F}$, the family $\mathscr{I}$ is said to be independent mod $\mathscr{F}$, if

$$F \cap \cap \mathscr{D} \cap_{E \in \mathscr{E}} N - E$$

is not empty for each $F \in \mathscr{F}$ and $\mathscr{D}$, $\mathscr{E}$ as earlier.

We will formulate and prove the Rudin–Shelah result in the language of topological dynamics. For example, as mentioned before, the structure of an independent family mod a filter $\mathscr{F}$ is dual to that of a closed subspace K of $\cap_{F \in \mathscr{F}} F^*$, which maps onto $2^{\mathscr{I}}$. The $\leq_{RK}$ ordering on N^* can be just thought of as the ordering induced by the inclusion ordering on the *orbits* of points under the family of *surjective* continuous self-maps of βN.

If S is a collection of continuous partial functions from X to itself, let, for $x \in X$,

$$O_s(x) = \{f(x) : f \in S \text{ and } x \in \text{dom}(f)\}.$$

A set Y is pairwise S-incomparable if, for distinct $x, y \in Y$, $x \notin O_s(y)$.

First we prove the following two theorems. (Actually we will be lazy and always assume, with no more apologies, that our spaces are all zero-dimensional.)

THEOREM 4.1: If X is any compact space of weight equal to a regular cardinal κ, which has no points of character less than κ, and if S is any family of at most κ continuous partial functions from X to X, then X contains a pairwise S-incomparable set of cardinality 2^κ.

THEOREM 4.2: If X is any compact space of weight κ, which has an irreducible mapping onto $[0, 1]^\kappa$ (hence no points of character less than κ), and if S is any family of at most κ continuous partial functions from X to X, then X contains a pairwise S-incomparable set of cardinality 2^κ.

Shelah observed that it suffices to find a set of size κ^+ because of the free set lemma. That is, each set $O_s(x)$ has cardinality at most κ and, by the Čech–Pospisil Theorem, X has cardinality 2^κ, hence if $2^\kappa > \kappa^+$, there is a $O_s(\cdot)$-free set of cardinality 2^κ.

The idea is standard. We have a compact space X. A subset of X is called a $G_{<\kappa}$-set if it is equal to the intersection of fewer than κ open subsets of X. By an obvious generalization of the Baire category theorem, when κ is regular the intersection of any family $\mathcal{U}$ consisting of at most κ $G_{<\kappa}$-dense and $G_{<\kappa}$-open sets is $G_{<\kappa}$-dense. Most of the time, an inductive construction can be reformulated as the statement that in some compact space X every member of a certain κ-sized family of sets are $G_{<\kappa}$-dense and $G_{<\kappa}$-open.

Let S be any collection of at most κ continuous functions from a compact subset of X into X. Suppose we have already chosen some collection Y of at most κ points in X. We want to find another point x in X so that

(1) For each $f \in S$ and $y \in Y \cap \text{dom}(f)$, $f(y) \neq x$.
(2) For each $f \in S$ and $y \in Y \cap \text{ran}(f)$, $x \notin f^{-1}(y)$.

Obviously, $X - \{f(y)\}$ is an open subset of X that is $G_{<\kappa}$-dense just as long as the character of $f(y)$ is not less than κ. Just as obviously, the set $X - f^{-1}(y)$ is $G_{<\kappa}$-dense just as long as $f^{-1}(y)$ does not contain a nonempty $G_{<\kappa}$-set—which in all likelihood it will. Therefore we must take care not to choose any such points; in particular, we must find a collection of no more than κ $G_{<\kappa}$-dense, $G_{<\kappa}$-open sets such that any point in the intersection will not be such a point. This is really the only hard part of the proof.

DEFINITION 4.2: Given a space X and a collection S of functions, let $B(S, X)$ denote the collection of points $x \in X$ such that $f^{-1}(x)$ contains a nonempty $G_{<w(X)}$-subset of X for some $f \in S$.

By adopting the topological approach, most of the difficulty is taken care of by the following two lemmas. We show that $X - B(S, X)$ contains the intersection of κ many $G_{<w(X)}$-dense, $G_{<w(X)}$-open sets.

LEMMA 4.3: Assume that X is a compact space in which every point has character at least κ, then no space of weight less than κ maps onto a $G_{<\kappa}$-dense subset of X.

Proof: Suppose that $g : Y \to X$, where Y has weight less than κ. Fix any x in the range of g and fix a base $\mathcal{A}$ for Y with cardinality less than κ. For each $a \in \mathcal{A}$, choose a neighborhood U_a of x such that, if possible, $U_a \cap g(a) = \emptyset$. Let $F = \cap\{U_a : a \in \mathcal{A}\}$. Now if C is any clopen subset of X that meets F but does not contain x, then $g^{-1}(C \cap F)$ is empty. Indeed, fix any $a \in \mathcal{A}$, which is contained in $g^{-1}(C)$ and which meets $g^{-1}(C \cap F)$. Therefore, U_a was chosen to miss $g(a)$, which contradicts that $\emptyset \neq g(a) \cap F \subset C \cap F \subset U_a$. □

LEMMA 4.4: Suppose that X is a compact space in which no point has character less than κ and suppose that f is a continuous function from some compact subset of X into X. Let B be any collection of fewer than κ regular-open subsets of X and define $U_{f,B}$ to be the set of points x such that $f^{-1}(x)$ does not contain any nonempty intersection of members of B. Then $U_{f,B}$ contains a $G_{<\kappa}$-dense, $G_{<\kappa}$-open subset.

Proof: To show that $U_{f,B}$ is as given earlier, it suffices to show that for any $G_{<\kappa}$-set K, $K - U_{f,B}$ is not $G_{<\kappa}$-dense in K. Since we may assume B is a Boolean subalgebra of $CO(X)$ (recall that we are being lazy), there is a canonical continuous function g from X onto $S(B)$. Let Y be the set of all points $y \in S(B)$ such that $f(g^{-1}(y))$ is a singleton in K, call it $h(y)$. Since no point of K has character less than κ and the weight of Y is less than κ, we are finished, by Lemma 4.3, if we show that h is a continuous function. Let C be a clopen subset of X and fix any $y \in Y$ such that $h(y) \in C$. Therefore, $g^{-1}(y) \subset f^{-1}(C)$. Since dom $(f) - f^{-1}(C)$ is compact and $g^{-1}(y)$ is the intersection of members of B, there is a $b \in B$ such that $g^{-1}(y) \subset b$ and $b \cap$ dom $(f) \subset f^{-1}(C)$. Therefore $b \in y$ (in the sense of $S(B)$) and $h(b) \subset C$. It follows that h is continuous. $\square$

COROLLARY 4.5: Let κ be a regular cardinal and let X be a compact space of weight κ such that no point of X has character less than κ. Then if S is any collection of no more than κ continuous partial functions on X and if $Y \subset X - B(S, X)$ has cardinality at most κ, there is a $G_{<\kappa}$-dense set of points $x \in X - B(S, X)$ such that $Y \cup \{x\}$ is pairwise S-incomparable.

It should be clear how to prove Theorem 4.2 for regular cardinals κ. The singular case is handled by expressing the system as an inverse limit. This decomposition is much easier algebraically. For example, the next lemma is trivial and is the only place we use the irreducible map onto $[0, 1]^\kappa$.

LEMMA 4.6: Suppose that $S_0 \subset S$ and $B_0 \subset CO(X)$ both have cardinality less than κ. Then there is a $B_1 \subset CO(X)$ containing B_0 of cardinality equal to $|B_0|$, such that for each $f \in S_0$ and $b \in B_1 f^{-1}(b) \in B^1$. In addition, if there is an irreducible mapping of X onto 2^κ, then B_1 can be chosen so that no point of $S(B_1)$ has character less than $|B_1|$.

Therefore, by applying the lemma repeatedly, we can express X as an inverse limit system, $\lim_\leftarrow \{X\alpha : \alpha < cf(\kappa)\}$ with bonding maps $\varphi_{\alpha,\beta}$ such that the weight of each X_α is a regular cardinal greater than $\Sigma_{\gamma < \alpha} w(X_\gamma)$ and such that for each $f \in S$ there is an $\alpha_f < cf(\kappa)$ such that

$$f_\beta = \varphi_\beta \circ f \circ \varphi_\beta^{-1}$$

is a continuous function on X_β for each $\beta \geq \alpha$, where φ_β is the function from X onto X_β. We can then express S as an increasing union of S_α with $|S_\alpha|$ at most the weight of X_α such that $\alpha_f \leq \alpha$ for each $f \in S_\alpha$.

Now let us complete the proof of 4.2. Suppose that we have chosen $Y \subset X$, which is pairwise S-incomparable, has cardinality at most κ, and for each $y \in Y$ and $\alpha < cf(\kappa)$, $\varphi(y) \notin B(S_\alpha, X_\alpha)$. Let $\{Y_\alpha : \alpha < cf(\kappa)\}$ be an increasing chain whose union is Y such that $|Y_\alpha| \leq w(X_\alpha)$ for each α. Now inductively choose x_α in the $G_{<w(X_\alpha)}$-set $\cap$ $\varphi_{\alpha,\gamma}^{-1}(x_\gamma)$ so that $\{x_\alpha\} \cup \{\varphi_\alpha(y) : y \in Y_\alpha\}$ is pairwise $\{f_\alpha : f \in S_\alpha\}$-incomparable. To see that this can be done, we notice that we just have to meet at most $w(X_\alpha)$ many $G_{<w(X_\alpha)}$-dense, $G_{<w(X_\alpha)}$-open subsets of X_α.

The referee observes that it is worthwhile making the following deduction from Theorem 4.2.

COROLLARY 4.7: Every infinite closed subset of N^* contains a family of 2^c Rudin–Keisler incomparable points.

Proof: Let Y be any closed subset of N^* and let D be any countable infinite discrete subset. Let φ be any function from D onto a dense subset of 2^c. It follows from Proposition 2.1 that $\overline{D}$ is homeomorphic to βN and that φ extends to a continuous map from $\overline{D}$ onto 2^c. As mentioned just before Theorem 3.5, there is a closed subset X of $\overline{D}$ such that $\varphi \upharpoonright X$ is irreducible. For each $f \in N^N$ there is an extension to a function, β*f*, from β*N* to β*N*. For each such *f*, let f_X denote the restriction β*f* ↾ $(X \cap \beta f^{-1}(X))$. Now apply Theorem 4.2 using the collection $\mathscr{S} = \{f_X : f \in N^N\}$. □

We are also grateful to Andreas Blass who supplied us with the following example. To motivate the example we remark that one is tempted to prove that every embedded copy of β*N* contains a large family of Rudin–Keisler incomparable points because the embedding should preserve the incomparability from the original Rudin and Shelah constructed family. However, this next result shows that, at least consistently, this is not the case.

PROPOSITION 4.8 (Blass): *If there are at least three $\leq_{RK}$-distinct $\leq_{RK}$-minimal free ultrafilters, then there are two $\leq_{RK}$-incomparable p, q and an embedding f of β*N* into itself such that $f(p) \leq_{RK} f(q)$.*

Proof: Given two ultrafilters *p, q* on *N*, let $p \otimes q$ be any ultrafilter that is a *p*-limit of a discrete collection $\{q_n : n \in N\}$ such that each q_n is $\leq_{RK}$-equivalent to *q*. Another way to think of this is to first construct the following ultrafilter $\mathscr{U}$ on $N \times N$: a set $U \in \mathscr{U}$ if and only if

$$\{m \in N : \{n \in N : (m, n) \in U\} \in q\} \in p.$$

Then any embedding of $N \times N$ into *N*, sends $\mathscr{U}$ to a suitable copy of $p \otimes q$. It is shown in [8] that this operation is associative.

If *p* and *q* are $\leq_{RK}$-minimal, then $\{p, q, p \otimes q\}$ is equal (up to equivalence) to $\{r \in N^* : r \leq_{RK} p \otimes q\}$. This is essentially proved in Theorem 4.2 when we proved there that *p* was the only -predecessor to *r*.

Now suppose that *p, q, r* are all distinct $\leq_{RK}$-minimals. By the preceding it follows that $p \otimes q$ and $p \otimes r$ are $\leq_{RK}$-incomparable. We will find an embedding that sends $p \otimes q$ to an ultrafilter that is $\leq_{RK}$ above that to which $p \otimes r$ is sent. Indeed, fix any ultrafilter *z* and send $p \otimes q$ to $p \otimes q \otimes r \otimes z$ and $p \otimes r$ to $p \otimes r \otimes z$. We can easily do this as follows: fix complementary $A, B \subset N$ such that $A \in p \otimes q$ and $B \in p \otimes r$. Fix a discrete sequence $\{d_n : n \in A\}$ of "copies" of $r \otimes z$ and another discrete sequence $\{d_n : n \in B\}$ of "copies" of *z* such that $\{d_n : n \in N\}$ is discrete. It should be clear that by extending this embedding, $p \otimes q$ is sent to $(p \otimes q) \otimes (r \otimes z) = p \otimes q \otimes r \otimes z$ while $p \otimes r$ is sent to $(p \otimes r) \otimes z = p \otimes r \otimes z$. Consider $p \otimes q \otimes r \otimes z$ as an ultrafilter on N^4. For each $(l, n, o) \in N^3$, let $A_{(l,n,o)} = \{l\} \times N \times \{(n, o)\}$. It is routine to check that $p \otimes q \otimes r \otimes z$ is the $p \otimes r \otimes z$-limit of the sequence $\{A_{(l,n,o)} : (l, n, o) \in N^3\}$. The result now follows by Proposition 2.2. □

5. NCF AND COMPOSANTS

A very important and exciting development in this subject in recent years is the formulation of NCF by Blass and the proof, by Shelah, that it is consistent. Mary Ellen's involvement in this comes from her paper in which she proves that it is consistent that there are 2^c distinct composants in the Stone–Čech remainder of the half-line ℍ (i.e., ℍ $= [0, \infty)$). NCF is equivalent to the statement that there is only one composant. NCF has a great deal of equally (or more) important applications (see [4, 5]). In addition, the study of the remainder of the real line has undergone something of a revival in recent years with the fine work of people such as Baldwin, Smith, Yu, and most of all Zhu. K. P. Hart has just finished a survey on the topic [14].

A pair x, y of a space X are in the same *composant* if there is a proper closed connected set C containing them both. A *continuum* is a compact connected space, and a continuum is called an *indecomposable continuum* if it cannot be written as the union of two proper subcontinua. If X is an indecomposable continuum, then the relation of being in the same composant is a transitive, hence an equivalence, relation. Mazurkiewicz proved that every metric indecomposable continuum contains a perfect set of points, each in a distinct composant. So after Woods and Bellamy had, independently, shown that $\mathbb{H}^*$ was an indecomposable continuum it was natural to want to estimate the number of composants.

Not surprisingly the structure of $\mathbb{H}^*$ is very tied up with the structure of N^* (although we still do not fully understand the relationship). For example, from [14], if we fix any sequence $[a_n, b_n]$ of disjoint intervals such that $b_n < a_{n+1}$ for each n, then, for any $p \in N^*$, the set of p-limits of the sequence $\{[a_n, b_n] : n \in N\}$ is a proper subcontinuum. In fact, it is not too difficult to see that any proper subcontinuum is contained in such a *standard* subcontinuum (any proper regular closed set containing the subcontinuum will induce the partition of $\mathbb{H}$ into intervals $\{[a_n, b_n] : n \in N\}$ and then each subset of N will correspond to a clopen subset of the remainder of the union of the intervals). Hence if two standard subcontinua are contained in a single proper subcontinuum there is a natural finite-to-one map induced on the intervals.

DEFINITION 5.1: Two filters on N are said to be *nearly coherent* if there is a finite-to-one function $f : N \to N$ that takes both the filters into a single filter. NCF is the assertion that all ultrafilters are nearly coherent.

THEOREM 5.1: $\mathbb{H}^*$ has only one composant if and only NCF holds.

Mary Ellen showed that if two ultrafilters on N when considered as points in $\mathbb{H}^*$ are in the same composant, then they are nearly coherent. She also showed that, assuming the continuum hypothesis, there is a set of 2^c ultrafilters on N, no two of which are nearly coherent (e.g., any family of $\leq_{RK}$-minimals).

Zhu has proved the remarkable fact that NCF is equivalent to the assertion that $\mathbb{H}^*$ can be covered by an increasing chain of nowhere dense P-sets. It is not immediately clear that it follows that N^* itself can then be covered by an increasing chain of such sets. However, Zhu has established this fact (it can also be easily deduced from [4, theorem 19]). Blass had already established that NCF implies that N^* can be covered by nowhere dense P-sets (even of a special kind). The question of whether a space can be covered by such sets arises in a variety of settings. One such setting is the question of whether the product of (κ-many) sequentially compact spaces is countably compact—the Scarborough–Stone problem. To give an example of how NCF can be applied we give a direct proof that it implies that N^* can be covered by a chain of nowhere dense P-sets (but we do not know if this is equivalent to NCF). Also, following a suggestion of J. Vaughan, we will modify one of the forcings that Shelah showed will produce a model of NCF to produce a model in which there is a noncountably compact product of $\mathfrak{h}$ (defined below) sequentially compact spaces (this answers question 343 of [23]).

LEMMA 5.2 ([4]): NCF implies that there is a P-point of character less than $\mathfrak{d}$. Hence every point is nearly coherent with a P-point, which in turn implies that N^* can be covered by nowhere dense P-sets.

Proof: By the definition of $\mathfrak{d}$ we can fix a family of strictly increasing functions $\{f_\alpha : \alpha < \mathfrak{d}\} \subset {}^N N$ of minimum cardinality such that for every function $g \in {}^N N$ there is an α such that $g(n) < f_\alpha(n)$ for all $n \in N$. Since every ultrafilter of character less than $\mathfrak{d}$ is a P-point [16], we need only show that there is some ultrafilter of character less than $\mathfrak{d}$. Assume there is not, fix any $p \in N^*$, and construct $q \in N^*$, which is not nearly coherent. For each $\alpha < \mathfrak{d}$ and $i < 4$, define $h_{i,\alpha} \in {}^N N$ by $h_{i,\alpha}([f_\alpha^{(4n+i)}(0), f_\alpha^{(4(n+1)+i)}(0))) = n$ and $h_{i,\alpha}([0, f_\alpha^i(0))) = 0$. We first show that an ultrafilter q is nearly coherent with p if and only if there is an $\alpha < \mathfrak{d}$ and an $i < 4$ such that $h_{i,\alpha}$ takes both to the same (ultra)filter. Fix any finite-to-one g that takes p, q to the same filter and fix an α such that for each k, $g^{-1}(g(k)) \subset [0, f_\alpha(k))$. Choose $i < 4$ so that $I' = \bigcup_n [f_\alpha^{(4n+i+1)}(0), f_\alpha^{(4n+i+2)}(0))$ is in p. Let $I \in p$ be arbitrary and $J \in q$. We must show that $h_{i,\alpha}(I) \cap h_{i,\alpha}(J) \neq \varnothing$. Indeed, choose any $k \in g(I \cap I') \cap g(J)$. Fix $l \in I \cap I'$ and $j \in J$ such that $g(l) = g(j) = k$. Since $l \in I'$, there is an n such that $f_\alpha^{(4n+i+1)}(0) \le l < f_\alpha^{(4n+i+2)}(0)$. It follows that $j < f_\alpha^{(4n+i+3)}(0)$ since $g^{-1}(g(j)) = g^{-1}(g(i)) \subset f_\alpha(i) \le f_\alpha(f_\alpha^{(4n+i+2)}(0))$. Similarly, $f_\alpha^{(4n+i)}(0) \le j$ since $g^{-1}(g(j)) \subset f_\alpha(j)$, while $i \in g^{-1}(g(j)) - f_\alpha(f_\alpha^{(4n+i)}(0))$. Therefore, $h_{i,\alpha}(j) = h_{i,\alpha}(l) = n \in h_{i,\alpha}(I) \cap h_{i,\alpha}(J)$.

Reenumerate the set $\{h_{i,\alpha} : i < 4 \text{ and } \alpha < \mathfrak{d}\}$ as $\{h_\alpha : \alpha < \mathfrak{d}\}$. Now inductively build an increasing sequence of filter bases $(q_\alpha : \alpha < \mathfrak{d})$, each of cardinality at most $|\alpha| \cdot \omega$ as follows. At stage α, $h_\alpha(p)$ is an ultrafilter and $h_\alpha(q_\alpha)$ is not (since it has character less than $\mathfrak{d}$). Therefore, there is some set Y_α such that $N - Y_\alpha \in h_\alpha(p)$ and Y_α meets each member of $h_\alpha(q_\alpha)$. Let $q_{\alpha+1}$ be generated by $q_\alpha \cup \{h_\alpha^{-1}(Y_\alpha)\}$. Any ultrafilter q that contains $\bigcup_{\alpha < \mathfrak{d}} q_\alpha$ will not be nearly coherent with p. □

PROPOSITION 5.3 (Zhu): NCF implies that N^* can be covered by an increasing chain of nowhere dense P-sets.

Proof: Fix a P-point p_0 with character less than $\mathfrak{d}$ and let f_0 be the identity function on N. We inductively construct P-point ultrafilters p_α and finite-to-one functions f_α for $\alpha < \mathfrak{d}$ so that $P_\alpha = (\beta f_\alpha)^{-1}(p_\alpha)$ is the desired chain. In addition, for each α there will be a function h_α such that $f_\alpha(n) = [h_\alpha^n(0), h_\alpha^{n+1}(0))$.

Let us first show that if $\{h_\alpha : \alpha < \mathfrak{d}\}$ is a dominating family of increasing functions, then each ultrafilter is in some P_α. Let q be any ultrafilter and let g be any finite-to-one function that witnesses that q and p_0 are nearly coherent. For each $X \in p_0$, define $h_X(n)$ to be large enough so that there are $j_0, j_1 \in X$ such that $\max(j_0 \cup g^{-1}(g(j_0))) < \min(j_1 \cup g^{-1}(j_1))$ and the interval $[n, h_X(n)]$ contains $\{j_0, j_1\} \cup g^{-1}(\{g(j_0), g(j_1)\})$. Find an α such that, for each X from a base for p_0, $h_\alpha(n) > h_X(n)$ for infinitely many n. We want to show that $q \in P_\alpha$. Let $Y = \{j : (\exists n)\ g^{-1}(g(j)) \subset f_\alpha^{-1}(n)\}$. It is sufficient to show that $Y \in p_0$, that is, that it meets every member of the base for p_0. To see this, suppose that it is true. Then $g^{-1}(g[Y])$ is in q and, for any $Z \in q$ with $g[Z] \subset g[Y]$, $f_\alpha(Z)$ is equal to $f_\alpha(g^{-1}(g[Z]))$. But now this is a member of p_α since $g^{-1}(g[Z]) \in p_0$ and f_α takes every member of p_0 to a member of p_α.

To show that $Y \in p_0$, let X be any member of the base for p_0. Fix any m such that $h_\alpha(m) > h_X(m)$ and fix $j_0, j_1 \in X$ as in the definition of $h_X(m)$. Let n be such that $m \in [h_\alpha^n(0), h_\alpha^{n+1}(0))$. Now either $g^{-1}(g(j_0)) \subset [h_\alpha^n(0), h_\alpha^{n+1}(0))$ (hence $j_0 \in X \cap Y$) or, if this is not the case, then $h_\alpha^{n+1}(0) \leq \max(g^{-1}(j_0)) < \min(g^{-1}(j_1))$. Since we also have that $h_\alpha^{n+2}(0) = h_\alpha(h^{n+1}(0)) \geq h_\alpha(m) > h_X(m)$, it follows that $j_1 \cup f^{-1}(j_1) \subset [h_\alpha^{n+1}(0), h_\alpha^{n+2}(0))$, hence $j_1 \in X \cap Y$.

Now suppose we have chosen p_α and h_α for each $\alpha < \beta < \mathfrak{d}$. Then in place of g and p_0 given earlier, do the same thing using f_α (hence the image of p_0 under f_α will be p_α). We can find an increasing function h (bigger than any given member of a dominating family) such that h "works" for each f_α (i.e., h is sufficiently large in the sense that the h_α given earlier was sufficiently large). This will be our h_β, and p_β is simply the image of p_0 under h_β. In the preceding paragraph, we can observe that we showed that any q that was taken by g to the same thing as p_0 was sent by f_α to the same thing as p_0. Therefore, in this case, $P_\alpha \subset P_\beta$ for all $\alpha < \beta$. $\square$

Now we answer question 343 in [23] by showing that there is a model in which there is a noncountably compact product of $\mathfrak{h}$ sequentially compact spaces. First some definitions. A *tower* is a maximal descending mod finite sequence of infinite subsets of N. The cardinal $\mathfrak{t}$ is the minimum length of a tower. The cardinal $\mathfrak{h}$ can be defined as the minimum cardinal such that there is a chain of that length consisting of dense open subsets of N^* whose intersection has empty interior. Of course, $\mathfrak{t}$ is just the minimum length of a chain of clopen subsets of N^* that has empty interior. It is easy to see that if "chain" is removed from the definition of $\mathfrak{h}$, then the value is unchanged, but it is not known if this is the case for $\mathfrak{t}$. The study of the cardinal $\mathfrak{h}$ (called the Baire number of N^*) was begun in [1].

It is straightforward to show that the product of $\mathfrak{t}$ sequentially compact spaces is countably compact, and it was shown by Simon that the product of fewer than $\mathfrak{h}$ such spaces is countably compact. It is known (see [26]) that if N^* can be covered by a family of κ nowhere dense sets that are each equal to the intersection of a decreasing chain of clopen sets, then there is a family of κ many sequentially compact spaces whose product is not countably compact. We produce a model in which N^* is covered by $\mathfrak{h}$-many such nowhere dense sets.

Let $\mathbb{Q}$ be the rational perfect poset as defined in [25] and let $\mathbb{M}$ be the Mathias poset as defined in [22]. Start with a model of CH and fix disjoint stationary subsets of ω_2, S_0, and S_1, consisting of ordinals of cofinality ω_1. Now perform an ω_2-length countable support iteration $\langle P_\alpha, Q_\alpha : \alpha < \omega_2 \rangle$ in which, at limits $\lambda \in S_0$ we use $Q_\lambda = \mathbb{M}$ and at all other α, we use $Q_\alpha = \mathbb{Q}$.

There is really nothing new to prove. Just collect together some published lemmas.

LEMMA 5.4: If $\langle P_\alpha, Q_\alpha : \alpha < \omega_2 \rangle$ is a countable support proper iteration of proper posets of cardinality $\aleph_1$ such that, Q_λ is $\mathbb{M}$, for a stationary set of λ in ω_2, each with cofinality ω_1, then $\mathfrak{h}$ is ω_2.

Proof: This lemma is essentially proven in [12], but we sketch the main idea. Recall that $\mathfrak{h}$ is equal to the minimum number of maximal antichains of $\mathscr{P}(N)/fin$ for which there is no subset of N that is almost contained in a member from each antichain. Suppose that we have a collection of fewer than ω_2 maximal antichains in the model $V(G)$ described in the statement of the lemma. We apply a straightforward reflection argument to observe that there is a cub subset C of ω_2 such that for every $\lambda \in C$ the intersection of each of the antichains will be maximal in

$$\mathscr{P}(N) \cap \bigcup_{\beta < \lambda} V[G_\beta].$$

This is only useful for λ with cofinality ω_1 since in this case $\cup_{\beta<\lambda} V[G_\beta] \cap \mathscr{P}(N) = V[G_\lambda] \cap \mathscr{P}(N)$. Now we can choose a $\lambda \in C$, such that Q_λ is Mathias forcing. It is well-known that $\mathbb{M}$ will introduce a set that is almost included in some member of the restriction of each of the maximal antichains.   □

The following lemma is one of the key steps in showing that NCF holds in the model obtained by iterating $\mathbb{Q}$. It can be proved, for example, by combining lemma 1 and theorem 3 of [6].

LEMMA 5.5 (CH): Forcing with $\mathbb{Q}$ introduces a finite-to-one function f and a P-point p such that the filter $\{f^{-1}(X) : X \in p\}$ is contained in the filter generated by every ground model ultrafilter.

The general iteration lemmas of Shelah in [31] imply that a proper iteration of posets that preserves all towers (i.e., all towers in the ground model are again maximal in the extension) will itself preserve all towers. Baumgartner and Dordal showed that Mathias forcing preserves all towers. By a straightforward generalization one can show that rational perfect forcing also does, but we do not prove it here. Hence the next lemma follows.

LEMMA 5.6: If T is a maximal descending mod finite family of subsets of N, then it remains so after any countable support iteration with iterands from $\{\mathbb{Q}, \mathbb{M}\}$.

We outline how to verify that the model $V[G]$ described earlier has the desired properties. Clearly $\mathfrak{h}$ will be ω_2 by Lemma 5.4. For each $\alpha \notin S_0$, consider the inner model $V[G \cap P_\alpha]$. By Lemma 5.4, we obtain the finite-to-one function f_α and a P-point p_α with the properties as stated. Hence, in the inner model $V[G \cap P_{\alpha+11}]$, $f_\alpha^{-1}(p_\alpha)$ contains a maximal descending mod finite tower. This tower remains maximal in $V[G]$ by Lemma 5.6. To finish the proof we have to show that every ultrafilter q in $V[G]$ contains one of these towers. By a standard reflection argument, we can fix a name for q and obtain a cub C in ω_2 such that for every $\lambda \in C \cap S_1$, $q_\lambda = q \cap V[G \cap P_\lambda]$ is a member of βN in the sense of $V[G \cap P_\lambda]$. Finally, we have, by Lemma 5.5, that $q \supset q_\lambda \supset f_\lambda^{-1}(p_\lambda)$.

REFERENCES

1. BALCAR, B., J. PELANT & P. SIMON. 1980. The space of ultrafilters on N covered by nowhere dense sets. Fundam. Math. **110**: 11–24.
2. BERNSTEIN, A. R. 1970. A new kind of compactness for topological spaces. Fundam. Math. **66**: 185–193.
3. BLASS, A. 1970. Orderings on ultrafilters. Ph.D. Dissertation, Harvard University, Cambridge, Mass.
4. ———. 1986. Near coherence of filters, I: Cofinal equivalence of models of arithmetic. Notre Dame J. Formal Logic **27**: 579–591.
5. ———. 1987. Near coherence of filters II, Applications to operator ideals, the Stone-Čech remainder of a half-line, order ideals of sequences, and slenderness of groups. Trans. Am. Math. Soc. **300**: 557–581.
6. ———. 1987. Applications of superperfect forcing and its relatives. *In* Set Theory and Its Applications. Lecture Notes in Mathematics **1401**: 18–40.
7. ———. 1993. Private communication. January.
8. BOOTH, D. D. 1970. Ultrafilters on a countable set. Ann. Math. Logic **2**: 1–24.

9. BUTKOVIČOVA, E. 1982. Ultrafilters without immediate predecessors in Rudin-Frolík order. Commun. Math. Univ. Carolinae **23:** 757–766.

10. ———. 1993. A remark on incomparable ultrafilters in the Rudin-Frolík order. Proc. Am. Math. Soc. In press.

11. COMFORT, W. W. & S. NEGREPONTIS. 1974. The Theory of Ultrafilters. Springer-Verlag. New York.

12. DOW, A. 1989. Tree pi-bases for N^* in various models. Topol. Appl. **33**(1): 3–19.

13. DOW, A., A. V. GUBBI & Z. SZYMANSKI. 1988. Rigid stone spaces within ZFC. Proc. Am. Math. Soc. **102:** 745–748.

14. HART, K. P. 1992. The Čech-Stone compactification of the Real line. *In* Reports on the Prague Conference on General Topology. North-Holland. Amsterdam, the Netherlands.

15. HINDMAN, N. 1988. Is there a point in ω^* that sees all others? Proc. Am. Math. Soc. **104:** 1235–1238.

16. KETONEN, J. 1976. On the existence of P-points in the Stone-Čech compactification of integers. Fundam. Math. **62:** 91–94.

17. KUNEN, K. 1970. On the compactification of the integers. Not. Am. Math. Soc. **17:** 299.

18. ———. 1976. Some points in βN. Proc. Cambridge Phil. Soc. **80:** 385–398.

19. ———. 1978. Weak P-points in N^*. *In* Society Colloquia Mathematica János Bolyai, Vol. 23: 741–749. Budapest.

20. ———. 1990. Large homogeneous compact spaces. *In* Open Problems in Topology: 261–270. North-Holland. Amsterdam, the Netherlands.

21. LAFLAMME, C. 1989. Forcing with filters and complete combinatorics. Ann. Pure Appl. Logic **42:** 125–163.

22. MATHIAS, A. 1977. Happy families. Ann. Math. Logic **12:** 59–111.

23. VAN MILL, J. & G. M. REED. 1990. Open Problems in Topology. North-Holland. Amsterdam, the Netherlands.

24. VAN MILL, J. 1984. An introduction to βN. *In* Handbook of Set-Theoretic Topology, K. Kunen and J. E. Vaughan, Eds. North-Holland. Amsterdam, the Netherlands.

25. MILLER, A. W. 1984. Rational perfect set forcing. *In* Axiomatic Set Theory. Contemporary Mathematics. American Mathematical Society. Providence, R. I.

26. NYIKOS, P. & J. E. VAUGHAN. 1987. Sequentially compact, Franklin-Rajagopalan spaces. Proc. Am. Math.Soc. **101:** 149–155.

27. RUDIN, M. E. 1966. Types of ultrafilters. Ann. Math. Studies **60:** 147–151.

28. ———. 1971. Partial orders on the types in βN. Trans. Am. Math. Soc. **155:** 353–362.

29. ———. 1975. Lectures on Set Theoretic Topology. CBMS Regional Conference Series in Mathematics, Vol. 23.

30. ———. 1977. Martin's Axiom. *In* Handbook of Mathematical Logic, J. Barwise, Ed.: 491–501. North-Holland. Amsterdam, the Netherlands.

31. SHELAH, S. 1984. On cardinal invariants of the continuum. *In* Axiomatic Set Theory. Contemporary Mathematics. American Mathematical Society. Providence. R. I.

32. SHELAH, S. & M. E. RUDIN. 1978. Unordered types of ultrafilters. Topol. Proc. **3:** 199–204.

33. SIMON, P. 1983. Applications of independent linked families. Colloquia Mathematica Society János Bolyai **41:** 561–580. Budapest.

Theorems from Measure Axioms, Counterexamples from $\diamondsuit^{++}$

WILLIAM FLEISSNER[a]

Department of Mathematics
University of Kansas
Lawrence, Kansas 66044

ABSTRACT: Daniels' proof that after adding supercompact many random reals, normal k'-spaces are collectionwise normal is simplified. Examples of how to use $\diamondsuit^{++}$ to construct spaces are given. Question 36 of Watson is answered.

1. INTRODUCTION

My assigned topic is the normal Moore space conjecture (NMSC). Although there has not been enough progress in this area since [12, 19] to justify an entire survey article, the recent work of Balogh and Daniels deserves attention and discussion, so that will be the topic of the first part of this article. The second part of this article introduces $\diamondsuit^{++}$ (diamond double plus) and illustrates its use. This axiom, a consequence of $V = L$, was first used in the context of NMSC, but soon after she saw $\diamondsuit^{++}$ Mary Ellen Rudin used it to construct a screenable Dowker space [18]. Later, collaborating with Bešlagić [3], she used $\diamondsuit^{++}$ to construct a normal such that every monotone increasing open cover shrinks, yet there is an open cover that does not shrink. Clearly Mary Ellen has used $\diamondsuit^{++}$ more skillfully and effectively than its formulator. After the details of Axiom $\diamondsuit^{++}$, I present a simpler construction of the Bešlagić–Rudin example. I hope that this exposition makes her techniques available to more researchers. The third part of this article is a correct construction of the space, Son of George [11].

We start the first part with some observations after reading [5]. Theorem 1.1 seems to be new; the proof here of Theorem 1.4 is much shorter than Daniels'.

We say that a topological space $(X, \mathcal{T})$ is *determined* by a family of subspaces $\mathcal{Z}$ when a set F is closed in X iff $F \cap Z$ is closed in Z for all $Z \in \mathcal{Z}$. Equivalently, a set U is open in X iff $U \cap Z$ is open in Z for all $Z \in \mathcal{Z}$. We list some examples.

1. X is sequential iff X is determined by $\{Z \subset X : Z \cong \omega + 1\}$.
2. X has countable tightness iff X is determined by $\{Z \subset X : Z$ is countable$\}$.
3. X is a *k-space* iff X is determined by $\mathcal{K}(X) = \{K \subset X : K$ is compact$\}$.

Mathematics Subject Classification. Primary 54A35, 03E35.
Key words and phrases. Mary Ellen Rudin, normal Moore space conjecture, Moore space, k'-space, normal, collectionwise normal, $V = L$, diamond double plus, shrinking.
[a]I gladly and enthusiastically express my thanks to Mary Ellen Rudin for her guidance and inspiration as a mathematician, and her warm acceptance of me as a friend. I am honored to contribute a paper to this volume dedicated to her.

A locally compact space is of point-countable type; a space of point-countable type is a k-space. Therefore the following conjecture would extend recent results of Balogh [1].

CONJECTURE: Let κ be supercompact, and let G be either Meas(κ)-generic or $Fn(\kappa, 2)$-generic over V. In $V[G]$, normal k-spaces are collectionwise normal and countably paracompact k-spaces are expandable.

A *weak base* for a space $(X, \mathcal{T})$ is a family $\mathbb{W} = \{\mathcal{W}(x) : x \in X\}$ satisfying:

(i) For each $x \in X$, $\mathcal{W}(x)$ is a filter base with $x \in \cap \mathcal{W}(x)$
(ii) For all $S \subset X$, S is open iff for all $x \in S$, there is $W_x \in \mathcal{W}(x)$ such that $W_x \subset S$.

We say that X has *weak character below* $\mathfrak{c}$ iff X has a weak base $\mathbb{W}$ such that $|\mathcal{W}(x)| < \mathfrak{c}$ for all $x \in X$.

The idea of using weak bases with the Product Measure Extension Axiom (PMEA [17]) seems to be due to Junnila (see [12, p. 738]), and we illustrate with the following theorem.

THEOREM 1.1: Assume PMEA. Let $(X, \mathcal{T})$ be a normal space with topology determined by $\mathcal{Z} = \{Z \subset X : Z$ has weak character below $\mathfrak{c}\}$. Then X is collectionwise normal.

Proof: Let $Y = \{Y_\alpha : \alpha < \lambda\}$ be a closed discrete family in X. Following Nyikos, for each $f \in 2^\lambda$ use normality to choose disjoint open sets $U(f, 0)$ and $U(f, 1)$ so that $Y_\alpha \subset U(f, f(\alpha))$. For $\alpha < \lambda$ and $x \in X$, set $A(\alpha, x) = \{f : x \in U(f, f(\alpha))\}$. Define

$$V_\alpha = \{x \in X : m(A(\alpha, x)) > \tfrac{7}{8}\}.$$

Clearly, $Y_\alpha \subset V_\alpha$. By Nyikos' argument, $\{V_\alpha : \alpha < \lambda\}$ is disjoint.

Fix $\alpha < \lambda$. To show that V_α is open, it suffices to show that $V_\alpha \cap Z$ is open in Z for all $Z \in \mathcal{Z}$. Fix $Z \in \mathcal{Z}$. Let $\{\mathcal{W}(x) : x \in Z\}$ be a weak base for Z with $|\mathcal{W}(x)| < \mathfrak{c}$ for all x. For $x \in V_\alpha \cap Z$. For each $f \in A(\alpha, x)$, there is $W_f \in \mathcal{W}(x)$ with $W_f \subset U(f, f_\alpha)$. Because $m(A(\alpha, x)) > \tfrac{7}{8}$, m is $\mathfrak{c}$-additive, and $\mathcal{W}(x)$ is a filter base, there is $W \in \mathcal{W}(x)$ so that $m(\{f : W \subset W_f\}) > \tfrac{7}{8}$. Therefore $W \subset V_\alpha$, hence $V_\alpha \cap Z$ is open in Z by the weak base property. $\square$

Let $(X, \mathcal{T})$ be a k-space. Set $\mathcal{T}_x = \{T \in \mathcal{T} : x \in T\}$. We define a weak base $\mathbb{W}_K$ for X. For $x \in X$, let $\Psi(x)$ be the set of functions $\psi : \mathcal{K}(X) \to \mathcal{T}_x$. For $\psi \in \Psi(X)$, define $W(\psi) = \cup\{K \cap \psi(K) : K \in \mathcal{K}(X)\}$. Set $\mathbb{W}_K(x) = \{W(\psi) : \psi \in \Psi(x)\}$. It is routine to verify that $\mathbb{W}_K = \{\mathcal{W}_K(x) : x \in X\}$ is a weak base. Call A a k-*neighborhood* of S if for each $x \in S$, there is $W \in \mathcal{W}_K(x)$ such that $W \subset A$. Alternately, A is a k-neighborhood of S if for all $K \in \mathcal{K}(X)$, if $x \in K \cap S$, then there is an open U such that $x \in U \cap K \subset A$. Note that if for all $n \in \omega$, S_{n+1} is a k-neighborhood of S_n, then $\cup\{S_n : n \in \omega\}$ is open. We say that $(X, \mathcal{T})$ is a k'-*space* if whenever A is a k-neighborhood of S, then S is in the interior of A.

We use the following lemmas in the proof of Theorem 1.4. The routine proofs are left to the reader.

LEMMA 1.2: Let $(X, \mathcal{T})$ be a k-space and let $\mathcal{Y} \subset \mathcal{P}(X)$. $\mathcal{Y}$ is discrete iff every $\mathcal{Y}' \in [\mathcal{Y}]^\omega$ is discrete.

LEMMA 1.3: Let Z be compact and let $\mathscr{U}$ be a family of open subsets of Z. If for all $U, U' \in \mathscr{U}$, there is $U'' \in \mathscr{U}$ such that $\overline{U''} \subset U \cap U'$, then $\mathscr{U}$ is an outer base for $\cap \mathscr{U}$.

THEOREM 1.4 (Daniels [5]): Let κ be supercompact, and let G be Meas(κ)-generic over V. In $V[G]$, normal k'-spaces are collectionwise normal.

Proof: In $V[G]$, let $\mathscr{Y} = \{Y_i : i \in I\}$ be a discrete family in a k'-space $(X, \mathscr{T})$. Let β be sufficiently large. Let $j : V \to M$ be a supercompact embedding so that $j \upharpoonright \kappa$ is the identity, $j(\kappa) > \beta$, and M contains a Meas(κ)-name for $(X, \mathscr{T}, \mathscr{Y})$. Let H be Meas($j(\kappa)\backslash\kappa$)-generic over $V[G]$. In $V[G][H]$ there are G^* and parameters defining j^* so that $j^* : V[G] \to M[G^*]$ is an elementary embedding extending j. Note that $(X, \mathscr{T}, \mathscr{Y}) \in M[G^*]$. (See [7, 2.1, 2.2, 2.3, 3.3] or [13, 9.4] for proofs, details, explanations, etc.)

Now we work in $M[G^*]$. For $x \in X$, define $\{x\}^* = \cap\{j^*(U) : U \in \mathscr{T}_x\}$. For $i \in I$, define $Y_i^* = \cup\{\{y\}^* : y \in Y_i\}$. Define $\mathscr{Y}^* = \{Y_i^* : i \in I\}$.

CLAIM: $\mathscr{Y}^*$ is discrete in $(j^*(X), j^*(T)) \in M[G^*]$.

Because $(j^*(X), j^*(\mathscr{T}))$ is a k-space, by Lemma 1.2 it suffices to show that if $I' \in [I]^\omega$, then $\{Y_i^* : i \in I'\}$ is discrete. Because X is normal, $\mathscr{Y}$ is discrete, and I' is countable, in $V[G]$, there is $\{O_i : i \in I'\}$ discrete open in $(X, \mathscr{T})$ with $Y_i \subset O_i$ for all $i \in I'$. Then, in $M[G^*]$, $\{j^*(O_i) : i \in I'\}$ is discrete open in $(j^*(X), j^*(\mathscr{T}))$, and $Y_i^* \subset j^*(O_i)$. Hence $\{Y_i^* : i \in I'\}$ is discrete in $(j^*(X), j^*(T))$.

From H define a "generic" function $h : I \to 2$. (Here's how. Let $\eta : I \to (j(\kappa)\backslash\kappa)$ be one-one. Set $h(i) = e$ iff the set $F(i, e) = \{f \in {}^{(j(\kappa)\backslash\kappa)}2 : f(\eta(i)) = e\}$ is in H. More precisely, iff the equivalence class of $F(i, e)$ mod the ideal of measure zero sets is in H.) Because $j^*(X)$ is normal and $\mathscr{Y}^*$ is discrete, there are disjoint open sets T^0, T^1 such that $Y_i^* \subset T^{h(i)}$ for all $i \in I$. This construction can be done for every generic H, so in $V[G]$, there are names τ^0, τ^1 so that $[\![\tau^0, \tau^1$ are disjoint, open, etc.$]\!] = \mathbf{1}$. For $i \in I$, let τ_i be a name so that $[\![\tau_i = \tau^{h(i)}]\!] = \mathbf{1}$.

Define $V_i = \{x \in X : m([\![j^*x \in \tau_i]\!]) > \frac{7}{8}\}$. Clearly, $Y_i \subset V_i$, and Nyikos' argument shows that $\{V_i : i \in I\}$ is disjoint. Fix $i \in I$. Toward showing that V_i is a k-neighborhood of Y_i, fix $y \in Y_i$ and $K \in \mathscr{K}(X)$ with $y \in K$. In $M[G^*]$, note that $j^*(K)$ is compact and that $j^*(U) \cap j^*(K) = j^*(U \cap K)$. We apply Lemma 1.3 with $Z = j^*(K)$ and $\mathscr{U} = \{j^*(U \cap K) : U \in \mathscr{T}_y\}$, and conclude that $\mathscr{U}$ is a base for $\cap \mathscr{U} = \{y\}^* \cap K$. Thus, there is $U \in \mathscr{T}_y$ so that $\{y\}^* \cap K \subset (j^*(U \cap K)) \subset T^{h(i)}$. Because this is valid for all generic H, there is in $V[G]$ a name σ such that $[\![\sigma \in T_y]\!] = \mathbf{1} = [\![j^*(\sigma \cap \hat{K}) \subset \tau_i]\!]$. Because $\mathscr{T}_y$ is a filter base, there is $O \in \mathscr{T}_y$ such that $m([\![j^*(\hat{O} \cap \hat{K}) \subset \tau_i]\!]) > \frac{7}{8}$. For all $x \in O \cap K$, $m([\![j^*x \in \tau_i]\!]) > \frac{7}{8}$; hence $O \cap K \subset V_i$. Because X is a k'-space, $\{$interior $V_i : i \in I\}$ is a disjoint family of open sets separating $\mathscr{Y}$. We conclude that X is collectionwise normal.

REMARKS AND VARIATIONS: Theorems 1.1, 1.4, and Lemma 1.2 are valid when countably paracompact replaces normal, expandable replaces collectionwise normal, and locally finite replaces discrete. Essentially the same proofs work. How to overcome the additional technical difficulties is illustrated in [1] and [13].

Note that in the proof of Theorem 1.4 we started with $m([\![\{y\}^* \subset \tau_i]\!]) = \mathbf{1}$, and ended with $m([\![j^*x \in \tau_i]\!]) > \frac{7}{8}$. The analogue of going from measure 1 to measure $\frac{7}{8}$

can be done in RegOpen(2^κ) using endowments. So Theorem 1.4 is valid with Meas(κ) replaced by Fn(κ, 2). The analogue of going from measure $\frac{7}{8}$ to measure $\frac{7}{8}$ cannot be done in RegOpen(2^κ), as discussed in [7, remark 3.4]. Thus the proof of Theorem 1.1, with Meas(κ) replaced by Fn(κ, 2), cannot be translated to an endowment proof. It is, of course, plausible that the analogue can be proved by a different proof.

It seems that all we need from j is that $(X, \mathcal{T}, \mathcal{Y}) \in M[G^*]$ (so that we can define $\{x\}^*, Y_i^*, \mathcal{Y}^*)$ and that $|j(\kappa)\backslash\kappa| > |I|$ (so that we can define the "generic" h). So a large cardinal assumption weaker than supercompact will suffice. Also, the conclusion of Theorem 1.1 can be proved assuming the Normal Measure Axiom (see [13]).

2. USING $\diamondsuit^{++}$ TO CONSTRUCT SPACES

We begin the second part by explaining how $\diamondsuit^{++}$ relates to NMSC. On the basis of [9], it was conjectured that $V = L$ implies the NMSC. A lot of effort was expended chasing this false scent. The first application of $\diamondsuit^{++}$ [11] was to construct a first-countable, normal, not collectionwise normal space. Of course, such spaces had been constructed assuming MA_{ω_1}—the point was to get one in L. This first application of $\diamondsuit^{++}$ was soon superceded by the construction of a normal nonmetrizable Moore space from the Continuum Hypothesis—a better conclusion from a weaker hypothesis!

After seeing $\diamondsuit^{++}$, Devlin [6] formulated and proved from $V = L$ the stronger axiom $\diamondsuit^\#$, which he describes as "simply a Π_2^1-reflecting sequence." More precisely, $\diamondsuit^\#$ asserts the existence of a sequence $(N_\alpha : \alpha < \omega_1)$ such that:

(i) N_α is a countable, transitive, primitively recursively closed set containing α.
(ii) If $X \subset \omega_1$, then there is a club set C such that $\alpha \in C \to X \cap \alpha, C \cap \alpha \in N_\alpha$.
(iii) $(N_\alpha : \alpha < \omega_1)$ is Π_2^1-reflecting.

To set theorists, and perhaps logicians in general, $\diamondsuit^\#$ may be simpler to express than $\diamondsuit^{++}$. However, the terms "Π_2^1-reflecting" and "primitively recursively closed set" need explanation to most mathematicians. Rather than explaining, we write down the specific instances of closure and reflection we need for the constructions in this paper and call the list $\diamondsuit^{++}$. The various published formulations of $\diamondsuit^{++}$ are very similar in spirit but different in detail because the list of specific instances needed differs. The version below is a long tedious list, but each item is combinatorial, rather than metamathematical.

The axiom $\diamondsuit^{++}$ asserts the existence of a triple $(\mathcal{A}, \Gamma, D)$ satisfying six clauses. Clause 0 simply state what types of objects $\mathcal{A}, \Gamma$, and D are. Clause 1 is essentially the statement of $\diamondsuit^+$ from [14, p. 83]. Clauses 2 and 3 give some properties that follow from the proof of $\diamondsuit^+$ from $V = L$. These properties show our concern about the notion of closed unbounded. The reflection aspects of $\diamondsuit^{++}$ are contained in clauses 4 and 5. These clauses should be compared to the following. Let κ be a weakly compact cardinal, and let D be the set of (strongly) inaccessible cardinals less than κ. Then D is stationary in κ, and for any χ, a set of less than κ many stationary subsets of κ, the following set is stationary in κ: $\{\delta \in D : (\forall X \in \chi)(X \cap \delta \text{ is stationary in } \delta)\}$. Indeed [2], $\diamondsuit^{++}$ holds in $M[G]$, where G is Lv(κ)-generic over M, a model of set theory, and $Lv(\kappa)$ is the Levy collapsing order defined and described in [14,chap. VII, 8.6–8.9].

DEFINITION: For δ an ordinal, let $Club(\delta)$ be the family of closed unbounded subsets of δ. If $\mathscr{F}$ is a filter base on a set I, then set $Stat(\mathscr{F}) = \{X \subset I : (\forall F \in \mathscr{F})(X \cap F \neq \emptyset)\}$.

DEFINITION: $\diamondsuit^{++}$ asserts that there are $\mathscr{A}$, Γ, and D, which satisfy:

(0.1) $\mathscr{A}$ is a sequence $(A_\alpha : \alpha < \omega_1)$.

(0.2) Each A_α is a countable subfamily of $\mathscr{P}(\alpha)$ containing all finite subset of α and all intervals (β, γ), where $\beta < \gamma \leq \alpha$.

(0.3) Each A_α is closed under union, intersection, and complement with respect to α.

(0.4) Γ is a function from $\mathscr{P}(\omega_1)$ to $Club(\omega_1)$.

(0.5) D is a stationary subset of ω_1.

(1) For all X in $\mathscr{P}(\omega_1)$, for all $\gamma \in \Gamma(X) : X \cap \gamma \in A_\gamma$ and $\Gamma(X) \cap \gamma \in A_\gamma$.

(2) For all X in $\mathscr{P}(\omega_1)$, for all $\gamma \in \Gamma(X) \cap D : sup(\gamma \cap \Gamma(X)) = \gamma$, and if $sup X = \omega_1$, then $sup(\gamma \cap X) = \gamma$.

(3) For all X in $Club(\omega_1) : \Gamma(X) \subset X$.

(4) For all $\delta \in D : Club(\delta) \cap A_\delta$ is a filter base.

(5) For all countable subfamilies χ of $Stat(Club(\omega_1))$: the set of $\delta \in D$ such that $\{X \cap \delta : X \in \chi\} \subset Stat(Club(\delta) \cap A_\delta)$ is stationary.

Perhaps the simplest illustration of $\diamondsuit^{++}$ is to construct a first-countable, normal topology finer than the order topology on ω_1 with the following two conflicting properties:

(a) Every club set has an uncountable clopen subset.

(b) There is a stationary set of nonisolated points.

The first item requires more open sets; it is satisfied by the discrete topology, but not by the order topology. The second item requires that we do not add too many new open sets; it is satisfied by the order topology but not the discrete topology. Assuming $\diamondsuit^{++}$, we construct a space Y with the two properties just given. Space Y is a continuous image of the Bešlagić–Rudin space Δ, and Y is the "tie-together space" of Son of George.

We present two constructions. The first is to simply declare open the additional sets we need to be open. This approach is conceptually clear, but working with a subbase is awkward. The second approach is to inductively define a topology, as Mary Ellen has done so often. The second topology might be finer than the first, but the verification of properties is essentially the same—just the construction is different.

Here is the first construction. We want final segments and sets of the form $\Gamma(X)$ to be open. To this end, set $\mathscr{C} = \{(\alpha, \omega_1) \subset \omega_1 : \alpha \in \omega_1\} \cup \{\Gamma(X) : X \in \mathscr{P}(\omega_1)\}$ and $U = [\mathscr{C}]^{<\omega}$. Sets C from $\mathscr{C}$ are closed (and unbounded) in the order topology. We create a new topological space Y with point set ω_1 by declaring sets C from $\mathscr{C}$ to be open and by isolating points α in $Y \backslash D$. Now we display a base $\mathscr{B}$ for Y. If $\alpha \notin D$, then $\{\alpha\} \in \mathscr{B}$. For $\delta \in D$, set $\mathscr{C}_\delta = \{C \in \mathscr{C} : \delta \in C\}$ (note that if $C \in \mathscr{C}_\delta$, then $C \cap \delta$ is club in δ) and set $U_\delta = [\mathscr{C}_\delta]^{<\omega}$. If $\delta \in D$, and $u \in U_\delta$, then $B(\delta, u) \in \mathscr{B}$ where

$$B(\delta, u) = (\delta + 1) \cap (\cap u).$$

In this context, if $u = \varnothing$, then $\bigcap u = \omega_1$. It is easy to check that $\mathscr{B}$ is a clopen base for a T_1, regular topology finer than the order topology. Observe that $\{B(\delta, u)\backslash\{\delta\} : u \in U_\delta\}$ is a subset of the countable set A_δ; hence Y is first countable even though U_δ is uncountable. Note that by (4) of $\diamondsuit^{++}$, a point $\delta \in D$ is not isolated. Further note that sets of the form $\Gamma(X)$ are clopen. The first construction is completed by Lemma 2.1.

LEMMA 2.1: Y is strongly zero-dimensional. To be more specific, for every two disjoint, closed subsets H and F of Y, there is a clopen set R such that $H \subset R \subset Y\backslash F$.

Proof: Toward a contradiction, we first assume that both H and F are stationary. By (5) of $\diamondsuit^{++}$, there is $\delta \in D \cap \overline{H} \cap \overline{F}$—contradiction! So, wlog we assume that $F \cap C = \varnothing$ for some club set C. Then $\Gamma(C)$ is clopen and $\Gamma(C) \cap F = \varnothing$. If $\alpha \in Y\backslash\Gamma(C)$, then $sup(\alpha \cap \Gamma(C)) < \alpha < min(\Gamma(C)\backslash(\alpha + 1))$. In this way, $\Gamma(C)$ partitions $Y\backslash\Gamma(C)$ into a discrete family $\mathscr{M}$ of countable open intervals. For $M \in \mathscr{M}$, let R_M be clopen in M such that $H \cap M \subset R_M \subset M\backslash F$. Then $R = \bigcup_{M\in\mathscr{M}}R_M \cup \Gamma(C)$ is as claimed. $\square$

In the second construction, we define $\mathscr{B}_\alpha$, a countable subfamily of $\mathscr{P}(\alpha + 1)$, by induction on $\alpha < \omega_1$. Having completed this induction, it is routine to verify that $\mathscr{B}_\alpha$ is a base for the point α in the topology on ω_1 generated by $\bigcup\{\mathscr{B}_\alpha : \alpha \in \omega_1\}$. Our induction hypothesis is:

If $\delta \in D \cap \Gamma(X)$, then there is $b \in \mathscr{B}_\delta$ such that $b \subset \Gamma(X)$.

If $\alpha \notin D$, set $\mathscr{B}_\alpha = \{\{\alpha\}\}$. Fix $\delta \in D$. List $A_\delta \cap Club(\delta)$ as $(c_i : i \in \omega)$, and list $Stat(Club(\delta) \cap A_\delta) = \{a \in A_\delta : (\forall i)\, a \cap c_i \neq \varnothing\}$ as $(a_i : i \in \omega)$, where each such a is listed ω times. By induction on n, choose $\beta_n \in a_n \cap (\bigcap\{c_i : i \leq n\})$. (This intersection is nonempty by (4) of $\diamondsuit^{++}$.) Further, choose $u_n \in \mathscr{B}_{\beta_n}$ so that $u_n \subset \bigcap\{c_i : i \leq n\}$. (This is possible by (2) of $\diamondsuit^{++}$ and the induction hypothesis. In constructions like this, we often require that $\beta_n, n \in \omega$ be strictly increasing and that the u_n be disjoint, but it does not seem to be necessary here.) Now set $\mathscr{B}_\delta = \{\{\bigcup\{u_n : n \geq m\} : m \in \omega\}$. It is routine to check that we have defined a locally countable, Hausdorff, zero-dimensional topology. A point $\delta \in D$ is not isolated by construction. If X is club, then $\Gamma(X)$ is an uncountable clopen subset. Normality is verified as in Lemma 2.1.

The Bešlagić–Rudin [3] example is called Δ because the point set is the "lower triangle," $\{(\alpha, \xi) \in \omega_1 \times \omega_1 : \xi < \alpha < \omega_1\}$. We repeat the second construction, adding a few "bells and whistles." The following notation will be convenient. Define $\pi : \Delta \to \omega_1$ by $\pi(\alpha, \xi) = \alpha$ and define $\Delta(\theta) = \{(\alpha, \xi) : \xi < \alpha \leq \theta\}$. We define $\mathscr{B}_{\alpha\xi}$, a countable family of countable sets containing the point (α, ξ), by induction on $\alpha < \omega_1$. Our induction hypothesis (on θ) is that $\bigcup\{\mathscr{B}_{\alpha\xi} : \xi < \alpha \leq \theta\}$ generates a topology on $\Delta(\theta)$ that satisfies:

(1) $\Delta(\theta)$ is locally countable, Hausdorff, and zero-dimensional.
(2) If $\gamma < \theta$, then $\Delta(\gamma)$ is closed in $\Delta(\theta)$.
(3) If $\xi < \alpha \leq \theta, b \in \mathscr{B}_{\alpha\xi}$, and $(\mu, \nu) \in b\backslash\{(\alpha\xi)\}$, then $\mu < \alpha$ and $\nu < \xi$.
(4) If $\xi < \delta$ and $\delta \in \Gamma(X)$, then there is $b \in \mathscr{B}_{\delta\xi}$ such that $\pi[b] \subset \Gamma(X)$.

If $\alpha \notin D$, then set $\mathscr{B}_{\alpha\xi} = \{\{(\alpha, \xi)\}\}$. Fix $\delta \in D$. Make the lists $(c_i : i \in \omega)$ and $(a_i : i \in \omega)$ as earlier. Let $p : \omega \to \delta \times \omega$ be a bijection so that for all $\xi \in \delta$ and $i < j < \omega, p^{-1}(\xi, i) < p^{-1}(\xi, j)$. By induction on $k \in \omega$, define $\beta(p(k)) \in \delta$ so that $\beta(p(k)) <$

$\beta(p(k+1))$, and for each $\xi \in \delta$, $\beta(\xi, n) \in a_n \cap (\cap\{c_i : i \leq n\})$. (This intersection is nonempty by (4) of $\diamondsuit^{++}$.) By (2) and (3), $\pi^{-1}(\text{range } \beta)$ is closed discrete, so by (1) there is a disjoint open family $\{u(\alpha, \zeta) : \zeta < \alpha \wedge \alpha \in \text{range } \beta\}$ such that $u(\alpha, \zeta) \in \mathscr{B}_{\alpha\zeta}$. By (4), we can additionally require that $\pi[u(\beta(\zeta, n), \zeta)] \subset \cap\{c_i : i \leq n\}$. Now set $\mathscr{B}_{\delta\xi} = \{\cup\{u(\beta(\xi, n), \zeta) : n \geq m \wedge \zeta < \xi\} : m \in \omega\}$.

It is routine to check that we have defined a locally countable, Hausdorff, zero-dimensional topology. We can now verify the properties of Δ as in [3], changing only the notation. (I repeat their proofs because I want to share my delight in the quick and easy verifications.)

LEMMA 2.2: If $\{Y_i : i \in \omega\}$ is a family of subsets of Δ such that each $\pi[Y_i]$ is stationary, then there are $\zeta < \delta < \omega_1$ such that $(\{\delta\} \times (\delta\backslash\zeta)) \subset \cap\{\overline{Y_i} : i \in \omega\}$.

Proof: Since $\pi[Y_i]$ is stationary and $\zeta < \alpha$ for all $(\alpha, \zeta) \in \Delta$, by the pressing down lemma we can assume that Y_i has the form $X_i \times \{\zeta_i\}$. Let $\zeta = sup\{\zeta_i : i \in \omega\}$ and choose $\delta \in D$ with $\zeta < \delta < \omega_1$ such that $X_i \cap c \neq \varnothing$ for all $i \in \omega$ and $c \in Club(\delta) \cap A_\delta$. (Possible by (5) of $\diamondsuit^{++}$.) If $\zeta < \xi < \delta$, then $\{n \in \omega : \beta(\xi, n) \in X_i\}$ is infinite and $\zeta_i < \xi$; therefore, $(\xi, \delta) \in \overline{Y_i}$. $\square$

LEMMA 2.3: Δ is strongly zero-dimensional; *a fortiori,* Δ is normal.

Proof: Let H and F be disjoint and closed in Δ. By Lemma 2.2, we can assume that $\pi[F]$ is not stationary in ω_1. So let C be club and disjoint from $\pi[F]$. Then $\pi^{-1}(\Gamma(C))$ is clopen in Δ and disjoint from F. Moreover, $\Delta\backslash\pi^{-1}(\Gamma(C))$ is the free union of countable spaces. We finish as in Lemma 2.1. $\square$

LEMMA 2.4: Every countable open cover of Δ has a clopen refinement.

Proof: Let $\mathscr{U} = \{U_n : n \in \omega\}$ be an open cover of Δ. Toward a contradiction, assume that for all $n \in \omega$, $\pi[\Delta\backslash U_n]$ is stationary. Applying Lemma 2.2, we get that $\mathscr{U}$ does not cover. Hence there is $n \in \omega$ and a club C such that $\pi^{-1}(C) \subset U_n$. Then $\pi^{-1}(\Gamma(C))$ is a clopen subset of U_n. Moreover, $\Delta\backslash\pi^{-1}(\Gamma(C))$ is the free union of countable spaces. We finish as in Lemma 2.1. $\square$

LEMMA 2.5: Δ is collectionwise normal.

Proof: Let $\mathscr{H}$ be a discrete family of closed subsets of Δ. By Lemma 2.4 and $|\Delta| = \omega_1$, we assume that $|\mathscr{H}| = \omega_1$. By Lemma 2.2, at most one $H \in \mathscr{H}$ has $\pi[H]$ stationary, and by Lemma 2.3 we can separate it from the others. So we assume that $\pi[H]$ is not stationary for all $H \in \mathscr{H}$.

Toward a contradiction, assume that $\pi[\cup\mathscr{H}]$ is stationary. Because ω_1 is not a measurable cardinal, there are disjoint subsets $\mathscr{H}'$ and $\mathscr{H}''$ of $\mathscr{H}$ such that both $\pi[\cup\mathscr{H}']$ and $\pi[\cup\mathscr{H}'']$ are stationary. Then Lemma 2.2, $\overline{\cup\mathscr{H}'}$ and $\overline{\cup\mathscr{H}''}$ are not disjoint, and $\mathscr{H}$ is not discrete. Hence there is a club C so that $\pi[\cup\mathscr{H}] \cap C = \varnothing$, and we finish as in Lemmas 2.1 and 2.4. $\square$

We say that $\{M_i : i \in I\}$ is a *shrinking* of a cover $\{U_i : i \in I\}$ if $\{M_i : i \in I\}$ also covers and $M_i \subset U_i$ for all $i \in I$. This notion is useful in expressing properties between normality and paracompactness. A regular space is normal iff every point-finite open cover has a closed shrinking. Hence, every open cover of a normal metacompact space has a closed shrinking. A regular space is paracompact iff every open cover has a locally finite closed shrinking. (See [8, 1.5.18, 5.1.12, and discussion after 7.2.4].)

LEMMA 2.6: Every increasing open cover $\mathscr{U}$ of Δ has a clopen shrinking.

Proof: Note that $|\Delta| = \omega_1$, that Lemma 2.4 does the countable case, and that it suffices to consider a cofinal sequence of $\mathscr{U}$. Thus we may assume that $\mathscr{U}$ has the form $(U_\alpha : \alpha \in \omega_1)$ where $\beta < \alpha$ implies that U_β is a proper subset of U_α. For $\theta < \omega_1$, $\Delta(\theta)$ is countable clopen subset of Δ, so we may construct a one-to-one map α from ω_1 to itself so that $\Delta(\theta) \subset U_{\alpha(\theta)}$. Define $M_\alpha = \Delta(\theta)$ if $\alpha = \alpha(\theta)$, and $M_\alpha = \varnothing$ otherwise. $\{M_\alpha : \alpha < \omega_1\}$ is a shrinking of $\mathscr{U}$. $\square$

LEMMA 2.7: For $\gamma < \omega_1$, let $V_\gamma = \{(\alpha, \xi) \in \Delta : \xi \le \gamma < \alpha\}$. Then $\{V_\gamma : \gamma < \omega_1\}$ is an open cover of Δ that has no closed shrinking.

Proof: It is easy to see that the V_γ are open and cover Δ. Toward a contradiction, assume that $\{M_\gamma : \gamma < \omega_1\}$ is a closed cover of Δ with $M_\gamma \subset V_\gamma$ for all $\gamma < \omega_1$. For each $\alpha \in \omega_1$, the point $(\alpha, 0)$ is in M_γ for some $\gamma < \alpha$. So by the pressing down lemma, there is some $\gamma < \omega_1$ such that $Y_\gamma = \{\alpha \in \omega_1 : (\alpha, 0) \in M_\gamma\}$ is stationary in ω_1. Applying Lemma 2.2 yields the contradiction that $M_\gamma = \overline{M_\gamma}$ is not a subset of V_γ. $\square$

3. SON OF GEORGE RETURNS

The axiom $\diamondsuit^{++}$ has been quite useful, but perhaps the space Son of George should have faded into obscurity because of the construction of a normal, nonmetrizable Moore space from the continuum hypothesis (CH). However, Watson noticed that the construction was incorrect. Basically flawed is a more accurate description— the topology was described in terms of a family of sets that did not form a basis. He asked, in effect, whether the construction could be repaired ([20, question 36], which is [16, question 104]). The third and final part of this article is a correct construction, which follows the basic ideas of [11], but the details are more involved. The presentation here is complete and self-contained. However, the longer discussions of tie-together spaces in [10, 11, sec. 2] may be helpful to some readers.

Recall the space Y, our first example of the use of $\diamondsuit^{++}$. The idea is to use final segments of Y to tie together small approximations to Bing's Example G (see, e.g., [4], [8, 5.1.23], or [12, 5.1]). Let the nonisolated points be p_η, $\eta < \omega_1$. Bing's G has point set $2^{\mathscr{P}(\omega_1)}$, and character 2^{ω_1}. George has point set $\bigcup\{2^{\mathscr{P}(\alpha)} : \alpha < \omega_1\}$ and character 2^ω. Son of George has point set $X = \bigcup\{2^{r(\delta)} : \delta \in D\}$ and character ω, because $r(\delta)$ is a subset of the countable set A_δ.

Since there are a number of definitions before we get to the topology on X, let's describe our plan. First consider the product of Bing's G with Y. The horizontal sections are too wide for first countability, so we replace the αth horizontal section by X_α, a version of G with nonisolated points p_η, $\eta < \alpha$. We imagine Bing's G on top of X (but *not* part of X). The sequence of points (p_η, α), $\eta < \alpha < \omega_1$, approaches the point p_η in G. The canonical subbasic set separating $\{p_\eta : \eta \in H\}$ from $\{p_\eta : \eta \in \omega_1 \backslash H\}$ reaches down to a tail of $\bigcup\{X_\alpha : \alpha \in \Gamma(H)\}$. Since all the action is with δ's in D, we will omit X_α for $\alpha \notin D$ rather than doing something arbitrary.

Now we begin the construction of Son of George. Let Z be the upper triangle of the product of ω_1 (with the discrete topology) and Y. Specifically, the point set of Z is $\{(\alpha, \beta) : \alpha < \beta < \omega_1\}$. Set $Z_\delta = \{(\alpha, \beta) : \beta \le \delta\}$. Set $Z' = \{(\alpha, \delta) \in Z : \delta \in D\}$. Points $(\alpha, \beta) \in Z \backslash Z'$ are isolated. A basic open (in Z) set for $(\alpha, \delta) \in Z'$ is

$Q(\alpha, \delta, u) = \{\alpha\} \times B(\delta, u)$, where $u \in U_\delta$ and $\alpha < min(\cap U)$. (The last clause is to obtain $Q \subset Z$.) Set

$$\mathscr{C} = \{Q(\alpha, \delta, u) : \delta \in D, u \in U_\delta, \text{ and } \alpha < min(\cap u)\}.$$

Z enjoys the property that two disjoint closed subsets can be separated by a clopen set because Y does. Let $\mathscr{R}$ be the family of clopen subsets of Z. Let $b : \omega_1 \times \omega_1 \to \omega_1$ be a bijection so that $b[\delta \times \delta] = \delta$ for all $\delta \in D$. Now we can define

$$r(\delta) = \{R \cap Z_\delta : R \in \mathscr{R} \wedge \delta \in \Gamma(b[R])\} \cup \{Q(\alpha, \delta, u) \in \mathscr{C} : u \in U_\delta\}.$$

Observe that $r(\delta)$ is countable. Next, set

$$X_\delta = 2^{r(\delta)} \quad \text{and} \quad X = \cup_{\delta \in D} X_\delta.$$

Define d from X to D so that $x \in X_{d(x)}$. Define p from Z' to X so that $p(\alpha, \delta)$ is the point of X_δ that satisfies

$$p(\alpha, \delta)(t) = 1 \quad \text{iff} \quad (\alpha, \delta) \in t \quad \text{for all } t \in r(\delta).$$

Let $P \subset X$ be the range of p. If $x \in X \backslash P$, then x is isolated. The basic open sets for a point $p(\alpha, \delta) \in P$ are $N(\alpha, \delta, u, a, b)$ for acceptable quintuples. We say that $(\alpha, \delta, u, a, b)$ is *acceptable* if

(1) $\alpha < \delta, \delta \in D, u \in U_\delta, a \in [\mathscr{R}]^{<\omega}$, and $b \in [\mathscr{C}]^{<\omega}$.
(2) If $R \in a$, then $\Gamma(b[R]) \in u$.
(3) If $Q(\alpha', \delta', u') \in b$, then $u' \subset u$.

For $(\alpha, \delta, u, a, b)$ acceptable, if $R \in a$ and $\delta' \in (\delta + 1) \cap (\cap u)$, then $R \cap Z'_\delta \in r(\delta')$. Similarly for "$Q \in b$" in place of "$R \in a$". Thus, for acceptable quintuples the following definition makes sense. Let $N(\alpha, \delta, u, a, b)$ be the set of $x \in X$ satisfying

(4) $d(x) \in (\delta + 1) \cap (\cap u)$.
(5) $x(t \cap Z_{d(x)}) = p(\alpha, \delta)(t \cap Z_\delta)$ for all $t \in a \cup b$.

We now verify that we have defined a base. Let $N = N(\alpha, \delta, u, a, b)$ and $N' = N(\alpha', \delta', u', a', b')$. If $p(\alpha'', \delta'') \in N \cap N'$, then $p(\alpha'', \delta'') \in N(\alpha'', \delta'', u \cup u', a \cup a', b \cup b') \subset N \cap N'$. It is easy to additionally verify that the N's are clopen and that the topology is first countable, T_1, and regular.

Our next goal is to show that the map p from Z' to P is a homeomorphism. In the forward direction, $p[Q(\alpha, \delta, u)] = N(\alpha, \delta, u, \varnothing, \{Q(\alpha, \delta, u)\}) \cap P$. In the inverse direction, $p^{-1}(N(\alpha, \delta, u, a, b))$ is the intersection of Z' with the following three clopen subsets of $Z : \{(\alpha, \gamma) \in Z : \gamma \in \cap u\}$, $\cap\{R \in a \cup b : (\alpha, \delta) \in R\}$, and $\cap\{Z \backslash R : R \in a \cup b \wedge (\alpha, \delta) \in R\}$.

The proof that X is normal parallels the proof of Lemma 2.1. Let H and F be disjoint closed sets. It suffices to consider the case where H and F have no isolated points; so we assume that $H \cup F \subset P$. Now $p^{-1}(H)$ and $p^{-1}(F)$ are disjoint closed in Z; let $R \in \mathscr{R}$ satisfy $p^{-1}(H) \subset R \subset Z \backslash p^{-1}(F)$. Set $W = \{x \in X : d(x) \in \Gamma(b[R])\}$; W is clopen in X. Define $U(H) = \cup\{N(\alpha, \delta, \{\Gamma(b[R])\}, \{R\}, \varnothing) : p(\alpha, \delta) \in H\}$ and $U(F) = \cup\{N(\alpha, \delta, \{\Gamma(b[R])\}, \{R\}, \varnothing) : p(\alpha, \delta) \in F\}$; then $H \subset U(H), F \subset U(F)$, and $U(H) \cap U(F) = \varnothing$.

As in Lemma 2.1, $p(b^{-1}(\Gamma(b(R))))$ partitions $X \backslash W$ into a relatively discrete family $\mathscr{M}$ of open sets M, each with countably many nonisolated points. Within each $M \in \mathscr{M}$ find disjoint open $U(H, M)$ and $U(F, M)$ with $H \cap M \subset U(H, M)$ and $F \cap M \subset U(F, M)$. Then $U(H) \cup (\bigcup_{M \in \mathscr{M}} U(H, M))$ and $U(F) \cup (\bigcup_{M \in \mathscr{M}} U(F, M))$ separate H and F.

Toward showing that X is not collectionwise normal, observe that $\{H_\alpha : \alpha \in \omega_1\}$ is a closed discrete family in X, where $H_\alpha = \{p(\alpha, \delta) \in P : \delta \in D\}$. For each $\alpha \in \omega_1$, let $H_\alpha \subset U_\alpha$, open. For each α, δ choose $N(\alpha, \delta, u_{\alpha\delta}, a_{\alpha\delta}, b_{\alpha\delta})$ so that

(1) $(\alpha, \delta, u_{\alpha\delta}, a_{\alpha\delta}, b_{\alpha\delta})$ is acceptable.
(2) $N(\alpha, \delta, u_{\alpha\delta}, a_{\alpha\delta}, b_{\alpha\delta}) \subset U_\alpha$.
(3) $|a_{\alpha\delta} \cup b_{\alpha\delta}|$ is minimal.

Having chosen $a_{\alpha\beta} \cup b_{\alpha\delta}$, enlarge $u_{\alpha\delta}$, if necessary, so that

(4) $p(\alpha, \delta)(t \cap Z_\delta) = p(\alpha, \gamma)(t \cap Z_\gamma)$ for all $t \in a_{\alpha\delta} \cup b_{\alpha\delta}$ and all $\gamma \in B(\delta, u_{\alpha\delta})$.

For $\alpha \in \omega_1$ and $n \in \omega$, set

$$J_{\alpha n} = \{\delta \in D : |a_{\alpha\delta} \cup b_{\alpha\delta}| = n\} \quad \text{and} \quad s_\alpha = \{n \in \omega : J_{\alpha n} \text{ is stationary}\}.$$

Toward a contradiction, suppose that s_α is infinite for some $\alpha \in \omega_1$. From (5) of $\Diamond^{++}$, there is $\gamma \in D$ such that $c \cap J_{\alpha n} \neq \varnothing$ for all $c \in Club(\gamma) \cap A_\gamma$ and $n \in s_\alpha$. Choose $k \in s_\alpha$ with $k > |a_{\alpha\gamma} \cup b_{\alpha\gamma}|$. Pick $\delta \in B(\gamma, u_{\alpha\gamma}) \cap J_{\alpha k}$. Then $N(\alpha, \delta, u_{\alpha\gamma}, a_{\alpha\gamma}, b_{\alpha\gamma})$ satisfies (1) and (2), but by (4) $|a_{\alpha\gamma} \cup b_{\alpha\gamma}| \leq k = |a_{\alpha\delta} \cup b_{\alpha\delta}|$, contradicting (3).

For all $\alpha \in \omega_1$, set $k_\alpha = \max s_\alpha$. By counting, we find $k \in \omega_1$ and $e \subset \omega_1$ such that $|e| = 2^k + 1$ and $k_\alpha = k$ for all $\alpha \in e$. Since the union of countably many nonstationary sets is nonstationary, there is $\delta \in D \setminus \bigcup \{J_{\alpha n} : \alpha \in e \wedge n > k\}$. Then $|a_{\alpha\delta} \cup b_{\alpha\delta}|$ for $\alpha \in e$, so each $U_\alpha \cap X_\delta$ has "measure at least 2^{-k}" in X_δ. We conclude that the U_α are not disjoint; hence X is not collectionwise normal.

REFERENCES

1. BALOGH, Z. 1991. On collectionwise normality of locally compact, normal spaces. Trans. Am. Math. Soc. **323:** 389–411.
2. BEŠLAGIĆ, A. & W. FLEISSNER. Forcing $\Diamond^{++}$. In preparation.
3. BEŠLAGIĆ, A. & M. E. RUDIN. 1985. Set-theoretic constructions of nonshrinking open covers. Topol. Appl. **20:** 167–177.
4. BING, R. H. 1951. Metrization of topological spaces. Can. J. Math. **3:** 175–186.
5. DANIELS, M. 1991. Normal $\mathscr{K}$-spaces can be collectionwise normal. Fundam. Math. **138:** 225–234.
6. DEVLIN, K. 1982. The combinatorial principle $\Diamond^{\#}$. J. Symb. Logic **47:** 888–899.
7. DOW, A., F. TALL & W. WEISS. 1990. New proofs of the consistency of the normal Moore space conjecture I. Topol. Appl. **37:** 33–51.
8. ENGELKING, R. 1989. General Topology. Heldermann Verlag. Berlin.
9. FLEISSNER, W. 1974. Normal Moore spaces in the constructible universe. Proc. Am. Math. Soc. **46:** 294–298.
10. ———. 1976. A normal, collectionwise Hausdorff, not collectionwise normal space. Gen. Topol. Appl. **6:** 57–71.
11. ———. 1983. Son of George and $V = L$. J. Symb. Logic **48:** 71–77.
12. ———. 1994. The normal Moore space conjecture and large cardinals. In Handbook of Self-theoretic Topology, K. Kunen and J. Vaughan, Eds. North-Holland. Amsterdam, the Netherlands.
13. ———. 1991. Normal measure axiom and Balogh's theorems. Topol. Appl. **39:** 123–143.

14. KUNEN, K. 1980. Set Theory. North-Holland. Amsterdam, the Netherlands.
15. KUNEN, K. & J. VAUGHAN, Eds. 1984. Handbook of Set-theoretic Topology. North-Holland. Amsterdam, the Netherlands.
16. VAN MILL, J. & G. M. REED, Eds. 1990. Open Problems in Topology. North-Holland. Amsterdam, the Netherlands.
17. NYIKOS, P. 1978. A provision solution to the normal Moore space problem. Proc. Am. Math. Soc. **78:** 429–435.
18. RUDIN, M. E. 1983. A normal, screenable, non-paracompact space. Topol. Appl. **15:** 313–322.
19. TALL, F. 1984. Normality versus collectionwise normality. *In* Handbook of Self-theoretic Topology, K. Kunen and J. Vaughan, Eds. North-Holland. Amsterdam, the Netherlands.
20. WATSON, S. 1990. Problems I wish I could solve. *In* Open Problems in Topology. North-Holland. Amsterdam, the Netherlands.

Toward a Theory of Normality and Paracompactness in Box Products

L. BRIAN LAWRENCE

Department of Mathematics
George Mason University
Fairfax, Virginia 22030-4444

ABSTRACT. An introduction to the subject in the title where the factors are copies of a separable metric space is given.

1. INTRODUCTION

History

A milestone in set-theoretic topology is Mary Ellen Rudin's ZFC example of a normal topological space whose product with the closed unit interval is nonnormal (1971, [25]). Rudin constructed her example as a subspace of a box product of ω-many copies of a singular cardinal with the order topology (see Szeptycki and Weiss's article in this volume). The issue of normality and paracompactness in full box products had been raised several years earlier by A. H. Stone (1964, [8]). A box product is a topological space that takes a Cartesian product of spaces for the point set, and takes an arbitrary Cartesian product of open subsets for a base element; note that in contrast to the Tychonoff topology of pointwise convergence, each factor of a basic open set is permitted to be a proper subset of the factor space. Stone's problem is to determine which box products are normal or paracompact. The problem was attacked in the 1960s without success. Having demonstrated the usefulness of the box topology in her solution of Dowker's problem (the counterexample just mentioned), Rudin entered the field in the early 1970s and achieved the first breakthroughs. These include results on three different types of factor spaces: metrizable, compact, and noncompact linearly ordered. In this paper we focus our attention on the case of metrizable factors.

Suppose M is a separable metric space (this assumption holds throughout the paper), and ν is an infinite cardinal number. Let $\Box^\nu(M)$ denote the box product where ν is the index set and each factor is M ($^\nu M$ denotes the point set consisting of all functions from ν into M).

STONE'S PROBLEM: Is $\Box^\nu(M)$ normal (paracompact)?

We first consider a countable index set. Suppose M is locally compact. Then under the Continuum Hypothesis (CH), $\Box^\omega(M)$ is paracompact (Rudin, [28], 1972). Abstracting from Rudin's proof and building on ideas of E. A. Michael [18, 19],

Mathematics Subject Classification: Primary 54D18; Secondary 54A35, 54B10, 54B20.
Key words and phrases: Normality, paracompactness box product, continuum hypothesis, Martin's axiom.

78

Kenneth Kunen developed a theoretical decomposition framework for determining whether $\square^\omega(M)$ is paracompact under alternative set-theoretic hypotheses ([10], 1978). Using Kunen's technique, the following improvement on Rudin's result was achieved collectively by Judy Roitman, E. K. van Douwen, and Scott Williams: Under either of the combinatorial statements $\mathfrak{b} = \mathfrak{d}$ or $\mathfrak{d} = \mathfrak{c}$ (and the assumption that M is locally compact), $\square^\omega(M)$ is paracompact (for $f, g \in {}^\omega\omega$, define $f \leq^* g$ iff $f(n) \leq g(n)$ for all but finitely many n; with respect to the relation $\leq^*$ on ${}^\omega\omega$, $\mathfrak{b}$ is the minimum cardinality of an unbounded family, and $\mathfrak{d}$ is the minimum cardinality of a cofinal family; $\mathfrak{c}$ denotes the cardinality of the continuum of real numbers; see the next section for the details). It is unknown if the failure of normality in $\square^\omega(\omega + 1)$ is consistent with ZFC; Roitman has shown that neither $\mathfrak{b} = \mathfrak{d}$ nor $\mathfrak{d} = \mathfrak{c}$ is necessary for paracompactness ([23, 24], 1978, 1979).

Now suppose we drop the assumption of local compactness on the factor space, in which case, Kunen's technique no longer applies. Let $\mathbb{P}$ and $\mathbb{Q}$ denote the irrationals and the rationals, respectively, with the subspace topology from the real line. Then $\square^\omega(\mathbb{P})$ is nonnormal in ZFC (Van Douwen, [2], 1975); whereas, under either $\mathfrak{b} = \mathfrak{d}$ or $\mathfrak{d} = \mathfrak{c}$, $\square^\omega(\mathbb{Q})$ is paracompact (Lawrence, [14], 1988). So metrizability in the factor space is not sufficient for paracompactness (or normality) in the box product, and local compactness in not necessary. Since the rationals and an arbitrary separable, locally compact metric space are both σ-compact, the results of Rudin and Lawrence are partially generalized (in the case of Lawrence, the hypothesis $\mathfrak{b} = \mathfrak{d}$ is lost; and in the case of Rudin, the factor space must now have a base consisting of open–closed sets) by the following theorem of Louis Wingers: Suppose $\mathfrak{d} = \mathfrak{c}$, and suppose M is σ-compact and zero-dimensional; then $\square^\omega(M)$ is paracompact ([35], first appearing as a preprint in 1991).

In showing that $\square^\omega(\mathbb{P})$ is nonnormal, van Douwen proved a stronger result: $\mathbb{P} \times \square^\omega(\omega + 1)$ is nonnormal. In his talk at the Summer Topology Conference in honor of Mary Ellen Rudin (Madison, Wisconsin, 1991), Kunen pointed out that $\mathbb{P}$ can be replaced with any $\leq^*$ cofinal subset $D \subseteq \mathbb{P}$ with the subspace topology (the proof uses results from [9], 1973). Wingers showed that (at the level of homeomorphs) cofinality with respect to $\leq^*$ is also a necessary condition for nonnormality: Suppose Λ is a subspace of $\mathbb{P}$; then $\Lambda \times \square^\omega(\omega + 1)$ is nonnormal in ZFC iff Λ is homeomorphic to a $\leq^*$ cofinal subset of $\mathbb{P}$ ([36], first appearing as a preprint in 1992).

We turn now to the case of uncountably many factors. In ZFC, $\square^{\omega_1}(\omega + 1)$ is nonnormal (Lawrence, [15], first appearing as a preprint in 1991; a new proof is given in [16]); moreover, if $2^\omega = 2^{\omega_1}$, then the closed subspace of $\square^{\omega_1}(\omega + 1)$ consisting of all functions with countable support is nonnormal (where a function f has countable support iff $\exists \alpha < \omega_1 \, \forall \beta \geq \alpha \, [f(\beta) = \omega]$). Note that $\square^{\omega_1}(\omega + 1)$ is homeomorphic to a closed subspace of $\square^\nu(M)$ whenever ν is uncountable and M is nondiscrete.

Outline of the Paper

This paper is intended to serve as an introduction to the field and not as a survey (two surveys already exist: [4], [34]); the many interesting ideas and unsolved problems concerning box products with nonmetrizable factors have been omitted. Our focus is on Rudin's Theorem: the Continuum Hypothesis implies that $\square^\omega(M)$ is paracompact if M is locally compact; and on the efforts to extend this theorem to obtain a complete solution to Stone's Problem: under what set-theoretic conditions, and for which separable metric spaces M and cardinal numbers ν, is $\square^\nu(M)$ normal (paracompact)? We divide the subject into three cases: the presence or absence of local compactness in the factor over a countable index set, and an uncountable index set. In each case, we give the principal theorem on what is known and the corresponding problem on what remains to be discovered.

Our emphasis is on the convergence properties that underlie the proofs of the separation and covering theorems. For each of our three main theorems, we introduce an auxiliary proposition on convergence that lies at the heart of the proof (see Section 4).

The two propositions concerning a countable index set are formulated in terms of the fibers of the equivalence relation for finite disagreement (see Section 2). It is important to distinguish sharply between convergence in the quotient space (which is relevant to the locally compact case, and is discussed in Section 2), and convergence in the product space in relation to the fibers of the decomposition but independent of the quotient space topology (which is relevant in the absence of local compactness in the factor, and is discussed in Section 4).

We use the following standard notation: for a function f, Dom f and Ran f denote the domain and range, respectively; vertical bars denote the cardinality operator; and a horizontal bar denotes the closure operator in the context of a topological space.

2. SET-THEORETIC AXIOMS AND THE NABLA PRODUCT

Partial Orders on the Irrationals

Recall that $\mathbb{P}$ can be identified with $^\omega\omega = \{f | f : \omega \to \omega\}$ with the product topology: for each $\xi \in {}^{<\omega}\omega = \{\eta | \eta : [0, n] \to \omega$ for some $n \in \omega\}$, $\{f \in {}^\omega\omega : \xi \subseteq f\}$ is a basic open set (see [21, p. 204]).

For all $f, g \in {}^\omega\omega$, define $f \le g$ if for each $n \in \omega, f(n) \le g(n)$ (in which case, g is said to strictly dominate f); and define $f \le^* g$ if there exists $m \in \omega$ such that for each $n \ge m, f(n) \le g(n)$ (in which case, g is said to eventually dominate f). A subcollection $D \subseteq {}^\omega\omega$ is strictly (respectively, eventually) dominant provided that D is cofinal with respect to $\le$ (respectively, $\le^*$); and for a subset of the index set, $w \subseteq \omega$, D is strictly dominant on w provided that for every $f \subseteq {}^\omega\omega$ there exists $g \in D$ such that for each $n \in w, g(n) \ge f(n)$.

Cardinal Numbers

With respect to $\le^*$, let $\mathfrak{b}$ be the minimum cardinality of an unbounded family, and let $\mathfrak{d}$ be the minimum cardinality of a dominant family. (In fact, the minimum cardinality of a cofinal subcollection is the same for strict domination as it is for eventual domination, as can be easily verified.)

First note that in ZFC, $\omega_1 \le \mathfrak{b} \le \mathfrak{d} \le \mathfrak{c}$ [a standard diagonal argument proves $\omega_1 \le \mathfrak{b}$ (see Fact 2)]; there is a $\le^*$ well-ordered unbounded family of order type $\mathfrak{b}$, so $\mathfrak{b}$ is regular; $\mathfrak{b} \le$ Cof $\mathfrak{d}$ (the cofinality of $\mathfrak{d}$); and $\mathfrak{b} = \mathfrak{d}$ iff there is a well-ordered eventually dominant family (a collection of this type is often called a *scale*).

It is consistent with ZFC to simultaneously change any of the preceding relations to either strict equality or strict inequality. In particular, Martin's Axiom implies $\mathfrak{b} = \mathfrak{d} = \mathfrak{c}$ (see [11, p. 87, example 8]), hence the Continuum Hypothesis implies $\omega_1 = \mathfrak{b} = \mathfrak{d} = \mathfrak{c}$. More generally, S. H. Hechler has shown in [7] that if N is a (countable transitive) model of ZFC, in which α and β are any two regular cardinals with $\omega_1 \leq \alpha \leq \beta \leq \mathfrak{c}$, then there is an extension N' of N in which $\mathfrak{b} = \alpha$, $\mathfrak{d} = \beta$, and $\mathfrak{c}$ is unchanged.

Nabla Product

Define $X = X(M)$ by $X(M) = \square^\omega(M)$ (recall that throughout this paper, M denotes a separable metric space). Define two points in X to be equivalent if they disagree at most a finite number of times, and for each f, let $E(f)$ be the equivalence class to which f belongs. For each $f \in X$ and each $m \in \omega$, let $F_m(f) = \{g \in X : \forall n \geq m[g(n) = f(n)]\}$; then $E(f) = \bigcup_{m \in \omega} F_m(f)$. Let $\nabla^\omega(M)$ denote the quotient space on X induced by E, and let σ be the quotient map [so $\sigma(f)$ denotes $E(f)$ as a point in the quotient space]; $\nabla^\omega(M)$ is called the *nabla product*. Note that σ is an open map, so $\{\sigma[U] : U$ is open in $X\}$ is a base for the quotient space.

Fact 1: Suppose $D \subseteq {}^\omega\omega$, and $f : \omega \to \omega + 1$ is constant with $\text{Ran} f = \{\omega\}$. Then D is strictly dominant (respectively, eventually dominant) iff f [respectively, $\sigma(f)$] is a limit point of D (respectively, $\sigma[D]$) in $\square^\omega(\omega + 1)$ [respectively, $\nabla^\omega(\omega + 1)$].

Fact 2: Each of the following statements is a corollary of the diagonal construction: $g(n) = f_n(n)$, where $\langle f_n : n \in \omega \rangle$ is a given sequence of functions in ${}^\omega\omega$.

(1) Every countable subcollection of ${}^\omega\omega$ has a $\leq^*$ upper bound (and therefore, $\mathfrak{b} \geq \omega_1$);
(2) If for each $n \in \omega$, $D_n \subseteq {}^\omega\omega$ such that $\bigcup_{n \in \omega} D_n$ is eventually dominant, then there exists $m \in \omega$ such that D_m is eventually dominant;
(3) [Rudin] $\nabla^\omega(M)$ is ω_1-open (i.e., every intersection of a countable collection of open sets is open).

REMARK: The following results provide topological characterizations of the combinatorial statements $\mathfrak{b} = \mathfrak{d}$ and $\mathfrak{d} = \mathfrak{c}$, the corollary to which plays a key role in proving that either statement implies that $\square^\omega(M)$ is paracompact if M is locally compact.

FIRST LEMMA ON THE NABLA PRODUCT (Van Douwen, [4]; Williams [32]): Suppose M is nondiscrete. Then $\mathfrak{b} = \mathfrak{d}$ iff there is a sequence $\langle \mathcal{U}_\alpha : \alpha \in \mathfrak{d} \rangle$ such that (1) each $\mathcal{U}_\alpha$ is a pairwise disjoint open cover of $\nabla^\omega(M)$; (2) for all α, $\beta \in \mathfrak{d}$ with $\alpha < \beta$, $\mathcal{U}_\beta$ refines $\mathcal{U}_\alpha$; and (3) $\bigcup\{\mathcal{U}_\alpha : \alpha \in \mathfrak{d}\}$ is a base for $\nabla^\omega(M)$ (in which case, $\nabla^\omega(M)$ is said to be a $\mathfrak{d}$-metrizable topological space).

Proof: We give a new proof. Suppose $\mathfrak{b} = \mathfrak{d}$. Then we can choose $\delta : \mathfrak{d} \to {}^\omega\omega$ such that (1) for all α, $\beta \in \mathfrak{d}$, with $\alpha < \beta$, $\delta_\alpha \leq^* \delta_\beta$; and (2) the range of δ is eventually dominant. We can also take each value of δ to be a strictly increasing sequence. Let ρ be a distance function for M. For each $\alpha \in \mathfrak{d}$ and each $f \in \square^\omega(M)$, define $\mu_\alpha(f) = \{g \in \square^\omega(M) : \langle \rho(f(n), g(n)) \cdot \delta_\alpha(n) : n \in \omega \rangle$ converges to zero$\}$. For each $\alpha \in \mathfrak{d}$, let $\mathcal{U}_\alpha = \{\sigma[\mu_\alpha(f)] : f \in \square^\omega(M)\}$.

For the converse, let $f : \omega \to M$ be constant with $\text{Ran} f = \{p\}$, where p is a nonisolated point; and for each $\alpha \in \mathfrak{d}$, let $U_\alpha \in \mathcal{U}_\alpha$ with $\sigma(f) \in U_\alpha$. Use $\langle U_\alpha : \alpha \in \mathfrak{d} \rangle$

to construct a scale as follows. Let $\langle R_n : n \in \omega \rangle$ be a local base at p in M, where for each n, R_{n+1} is a proper subset of R_n. For each $g \in {}^\omega\omega$, let $V(g) = \Pi_{n \in \omega} R_{g(n)}$. Note that $D \subseteq {}^\omega\omega$ is eventually dominant iff $\{\sigma[V(g)] : g \in D\}$ is a local base at $\sigma(f)$ in $\nabla^\omega(M)$. Let λ: Cof $\mathfrak{d} \to \mathfrak{d}$ be strictly increasing with Ran λ cofinal in $\mathfrak{d}$. Recursively define δ : Cof $\mathfrak{d} \to {}^\omega\omega$ so that for each $\alpha \in$ Cof $\mathfrak{d}$, there exists $\gamma \geq \lambda_\alpha$ satisfying (1) $\sigma[V(\delta_\alpha)] \subseteq U_\gamma$, and (2) for each $\beta < \alpha$, U_γ is a proper subset of $\sigma[V(\delta_\beta)]$. Then Ran δ is a scale. (This implies $\mathfrak{b} = $ Cof $\mathfrak{d} = \mathfrak{d}$.) □

ROITMAN'S LEMMA ON DOMINATION ([24]): Suppose $\mathcal{F} \subseteq {}^\omega\omega$ and $\mathcal{A} \subseteq [\omega]^\omega$ such that $|\mathcal{F}|, |\mathcal{A}| < \mathfrak{d}$. Then there exists $g \in {}^\omega\omega$ such that for each $f \in \mathcal{F}$ and each $A \in \mathcal{A}$, $\{n \in A : g(n) \geq f(n)\}$ is infinite.

Proof: We can assume that each function in $\mathcal{F}$ is strictly increasing. For each f and $A = \{a_0, a_1, \dots\}$ (where A has been enumerated in increasing order), define $h = h(f, A)$ by $h(m) = f(a_{n+1})$ for each m in the interval $[a_n, a_{n+1})$. Let g be a strictly increasing function that is not eventually dominated by any function in the range of h. □

SECOND LEMMA ON THE NABLA PRODUCT (Roitman, [24]): Suppose M is nondiscrete. Then $\mathfrak{d} = \mathfrak{c}$ iff there is a base for the nabla product where the union of each subcollection of cardinality less than $\mathfrak{c}$ is closed. (For sufficiency, use Fact 1; and for necessity, use Roitman's Lemma.)

COROLLARY TO THE LEMMAS: Suppose $\mathfrak{b} = \mathfrak{d}$ or $\mathfrak{d} = \mathfrak{c}$. Then $\nabla^\omega(M)$ is ultraparacompact (i.e., every open cover has an open pairwise disjoint covering refinement).

Order Hypothesis

Let $Y = \nabla^\omega(M)$, and let $OH(M)$ be the following combinatorial statement: There exists a partial ordering $\leqslant$ of Y such that $\langle Y, \leqslant \rangle$ is a tree (i.e., $\leqslant$ is a reflexive, antisymmetric, and transitive relation for which the set of all predecessors of any given point is well-ordered by $<$), where (1) the height of the tree has order type $\leq \mathfrak{d}$ (i.e., every chain has order type $\leq \mathfrak{d}$), and (2) for each $y \in Y$, $\{y' : y \leqslant y'\}$ is open in the nabla product.

This concept was introduced in [14] to prove the following: If M is countable (e.g., $\omega + 1$ or $\mathbb{Q}$), then $OH(M)$ follows from either $\mathfrak{b} = \mathfrak{d}$ or $\mathfrak{d} = \mathfrak{c}$, and implies $\square^\omega(M)$ is paracompact. Since $\mathfrak{b} = \mathfrak{d}$ and $\mathfrak{d} = \mathfrak{c}$ are independent of one another, $OH(M)$ is a strictly weaker hypothesis. It is unknown if $OH(\omega + 1)$ is provable in ZFC.

If M is locally compact, then $OH(M)$ follows from $\mathfrak{d} = \mathfrak{c}$, and implies $\square^\omega(M)$ is paracompact. Alan Dow pointed out to the author a proof of the consistency of the conjunction of $\mathfrak{b} = \mathfrak{d}$ and the negation of OH(Cantor Set). This result is a corollary of [5].

3. STANDARD CHARACTERIZATIONS OF A CLOSED MAP

Hypothesis

For this section, suppose that X and Y are abstract Hausdorff regular spaces where Y is a quotient space on X and $\sigma : X \to Y$ is the quotient map; so the point set of Y is a partition of X, and for each $x \in X$, $\sigma(x)$ is the partition set to which x belongs.

Decompositions

The quotient map $\sigma : X \to Y$ is closed iff Y is an upper semicontinuous decomposition of X; that is, for each $x \in X$ and each open set U in X with $\sigma(x) \subseteq U$, there exists an open set V in Y such that $\sigma(x) \in V$ and $\bigcup V \subseteq U$.

Limit Points

The quotient map $\sigma : X \to Y$ is closed iff for every $A \subseteq X$, σ maps the limit points of A onto a set that contains the limit points of $\sigma[A]$.

We can visualize this condition as follows. Let S be a subspace of Y and let $T \subseteq X$ be a transversal of the corresponding point-inverse sets; that is, given $S \subseteq Y$, construct T by arbitrarily choosing a point from $\sigma^{-1}[y]$ for each $y \in S$. Then for each limit point y of S, there is a limit point x of T with $x \in \sigma^{-1}[y]$.

4. CONVERGENCE

KUNEN'S LEMMA ON DOMINATION ([9]): If $D \subseteq {}^\omega\omega$ is eventually dominant, then there is a finite subset $u \subseteq \omega$ and a function $\eta : u \to \omega$ such that $\{f \in D : f|u = \eta\}$ is strictly dominant on $\omega \backslash u$.

Proof: Apply Fact 3 below to obtain a suitable initial segment, and then apply Fact 4 successively to each index of the initial segment. □

Fact 3: If $D \subseteq {}^\omega\omega$ is eventually dominant, then there is an $m \in \omega$ such that D is strictly dominant on $\omega \backslash \{0, 1, 2, \ldots, m - 1\}$.

Fact 4: If $D \subseteq {}^\omega\omega$ is strictly dominant on $w \subseteq \omega$ and $n \notin w$, then either:

(1) D is strictly dominant on $w \cup \{n\}$, or,
(2) For some $i \in \omega$, $\{f \in D : f(n) = i\}$ is strictly dominant on w.

REMARK: Note that Kunen's Lemma is an immediate consequence of the fact that the quotient map $\sigma : \square^\omega(\omega + 1) \to \nabla^\omega(\omega + 1)$ is closed (use the limit point characterization of a closed map and Fact 1). Conversely, the ideas in Kunen's proof (Facts 3 and 4) can also be used to prove the general theorem that the quotient map is closed whenever the factor space is locally compact. This theorem was first established (implicitly) by Rudin. Alternative proofs are given in [4] and [10]. We give yet another proof below to emphasize the connection with Kunen's Lemma, but first a corollary.

COROLLARY: Suppose $D \subseteq {}^\omega\omega$ is eventually dominant. Then as a subspace of $\square^\omega(\omega + 1)$, D cannot be written as a countable union of closed sets. (Use Fact 2(2) and Kunen's Lemma.)

LEMMA ON THE QUOTIENT MAP: Suppose M is locally compact. Then the quotient map $\sigma : \square^\omega(M) \to \nabla^\omega(M)$ is closed.

Proof: We'll use the limit point characterization of a closed map. Suppose $f \in \square^\omega(M)$ and $S \subseteq \square^\omega(M)$ such that $S \cap E(f) = \varnothing$ and $\sigma(f)$ is a limit point of $\sigma[S]$ in the quotient space. Let $\mathcal{U}$ be a countable open cover of M where each member has a compact closure. For each $n \in \omega$, let $U_n \in \mathcal{U}$ with $f(n) \in U_n$. Let $\mathcal{V} = \{\Pi_{n\in\omega}V_n : \forall n \in \omega[V_n \in \mathcal{U}] \ \& \ \exists m \in \omega \ \forall n \geq m \ [V_n = U_n]\}$. Then $\mathcal{V}$ is a countable open cover of $E(f)$, and $\bigcup \mathcal{V}$ is an open inverse set (with respect to σ). Since the quotient space is ω_1-open (Fact 2(3)), there exists $\mathcal{O} = \Pi_{n\in\omega}V_n \in \mathcal{V}$ such that $\sigma(f)$ is a limit point of $\sigma[S \cap \mathcal{O}]$.

Let $g \in \mathcal{O} \cap E(f)$. For each $n \in \omega$, let $\langle R_{n,0}, R_{n,1}, \dots \rangle$ be a nested local base at $g(n)$ with $R_{n,0} = V_n$. Let $D = \{k \in {}^\omega\omega : S \cap \Pi_{n\in\omega}R_{n,k(n)} \neq \varnothing\}$. Then D is eventually dominant. By Fact 3, there is an initial segment $\{0, 1, \dots, m-1\}$ such that D is strictly dominant on the complement.

The following claim is the analogue of Fact 4 in the current setting. There is a point h in the closure of $\mathcal{O}$ such that $h(n) = g(n)$ for each $n \geq m$, and, h is a limit point of S in the product space. Otherwise, by the compactness of the set $H = \overline{V_0} \times \cdots \times \overline{V_{m-1}} \times \{g(m)\} \times \{g(m+1)\} \times \cdots$, there is an open set W with $H \subseteq W$ and $W \cap S = \varnothing$, which contradicts the choice of m. $\quad\square$

CLASSIFICATION OF LIMIT POINTS: Suppose $f \in \square^\omega(M)$ and $S \subseteq \square^\omega(M)$, where f is a limit point of S. Then define f to be an essential limit point of S iff f is a limit point of $S \cap E(f)$; otherwise, f is a nonessential limit point.

Fact 5: Suppose $f \in \square^\omega(M)$ and $S \subseteq \square^\omega(M)$. Then:

(1) f is an essential limit point of S iff f is a limit point of a countable subset of S;
(2) There is a sequence in $S \setminus \{f\}$ converging to f iff there exists $m \in \omega$ such that f is a limit point of $S \cap F_m(f)$.

LEMMA ON SEQUENTIAL CONVERGENCE (Lawrence, [13]): Suppose M is locally compact, and $C \subseteq \square^\omega(M)$ is closed. Then for every limit point f of C, there is a sequence in $C \setminus \{f\}$ converging to f; it follows (by Fact 5) that if $C \cap E(f) = \{f\}$, then f is an isolated point of C. (See [13] for the proof.)

COROLLARY: Suppose $D \subseteq {}^\omega\omega$ is eventually dominant, and let $f : \omega \to \omega + 1$ be constant with $\operatorname{Ran} f = \{\omega\}$. Then as a subspace of $\square^\omega(\omega + 1)$, the intersection of the derived set of D (i.e., the set of all limit points) with $E(f)$ is dense in itself.

LEMMA ON NONESSENTIAL LIMIT POINTS (Lawrence, [13]): Suppose M is not locally compact, and choose $p \in M$, where p does not have a compact neighborhood. Let $f: \omega \to M$ be constant with $\operatorname{Ran} f = \{p\}$. Then there is a closed set $C \subseteq \square^\omega(M)$ containing f such that:

(1) $C \cap E(f) = \{f\}$;
(2) f is a limit point of C;
(3) f is the only limit point of C.

So by Fact 5, there is a closed set C and a point f where f is isolated from each countable subset but is nevertheless a limit point, that is, f is a nonessential limit point.

Proof: Let $\{R_n : n \in \omega\}$ be a nested local base at p, and let $q: \omega \times \omega \to M$ be a 1–1 function, where for each $m \in \omega$, $\{q(m, n) : n \in \omega\}$ is closed and discrete in M and is a subset of $R_m \setminus \overline{R_{m+1}}$.

Let $\Phi : {}^\omega\omega \to \square^\omega(M)$ defined by $\Phi(k)(n) = q(k(n), k(n+1))$. Let $D \subseteq {}^\omega\omega$ be a strictly dominant family such that each pair of distinct functions in D disagrees on an infinite number of indices. Let $C = C' \cup \{f\}$, where C' is the range of Φ restricted to D. Note that $C' \cap E(f) = \varnothing$, and that f is a limit point of C'.

The following argument, due to the referee, shows that f is the only limit point of C'. Suppose g is a limit point of C' distinct from f. Then $\mathrm{Ran}\, g \subseteq \mathrm{Ran}\, q \cup \{p\}$. Let $A = \{n \in \omega : g(n) = p\}$. Choose a sequence of open sets in M, $\langle U_n : n \in \omega \rangle$, so that $U_n = M$ if $n \in A$, and $U_n \cap \mathrm{Ran}\, q = \{g(n)\}$ if $n \notin A$.

Case 1: A is finite. Then $\Pi_{n\in\omega}U_n$ is a neighborhood of g disjoint from $C'\backslash\{g\}$, contradicting the choice of g.

Case 2: A is infinite. Now, since $g \neq f$, we can choose $m \notin A$ with $m + 1 \in A$. Let $V_n = U_n$ for $n \neq m + 1$, and let $V_{m+1} = R_{j+1}$, where $g(m) = q(i,j)$. Then $\Pi_{n\in\omega}V_n$ is a neighborhood of g disjoint from $C'\backslash\{g\}$, again contradicting the choice of g. $\square$

FIRST COROLLARY (Van Douwen, [4]): If the quotient map $\sigma : \square^\omega(M) \to \nabla^\omega(M)$ is closed, then M is locally compact.

Proof: Suppose M is not locally compact and define f and Φ as in the proof of the lemma. Let $S = \{\Phi(k) : k \in {}^\omega\omega \text{ with } k(0) = 0\}$. Then S is a closed discrete subset of the product space, while $\sigma(f)$ is a limit point of $\sigma[S]$ in the quotient space. By the limit point characterization of a closed map, σ cannot be closed. $\square$

SECONDARY COROLLARY: If M is not locally compact, then there exists $f \in \square^\omega(M)$ and $S \subseteq \square^\omega(M)$ such that:

(1) f is a nonessential limit point of S;
(2) S is a countable union of closed sets.

REMARK: The method of proof in the preceding lemma can be used to construct a closed discrete subset of $\square^\omega(\mathbb{Q})$ whose image under the quotient map is dense in the quotient space.

PROPOSITION 1: Let $X = \square^\omega(M)$ and let $Y = \nabla^\omega(M)$. Then the following are equivalent:

(1) M is locally compact;
(2) The quotient map $\sigma : X \to Y$ is closed;
(3) For every limit point f of a closed set $C \subseteq X$, there is a sequence in $C\backslash\{f\}$ converging to f;
(4) Every limit point of a closed set $C \subseteq X$ is essential;
(5) If $S \subseteq X$ has a nonessential limit point, then S cannot be written as a countable union of closed sets.

PROPOSITION 2 (Lawrence, [14]): Suppose M is zero-dimensional (i.e., M has a base consisting of sets that are simultaneously open and closed), and let $\mathscr{B}$ be a base for $\square^\omega(M)$, where each member is a Cartesian product of open–closed sets. Suppose $\mathscr{A} \subseteq \mathscr{B}$ with $|\mathscr{A}| < \mathfrak{d}$. Then every limit point of $\cup \mathscr{A}$ is essential. (Use Roitman's Lemma for the proof.)

COROLLARY: Suppose $S \subseteq \square^\omega(M)$ (M need not be zero-dimensional) with $|S| < \mathfrak{d}$. Then every limit point of S is essential; it follows that $\cup \{E(f) : f \in S\}$ is closed.

PROPOSITION 3: Suppose $f \in \square^{\omega_1}(\omega + 1)$, where $|f^{-1}[\omega]| = \omega_1$. Then there is a closed set $C \subseteq \square^{\omega_1}(\omega + 1)$ such that:

(1) Each point in $C\backslash\{f\}$ disagrees with f uncountably many times;

(2) f is a limit point of C;

(3) f is the only limit point of C.

It follows that $C \setminus \{f\}$ can be written as a countable union of pairwise disjoint closed sets: let $A_0 = \{g \in C : g$ does not extend the function obtained by restricting f to $f^{-1}[[0, \omega)]\}$; and for $n \in \omega$ with $n > 0$, let $A_n = \{g \in C : g$ extends $f|f^{-1}[[0, \omega)]$ & $\forall \alpha \in f^{-1}[\omega]\ (g(\alpha) \geq n - 1)$ & $\exists \alpha \in f^{-1}[\omega]\ (g(\alpha) = n - 1)\}$.

Proof: For each $n \in \omega$ and each $u \subseteq f^{-1}[\omega]$ with $|u| = \omega_1$, define $h = h(n, u) \in \square^{\omega_1}(\omega + 1)$ by $h(\alpha) = f(\alpha)$ if $\alpha \notin u$, and otherwise, by $h(\alpha) = n$. Let $C = \{f\} \cup \mathrm{Ran}\ h$. $\square$

5. A LOCALLY COMPACT FACTOR OVER A COUNTABLE INDEX SET

Rudin's Proof of Rudin's Theorem

By recursion on $\alpha < \omega_1$, we show that under the Continuum Hypothesis, $X = \square^\omega(M)$ is paracompact if M is locally compact. We need two lemmas.

The first lemma asserts that if $\mathscr{U}$ is an open cover in X of an equivalence class $E(f)$ as a subspace of X, then $\mathscr{U}$ has an open refinement $\mathscr{V}$ such that (1) $\mathscr{V}$ covers $E(f)$; (2) if for some $g \in X$, $\cup\ \mathscr{V}$ intersects $E(g)$, then $E(g) \subseteq \cup\ \mathscr{V}$; (3) $\cup\ \mathscr{V}$ is closed in X; and, (4) $\mathscr{V}$ is locally finite in X. Note that the conjunction of the first two conditions is equivalent to the decomposition version of the statement that the quotient map is closed, which follows from Proposition 1.

The second lemma asserts that if $\mathscr{A}$ is a countable collection of closed sets in X where each set is a union of equivalence classes, then $\cup\ \mathscr{A}$ is also closed in X. Note that the second lemma is equivalent to Fact 2(3): the quotient space is ω_1-open.

We can now prove the theorem. Let $\langle k_\alpha : \alpha < \omega_1 \rangle$ be an enumeration of the equivalence classes, and let $\mathscr{U}$ be a given open cover of X. For each $\alpha < \omega_1$, choose (according to the first lemma) a locally finite refinement $\mathscr{V}_\alpha$ that covers k_α; and then define $\mathscr{W}_\alpha = \{V \setminus \cup_{\beta < \alpha}(\cup\ \mathscr{V}_\beta) : V \in \mathscr{V}_\alpha\}$. By the second lemma, each member of $\mathscr{W}_\alpha$ is open. It follows that $\cup_{\alpha < \omega_1} \mathscr{W}_\alpha$ is an open locally finite refinement of $\mathscr{U}$ that covers X. $\square$

Kunen's Theoretical Framework

Underlying Rudin's proof is an abstract decomposition technique for proving paracompactness. This was first recognized by Kunen in [10], where he set forth the basic principles, which include: isolating the nabla product and the closed quotient map, and establishing the following theorem extending results of E. A. Michael.

DECOMPOSITION LEMMA ON PARACOMPACTNESS: For the moment, suppose X and Y are arbitrary Hausdorff regular spaces and σ is a closed continuous map of X onto Y. Then:

(1) (Michael, [19]) If X is paracompact, then so is Y;

(2) (Kunen, [10]) If Y is paracompact and each point-inverse set ($\sigma^{-1}[y]$ for each $y \in Y$) is Lindelöf, then X is paracompact (Kunen proves this result as a corollary of [18]).

COROLLARY: If M is locally compact, then $\square^\omega(M)$ is paracompact iff $\nabla^\omega(M)$ is paracompact.

Proof: By Proposition 1, the quotient map is closed; and since for each $f \in \square^\omega(M)$, $E(f) = \bigcup_{m \in \omega} F_m(f)$ is a countable union of separable metric spaces, each point-inverse set is Lindelöf. $\square$

THEOREM 1: Suppose $\mathfrak{b} = \mathfrak{d}$ or $\mathfrak{d} = \mathfrak{c}$. Suppose further that M is locally compact. Then $\square^\omega(M)$ is paracompact.

Proof: Use the preceding corollary and the corollary to our lemmas on the nabla product. $\square$

PROBLEM 1 (Rudin): Is $\square^\omega(\omega + 1)$ paracompact (normal) in ZFC? (It is unknown if a box product of metric spaces can be normal without being paracompact; this issue was first raised some twenty years ago by Kunen.) See the papers of Roitman (or the survey paper of Williams where her work is discussed in detail) for some interesting partial results on Rudin's problem.

6. IN THE ABSENCE OF LOCAL COMPACTNESS

THEOREM 2: In the absence of local compactness in the factor space, we have the following partial results.

(1) (Lawrence, [14]) Suppose $\mathfrak{b} = \mathfrak{d}$ or $\mathfrak{d} = \mathfrak{c}$. Then $\square^\omega(\mathbb{Q})$ is paracompact.
(2) (Wingers, [35]) Suppose $\mathfrak{d} = \mathfrak{c}$. Suppose further that M is σ-compact and zero-dimensional. Then $\square^\omega(M)$ is paracompact.
(3) (Van Douwen, [2]) In ZFC, $\square^\omega(\mathbb{P})$ is nonnormal.
(4) (Kunen, Unpublished) In ZFC, $\square^\omega(D)$ is nonnormal for every eventually dominant $D \subseteq \mathbb{P}$ with the subspace topology.
(5) (Wingers, [36]) For every subject $\Lambda \subseteq \mathbb{P}$, $\Lambda \times \square^\omega(\omega + 1)$ is nonnormal in ZFC iff Λ is homeomorphic to an eventually dominant subspace of $\mathbb{P}$.

Outline of Proof of Theorem 2(1), (2): Suppose $\mathfrak{d} = \mathfrak{c}$. Let $\mathcal{U}$ be an open cover of $\square^\omega(\mathbb{Q})$, and let $\langle k_\alpha : \alpha < d \rangle$ be an enumeration of $\{E(f) : f \in \square^\omega(\mathbb{Q})\}$. As in the Rudin proof, the refinement is constructed by recursion on α; so as before, the problem is to guarantee that at each stage, the union of the sets chosen at preceding stages is closed. However, we no longer have Proposition 1 at our disposal (i.e., the decomposition is not upper semicontinuous) to ensure that whenever $E(f)$, for some f, is intersected at stage α, it is also covered at stage α. As an alternative strategy, we exploit the difference between essential and nonessential limit points.

At stage α, in the process of covering those points of k_α that were not covered at some previous stage, we introduce at most countably many new basic open sets (where a basic open set is a Cartesian product of open–closed sets in $\mathbb{Q}$); so it follows by Proposition 2 that the union of an initial segment of the refinement is closed if it contains all of its essential limit points. We are thus led to the following auxiliary concept. For each basic open set $\Pi_{m \in \omega} U_m$ (where U is an open–closed set-valued function) in the box product and each $n \in \omega$, define $\theta(U, n) = \{f : \forall m \geq n[f(m) \in U_m]\}$. A collection $\mathscr{A}$ of basic open sets is θ-closed iff $\forall \; \Pi_{m \in \omega} U_m \in \mathscr{A} \; \forall n \in \omega \; [\theta(U, n) \subseteq \cup \mathscr{A}]$. If $\mathscr{A}$ is θ-closed, then $\cup \mathscr{A}$ contains all of its essential limit points, and is therefore topologically closed whenever $|\mathscr{A}| < d$. So at stage α, we assume that we have a θ-closed refinement that covers k_β for each $\beta < \alpha$; we then extend the collection to a new θ-closed refinement that in addition covers k_α.

Now suppose in the factor space we replace each rational number with a compact set. At the strategic level, the proof of paracompactness in $\square^\omega(M)$ is similar to the proof for $\square^\omega(\mathbb{Q})$. However, there are substantial differences at the tactical level of developing the necessary machinery for extending θ-closed collections. The difficulty of passing from the rationals to the general case of a (zero-dimensional) σ-compact metric space is indicated in the set-theoretic hypotheses: $\square^\omega(\mathbb{Q})$ is paracompact under $OH(\mathbb{Q})$ (see Section 2 for the definition), which in turn follows from either $\mathfrak{b} = \mathfrak{d}$ or $\mathfrak{d} = \mathfrak{c}$; but in the general case, the implication $OH(M)$ from $\mathfrak{b} = \mathfrak{d}$ is known to fail, and whether paracompactness in $\square^\omega(M)$ follows from $\mathfrak{b} = \mathfrak{d}$ (or $OH(M)$) is an open question. $\square$

Proof of Theorem 2(3), (4) (Kunen): Suppose $D \subseteq \mathbb{P}$ is eventually dominant and has the subspace topology. We'll show that $D \times \square^\omega(\omega + 1)$ is nonnormal in ZFC. Let $A = \{\langle f, f\rangle : f \in D\}$ [note that $D \subseteq {}^\omega(\omega + 1)$], and let $B = \{\langle f, g\rangle : f \in D$, and g is eventually constant taking the value ω all but finitely many times$\}$. Then A and B are disjoint closed sets. (We remark that the existence of disjoint closed sets where one is the diagonal and the other is an equivalence class under finite disagreement is a manifestation of a nonclosed quotient map.) Let U be an open set containing A. For each $f \in D$, choose $\eta = \eta(f)$ so that for some $n \in \omega$, $\eta: [0, n] \to \omega$ satisfies: (1) $f \supseteq \eta$, and, (2) $\{\langle h, f\rangle: h \supseteq \eta\} \subseteq U$. By Fact 2(2), some η has an eventually dominant point-inverse set; so by Kunen's Lemma, $\overline{U}$ intersects B. $\square$

Outline of Proof of Theorem 2(5): A Hausdorff regular space Λ is Hurewicz if it possesses the following covering property introduced by Menger in 1924: for every sequence $\langle \mathscr{U}_n : n \in \omega\rangle$ of open covers of Λ, there is a sequence $\langle \mathscr{V}_n : n \in \omega\rangle$ such that (1) for each $n \in \omega$, $\mathscr{V}_n \in [\mathscr{U}_n]^{<\omega}$; and, (2) $\Lambda = \cup_{n \in \omega}(\cup \mathscr{V}_n)$. Note that σ-compact spaces are Hurewicz, and Hurewicz spaces are Lindelöf. Wingers proves the following two theorems which, together with Kunen's proof of 2(4), establish Theorem 2(5) as a corollary. (The proof of the first theorem uses ideas from Miller's paper [22].)

(1) A zero-dimensional, separable metric space is Hurewicz iff it is not homeomorphic to an eventually dominant subspace of $\mathbb{P}$. (It follows that $\mathbb{P}$ is an example of a Lindelöf space that is not Hurewicz; and a subspace $\Lambda \subseteq \mathbb{P}$ is an example of a Hurewicz space that is not σ-compact if Λ is $\leq^*$ unbounded but not homeomorphic to a $\leq^*$ cofinal subspace.)

(2) Suppose $\mathfrak{b} = \mathfrak{d}$ or $\mathfrak{d} = \mathfrak{c}$. Suppose further that Λ is a Hurewicz subspace of $\mathbb{P}$. Then $\Lambda \times \square^\omega (\omega + 1)$ is paracompact. (The key idea in the proof is that the quotient map (for the equivalence relation of finite disagreement) is closed (even though Λ is not locally compact) and has Lindelöf fibers; therefore, we can proceed, as in the proof of Theorem 1, to use the Decomposition Lemma on Paracompactness and the corollary to the first and second lemmas on the nabla product.) $\square$

REMARK: Also in [36] Wingers gives an example under Martin's Axiom of a Hurewicz subspace Λ of the real line such that $\Lambda \times \square^\omega (\omega + 1)$ is paracompact, while $\Lambda \times \Lambda \times \square^\omega (\omega + 1)$ is nonnormal.

PROBLEM 2 (van Douwen): Characterize the class of all separable metric spaces M for which $\square^\omega(M)$ is consistently paracompact. In particular, for which M is $\square^\omega(M)$ paracompact under the Continuum Hypothesis? (Solving this problem may involve developing a theoretical framework for proving paracompactness in $\square^\omega(M)$, where M is not locally compact.)

7. UNCOUNTABLY MANY FACTORS

THEOREM 3 [Lawrence, 15, 16]: In ZFC, $\square^{\omega_1}(\omega + 1)$ is nonnormal; under the assumption $2^\omega = 2^{\omega_1}$, the closed subspace consisting of all functions with countable support, $\{f \in \square^{\omega_1}(\omega + 1): \exists \alpha < \omega_1 \forall \beta \geq \alpha [f(\beta) = \omega]\}$, is nonnormal.

Outline of Proof: (See [16] for the details.) We first define an abstract topological space Λ, and then prove that Λ is both nonnormal and can be embedded as a closed subspace in $\square^{\omega_1}(\omega + 1)$.

Let $\mathscr{F}$ denote the set of all partial functions from ω_1 into ω : $\mathscr{F} = \{f: f$ is a function & Dom $f \subseteq \omega_1$ & Ran $f \subseteq \omega\}$. We partially order $\mathscr{F}$ by extension and domination (where a function g dominates a function f iff Dom $g \subseteq$ Dom f, and for every $\alpha \in$ Dom $g, g(\alpha) \geq f(\alpha)$). Yet a third partial order, $\preccurlyeq$, is defined by recursion on a subcollection of $\mathscr{F}$.

Our space Λ is defined using a tree $\langle T, \preccurlyeq \rangle$, where T is a subcollection of $\mathscr{F}$. As usual we mean that $\preccurlyeq$ is a reflexive, antisymmetric, and transitive relation such that for each $f \in T$, $\{g \in T : g < f\}$ is well-ordered by $<$; the height of f, denoted by Ht(f), is the order type of $\{g \in T : g < f\}$; for each ordinal δ, Lev$_\delta(T) = \{f \in T : $ Ht(f) $= \delta\}$; and Ht(T) is the least δ with Lev$_\delta(T) = \varnothing$. A branch of T is the range of 1–1 order-preserving function $B : \delta \to T$, where $\delta \leq$ Ht(T) and for every $\epsilon < \delta, B(\epsilon) \in$ Lev$_\epsilon(T)$; a branch is maximal in T iff there are no proper extensions contained in T.

For our construction, Ht(T) $= \omega_1$, and $\langle$Lev$_\delta(T) : \delta < \omega_1 \rangle$ is defined by recursion on δ (we remark that $f \preccurlyeq g$ implies $f \subseteq g$, but not conversely). Every point has 2^{ω_1} immediate successors, and every branch is countable.

Let $T' = \{f \in \mathscr{F} : f$ is the union of a maximal branch in $T\}$. Since every point in T has a successor, $T \cap T' = \varnothing$. Extend $\preccurlyeq$ to $T \cup T'$ in the obvious way: if f is the union of a maximal branch, then $g < f$ iff g is a point of the branch. Define a cut in $T \cup T'$ to be an ordered pair $\langle h, k \rangle$ where $h \in T, k \in T'$, and $h < k$. Let $\mathscr{C}$ be the collection of all cuts. The point set of our space Λ is the pairwise disjoint union $T \cup T' \cup \mathscr{C}$.

We define the topology of Λ to be the collection of open sets generated by the following basic neighborhoods. To begin, we declare each cut to be an isolated point. Suppose $f \in T$ and $g : \omega_1 \to \omega$ is a total function extending f. Then define the g-neighborhood of f by $N(f, g) = \{f\} \cup \{\langle h, k \rangle \in \mathscr{C}: h = f$ & k dominates $g\}$. Suppose $f \in T'$ with Ht(f) $= \delta$. Choose $B : \delta \to T$ so that Ran B is the maximal branch of T of which f is the union. Inherent in the construction of the tree is a particular cofinal subsequence of B of order type ω. We call this sequence the canonical sequence of f, and denote it by Seq (f). Suppose $n \in \omega$. Then define the n-neighborhood of f by

$N(f, n) = \{f\} \cup \{\langle h, k \rangle : k = f \,\&\, \exists m \geq n[h = \text{Seq } (f)(m)]\}$; so the local topology of Λ at f is completely described by a simple sequence of cuts converging to f.

Note that (1) the isolated points coincide with the cuts, so $T \cup T'$ is closed; (2) for each $f \in T \cup T'$, each basic neighborhood of f is the union of $\{f\}$ and a subcollection of the cuts, so $T \cup T'$ is discrete; (3) any two basic neighborhoods about distinct points in T, or distinct points in T', are disjoint, so each of T and T' individually can be pointwise separated by a pairwise disjoint collection of open sets; and, (4) the intersection of a basic neighborhood about a point $h \in T$ and a basic neighborhood about a point $k \in T'$ is either empty or the one-point set whose only element is the cut $\langle h, k \rangle$.

We claim that T and T' cannot be separated by disjoint open sets, and therefore X is nonnormal. We further claim that in ZFC, X can be embedded as a closed subspace of $\square^{\omega_1}(\omega + 1)$; and if $2^\omega = 2^{\omega_1}$, then we can modify X to obtain a nonnormal space that can be embedded as a closed subspace where each point has countable support. We remark that (1) Proposition 3 plays a key role in our construction of the embedding; and, (2) our proof that T and T' cannot be separated requires that the point set of Λ and a base for the topology have the same cardinality (which explains the role of $2^\omega = 2^{\omega_1}$ in the hypothesis of the second theorem). (Reference [17] is an announcement that contains the construction of the space X, and the proof that X is nonnormal; the embedding (along with the construction and the nonnormality proof) can be found in [16].) $\square$

PROBLEM 3 (Williams): Is the closed subspace of $\square^{\omega_1}(\omega + 1)$ consisting of all functions with countable support nonnormal in ZFC? In particular, what happens under the Continuum Hypothesis?

ACKNOWLEDGMENT

The author thanks the referee for a number of helpful comments.

REFERENCES

1. DANIEL, T. 1991. Normality in box products and Σ-spaces. Ph.D. Thesis, Univ. of Wisconsin, Madison.
2. VAN DOUWEN, E. K. 1975. The box product of countably many metrizable spaces need not be normal. Fundam. Math. **88:** 127–132.
3. ———. 1977. Another nonnormal box product. Gen. Topol. Appl. **7:** 71–76.
4. ———. 1980. Covering and separation in box products. Surveys in General Topology: 55–130. Academic Press. New York.
5. DOW, A. 1989. A separable space with no remote points. Am. Math. Soc. Trans. **312:** 335–353.
6. ERDÖS, P. & M. E. RUDIN. 1973. A non-normal box product. Colloq. Math. Soc. János Bolyai **10:** 629–631. Keszthely.
7. HECHLER, S. H. 1974. On the existence of certain cofinal subsets of $^\omega\omega$. In Proceedings of the Symposium on Pure Mathematics **10:** 155–174. American Mathematical Society. Providence, R.I.
8. KNIGHT, C. J. 1964. Box topologies. Q. J. Math. **15:** 41–54.

9. KUNEN, K 1973. Some comments on box products. Colloq. Math. Soc. János Bolyai **10:** 1011–1016. Keszthely.
10. ———. 1978. On paracompactness of box products of compact spaces. Am. Math. Soc. Trans. **240:** 307–316.
11. ———. 1980. Set Theory: An Introduction to Independence Proofs. North-Holland. Amsterdam, the Netherlands.
12. ———. Box products of ordered spaces. Topol. Appl. **20:** 245–250.
13. LAWRENCE, L. B. 1987. Convergence in the box product of countably many metric spaces. Topol. Proc. **12:** 85–92.
14. ———. 1988. The box product of countably many copies of the rationals is consistently paracompact. Am. Math. Soc. Trans. **309:** 787–796.
15. ———. Failure of normality in the box product of uncountably many real lines. To be published in Am. Math. Soc. Trans.
16. ———. Embedding a nonnormal space constructed from a tree in a box product of convergent sequences. Preprint.
17. ———. Failure of normality in the box product of uncountably many copies of a convergent sequence. Preprint.
18. MICHAEL, E. A. 1953. A note on paracompact spaces. Am. Math. Soc. Proc. **4:** 831–838.
19. ———. 1957. Another note on paracompact spaces. Am. Math. Soc. Proc. **8:** 822–828.
20. MILLER, A. W. 1982. On box products. Topol. Appl. **14:** 313–317.
21. ———. 1984. Special subsets of the real line. In Handbook of Set-Theoretic Topology, K. Kunen and J. E. Vaughan, Eds.: 201–234. North-Holland. Amsterdam, the Netherlands.
22. ———. 1988. On some properties of Hurewicz, Menger, and Rothberger. Fundam. Math. **129:** 17–33.
23. ROITMAN, J. 1978. Paracompact box products in forcing extensions. Fundam. Math. **102:** 219–228.
24. ———. 1979. More paracompact box products. Am. Math. Soc. Proc. **74:** 171–176.
25. RUDIN, M. E. 1971. A normal space X for which $X \times I$ is not normal. Fundam. Math. **73:** 179–186.
26. ———. 1971. Announcement on box products of ordinals. Prague Symposium: 385.
27. ———. 1971. The box topology. In Proceedings of the Univ. of Houston Point-Set Topology Conference: 191–199 (Houston, Tex.).
28. ———. 1972. The box product of countably many compact metric spaces. Gen. Topol. Appl. **2:** 293–298.
29. ———. 1972. Box products and extremal disconnectedness. In Proceedings of the Univ. of Oklahoma Topology Conference: 274–284 (Norman).
30. ———. 1974. Countable box products of ordinals. Am. Math. Soc. Trans. **192:** 121–128.
31. ———. 1975. Lectures on set-theoretic topology. In Regional Conference Series in Mathematics **23**. American Mathematical Society. Providence, R.I.
32. WILLIAMS, S. W. 1976. Is $\square^{\omega}(\omega + 1)$ paracompact? Topol. Proc. **1:** 141–146.
33. ———. 1977. Boxes of compact ordinals. Topol. Proc. **2:** 631–642.
34. ———. 1984. Box products. In Handbook of Set-Theoretic Topology, K. Kunen and J. E. Vaughan, Eds.: 169–200. North-Holland. Amsterdam, the Netherlands.
35. WINGERS, L. Box products of σ-compact spaces. To be published in Topol. Appl.
36. ———. Box products and Hurewicz spaces. Preprint.

Mary Ellen Rudin's Contributions to the Theory of Nonmetrizable Manifolds

PETER J. NYIKOS

Department of Mathematics
University of South Carolina
Columbia, South Carolina 29208

1. INTRODUCTION

There are still any number of books and articles that define "manifold" to mean "metrizable space, in which every point has a neighborhood homeomorphic to $\mathbb{R}^n$ for some finite n," or even require that the space be compact. But general topologists for at least the last fifteen years have mentally substituted "Hausdorff" for "metrizable." This change was wrought by a single unexpected result: in 1975, Mary Ellen Rudin provided consistent solutions to a problem posed in 1949 by Wilder [47].

The problem was whether every perfectly normal generalized manifold is metrizable, and Mary Ellen showed that under the axiom $\diamondsuit$ (a consequence of $V = L$, Gödel's Axiom of Constructibility) there is a perfectly normal, hereditarily separable, countably compact, nonmetrizable 2-manifold. By sacrificing countable compactness, she built an example using the continuum hypothesis (CH) alone, which P. L. Zenor helped to simplify [42].

Then, in 1978, Mary Ellen showed that the existence of a perfectly normal nonmetrizable manifold is independent of the usual axioms of set theory. Specifically, she showed $MA(\omega_1)$ implies all perfectly normal manifolds are metrizable.

This did not, as has generally been believed until now, solve completely the Wilder problem as originally stated, although Wilder was quite satisfied with the solution and perhaps no longer cared about his original problem. Nor did either discovery really touch a 1935 problem of Alexandroff. (For details, see Section 5). This is not to take away from the magnitude of Mary Ellen's discovery. In fact, the first result takes on added luster for answering Wilder's question (at least consistently) in an especially strong way: with an example that is a manifold and not just a "generalized" manifold. More importantly, her discoveries have inspired a spate of research in the topic of nonmetrizable manifolds by several topologists, myself included. In fact, the modern study of nonmetrizable manifolds can be said to begin with her results, in the same basic sense that the modern study of set-theoretic topology can be said to begin with the 1967 realization that $MA(\omega_1)$ implies the existence of separable nonmetrizable normal Moore spaces.

Elaborating a bit on the last theme, there are still large areas of the theory of nonmetrizable generalized manifolds that are "stuck in the classical period." This represents a challenge to future research that I will make more explicit in the final section, which gives not just open problems but also whole research topics.

In between, I will be mostly expounding on Mary Ellen Rudin's research into this area and the results leading to them. I will also make some effort to bring the reader up to date on the research into nonmetrizable manifolds that has been done since the *Handbook* article on manifolds [24], although space does not permit much more than a summary of results.

For the rest of this article, manifolds are assumed to be Hausdorff and connected.

2. PRECURSORS OF THE 1975 CONSTRUCTIONS

One of the streams of research relevant to Mary Ellen's results can be traced back to F.B. Jones's 1937 paper on normality in Moore spaces [15]. On the one hand, Jones shows $2^{\aleph_0} < 2^{\aleph_1}$ implies every separable normal Moore space is metrizable. On the other hand, he uses an uncountable λ-set to construct what is essentially [see [26] for the connection] a separable, pseudonormal, nonmetrizable Moore space. This latter construction was a big step toward the 1967 realization mentioned in the Introduction.

The next step was taken by Rothberger, who showed [33] that an axiom now known as "$\mathfrak{p} > \omega_1$" implies the existence of a Q-set. [For "Q-set" we can read "uncountable subset Q of $\mathbb{R}$ in which every subset is an F_σ in the relative topology of Q."] Bing then gave a construction [6, example E] of a separable nonmetrizable normal Moore space assuming the existence of a Q-set. His construction is closely related to that of Jones's pseudonormal example, and the proof of normality, which Bing omitted, uses the same ideas. Booth [49], without knowing of Rothberger's results, showed that $\mathfrak{p} > \omega_1$ follows from $MA(\omega_1)$. It remained for Silver, at the instigation of Frank Tall, to provide the missing link in the initial realization. He showed that a consequence of $MA(\omega_1)$ due to Solovay (later shown equivalent to $\mathfrak{p} > \omega_1$) implies the existence of a Q-set [36, p. 21].

There are a number of related constructions of separable Moore spaces, some of which dovetail nicely with Q-sets to produce nonmetrizable normal Moore spaces. Bing's example E was the Niemytzki tangent disk space (a.k.a. the Moore plane) with all but a Q-set of points on the x-axis thrown out.

In a 1974 expository article [35], Mary Ellen Rudin described a number of related constructions of separable nonmetrizable Moore spaces using the full binary tree of height $\omega + 1$. This tree is also known as the Cantor tree since there is a natural way of identifying the points at the top level of the tree with the Cantor set. In fact, there is a natural way of embedding the whole tree into $\mathbb{R}^2$ with a topology that is a rotation of the Sorgenfrey plane; details may be found in [26].

The Cantor tree itself is locally compact in the interval topology. If one inserts line segments to connect each node to its immediate successors, one obtains local connectedness at the cost of local compactness. The resulting "road space" is described in [35], followed by a construction of a "wide road space" in which the earlier one is "its center line."

Actually, these constructions are done in greater generality, using any tree of height $\leq \omega_1$, all of whose levels (except the topmost) are countable. If the tree has no maximal elements, the associated wide road space is a 2-manifold. Otherwise it is a 2-manifold-with-boundary, the boundary being a discrete set of copies of the open unit interval.

Mary Ellen goes on to tell us that if one throws out all but $\aleph_1$ points at the top of the Cantor tree or the associated (narrow) road space, the resulting space, which is nonmetrizable in ZFC, is normal under MA + ¬CH; whereas the analogous pruning does not give normality in the wide road space, for if one picks points one apiece from the boundary intervals, then repeats this, the two resulting closed discrete subspaces cannot be put into disjoint open sets.

All this was known for quite some time before this 1974 survey paper, but in the same volume there was a new theorem (also mentioned by Mary Ellen) due to Reed and Zenor, showing that nonnormality of the wide-road space is no accident.

THEOREM 2.1 [31]: Every locally compact, locally connected, normal Moore space is metrizable.

Thus, in particular, there are no nonmetrizable normal Moore manifolds.

There are a number of natural unsolved problems related to this theorem.

PROBLEM 1: Is every locally compact, locally connected, countably paracompact Moore space metrizable?

Nowadays it is a conditioned reflex among set-theoretic topologists, when confronted with a theorem like Theorem 2.1, to ask whether "normal" can be replaced by "countably paracompact," but Theorem 2.1 came early enough so that not much attention has been paid to Problem 1 to date. It remains open even for manifolds.

We do have some consistency results for Problem 1, including of course Burke's results that the product measure extension axiom (PMEA) implies that all countably paracompact Moore spaces are metrizable. Also one does not need large cardinals: Mary Ellen's argument in [36, p. 46] that $V = L$ implies all locally compact normal Moore spaces are metrizable, works equally well for "countably paracompact" in place of "normal" if one substitutes W. S. Watson's theorem in [46] for the theorem of Fleissner used by Mary Ellen.

Even more germane is the following problem, first posed by Mary Ellen in the form, "Does MA + ¬CH imply . . . ?"

PROBLEM 2: Is it consistent that every perfectly normal, locally compact, locally connected space is metrizable?

In other words, can "normal Moore" in Theorem 2.1 be weakened to "perfectly normal"? (Recall that every normal Moore space is perfectly normal.) The opening phrase in Problem 2 is stated the way it is because we know of consistent examples: Mary Ellen's manifolds, but also the much earlier Souslin lines, are examples of perfectly normal, locally compact, locally connected spaces.

We shall return to Problem 2 in Section 5. In Section 6 we shall look at the following variant, also due to Mary Ellen.

PROBLEM 3: Is every perfectly normal, locally compact, locally connected space collectionwise normal?

This question would be a consolation prize for Problem 2, were it not for the fact that the known consistency results go in the opposite direction: Diane Lane [19], generalizing Mary Ellen's 1978 result, showed that $MA(\omega_1)$ implies that every perfectly normal, locally compact, locally connected space is paracompact.

But let's take one more quick look at 1974, when nothing was known about perfectly normal nonmetrizable manifolds.

As part of a follow-up to Theorem 2.1, Reed and Zenor called general topologists' attention to the problem at the end of Wilder's book [47].

PROBLEM 4: Is there a perfectly normal, nonmetrizable generalized manifold?

They did not care much about the word "generalized" and even restricted its meaning in [32] to "not necessarily metrizable," so that Mary Ellen specifically aimed for manifolds in her 1975 constructions.

3. THE 1975 CONSTRUCTION AND RELATED EXAMPLES

One inspiration for Mary Ellen's 1975 perfectly normal manifolds was the second of a pair of Moore manifolds described by R. L. Moore in [21] and at the end of the

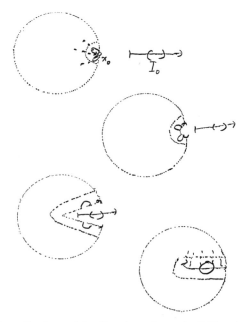

FIGURE 1. A disk swallowing a sticker "at a boundary point." (Drawing by Peg Daniels.)

second edition of his book on point set theory [22]. This example can be produced by replacing each point on the x-axis of the tangent disk space by a copy of $[0, 1)$ in a certain way. It is the method, rather than the example itself, that Mary Ellen adapted to her needs. It is illustrated by FIGURE 1, in which the "nicked disk" that engulfs the half-open segment can be thought of as a closed tangent disk at the point $\langle x, 0 \rangle$, with the point itself removed. The R. L. Moore example does this in a simple geometric fashion at every point, explained in detail in [24, examples 3.6 and 3.9]. (Moore's description was completely different, reminiscent of the description of the projective plane as being the set of all lines through the origin in $\mathbb{R}^3$ with a certain natural topology.)

A feature of Moore's example that Mary Ellen did not adopt is that the individual additions do not interact: none of the tangent disks meets more than one added copy of [0, 1), and these copies form a discrete collection of closed subsets of Moore's example. In the CH example of [42], the added intervals interact very much as though each were replacing a point of the Kunen "line" [16]; in the $\diamond$ example, as though each were replacing a point in the Ostaszewski space [29]. For the latter example, see [24].

The constructions both begin with a copy of the open unit disk, and proceed by adding, one at a time, copies of [0, 1) indexed by the countable ordinals (FIG. 2). As can be seen from FIGURE 1 (modified by removing the boundary of the disk), the space resulting after the first interval is added is still a copy of the disk. One might think that, after infinitely many such steps, the resulting space is no longer homeomor-

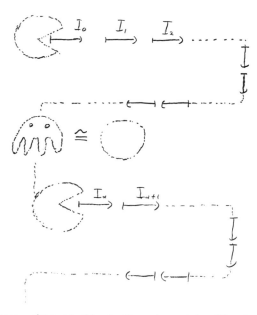

FIGURE 2. Repeat ω_1 times, attaching I_α at boundary point x_α. (Drawing by Peg Daniels.)

phic to the open disk (middle of FIG. 2), but the title of a paper by Morton Brown [7] tells us otherwise: "The [countable] monotone union of open n-cells is an n-cell." And so the process can be continued through all the countable ordinals (rest of FIG. 2 and beyond).

Now, one cannot just go mindlessly gobbling up intervals and expect the resulting manifold to have the properties one is looking for. Rudin and Zenor use CH to list all the countable subsets of the underlying set of the final space—not just the starting disk but its union with the $\aleph_1$ added intervals. Then at the αth stage they manipulate the closure of the αth countable set on their list, much as Kunen did in [16]. Kunen's construction gave a locally countable space, as did the space of Ostaszewski that guides the construction of the $\diamondsuit$ example [24, example 3.11], yet the modifications in the manifold constructions are quite simple.

Manipulating the closures of a countable set is done through a sequence of homeomorphisms that move the points around to strategic locations prior to the engulfing of the αth interval. The method is outlined in [24, example 3.9] and done in some detail in [42]. The most detailed description of all is in the paper of Kozlowski and Zenor [18], which produces, in four careful steps, a C^∞ transformation of $\mathbb{R}^n$ to its union with a copy of $[0, 1)$, while moving the points of a closed discrete subspace to appropriate locations. En route to their end product, a smooth version of the main example of [42], they prove the C^∞ modification of Morton Brown's theorem—which, in view of all the exotic smoothings of $\mathbb{R}^4$ subsequently discovered by S. K. Donaldson and others, is no mean feat.

Other, more recent extensions of the 1975 results have to do with weakening the set-theoretic hypotheses employed. An unpublished 1986 result of mine is that a hereditarily separable, perfectly normal, countably compact, nonmetrizable manifold exists in any model where $\aleph_1$ Cohen or random reals are added to any model of CH. Balogh [5] gives an example of such a manifold in L that is "preserved" by the addition of any number of Cohen reals. (Points are added to it just as to $\mathbb{R}$ itself, but a general description, using Borel codes, and good enough to give the relevant properties, emerges unscathed from the forcing.) In the same preprint, Gruenhage shows that a manifold essentially like the main example of [42] can be produced from any Luzin set.

But, of course, extra set-theoretic hypotheses are unavoidable. This is the theme of the next section.

4. THE 1978 INDEPENDENCE RESULT

At the 1978 ICM in Helsinki, in a lull between sessions, I noticed Mary Ellen busily explaining a proof to Teodor Przymusiński. On learning that it showed MA(ω_1) implies perfectly normal manifolds are metrizable, I exclaimed, "That's a major new result!" In an offhand manner she responded, "Well, I worked long enough on it," and resumed her explanation.

The proof and various generalizations have appeared in several easily accessible places [37], [19], [12], [24], so I will only comment on the main features here. It uses MA(ω_1) twice. The first use is a short and straightforward application of Szentmiklóssy's result that every locally compact space of countable spread is hereditarily Lindelöf. The second use of MA(ω_1) is a characteristic dirty-hands proof. In [24] this is further divided into two applications. In the first, finite families of open sets with compact closures are the members of a countable chain condition partially ordered set (ccc poset) P of cardinality $\aleph_1$, and every centered subset of P unions up to a discrete collection of closed sets. In the second, use is made of the result that MA(ω_1) implies every ccc poset of cardinality $\aleph_1$ is σ-centered. These two applications give a σ-discrete cover by compact sets, which in a manifold is equivalent to being a Moore space [Theorem 8.3 below]. An application of the Reed–Zenor theorem 1.1 completes the proof.

Szentmiklóssy's theorem is used to show that MA(ω_1) implies that every perfectly normal manifold is what I call "of Type I." No better name has come up since this term was introduced in [23], nor is it clear what might be a good definition of "Type II," and so on.

DEFINITION 4.1: A manifold is of *Type I* if every separable subset is Lindelöf (equivalently, second countable).

A more complicated but structurally revealing definition, embodied in the following proposition, was used in [24]. It seems more suitable for topological spaces in general.

PROPOSITION 4.2: A manifold is of Type I if, and only if, it is the union of an ω_1-sequence $\{U_\alpha : \alpha < \omega_1\}$ of open subspaces such as that $\overline{U_\beta} \subset U_\alpha$ whenever $\beta < \alpha$, and such that $\overline{U_\alpha}$ is Lindelöf for all α.

Proof: Sufficiency is easy. To show necessity, begin with a Euclidean open set U_0. With U_α defined and separable, cover the boundary of U_α with countably many Euclidean open sets and let their union be $U_{\alpha+1}$. At limit ordinals, take unions. By induction, $\overline{U_\alpha}$ is Lindelöf for all $\alpha < \omega_1$ and $\overline{U_\beta} \subset U_\alpha$ whenever $\beta < \alpha$. By connectedness, $\cup\{U_\alpha : \alpha < \omega_1\}$ is the whole manifold: it is clearly open, and it is also closed because of first countability. $\square$

Type I spaces have a number of nice structural properties, exploited in the second half of the MA(ω_1) proof and in several generalizations recounted in [24, sec. 4]. One of the problems at the end of that section remains unsolved.

PROBLEM 5: Is it consistent that every hereditarily normal manifold of dim > 1 is metrizable?

Since the long ray and long line, the only nonmetrizable manifolds of dimension 1, are not perfectly normal, an affirmative solution under MA(ω_1) would be a major generalization of Mary Ellen's 1978 theorem. This does not seem very likely; however, it does seem likely that the PFA implies that every hereditarily normal, hereditarily collectionwise Hausdorff manifold of dim > 1 is metrizable.

5. HISTORICAL REVISIONISM

After Mary Ellen obtained her 1978 independence result, Wilder (and apparently everyone else aware of her results) agreed that she had answered his 1949 problem, and it was also assumed that she had solved a 1935 problem of Alexandroff [1], whether there could be a "manifold in the sense of Čech" that is not metrizable. But in February of 1992, while I was trying to determine exactly which paper of Čech Alexandroff was referring to, an altogether different picture emerged.

To begin with, note the word "generalized" in Problem 4 (end of Section 3). To Wilder in 1949 it had a specific meaning, defined on page 244 of his text [47]. He refers to a space S as a *generalized n-manifold* if it is of covering dimension n and:

(i) Locally coconnected in all dimensions, $0, \ldots, n-1$;
(ii) Of n-dimensional local co-Betti number 1 at each point.

These concepts are defined in [47, pp. 189–191]. They imply local compactness and local connectedness but not local metrizability. There are generalized manifolds in Wilder's sense that are metrizable but not manifolds, and others that are neither.

Also, every manifold in our sense is a generalized manifold in Wilder's sense. So Mary Ellen's 1975 constructions certainly showed that a positive answer (a particularly nice one) to Wilder's problem is consistent. Her 1978 $MA(\omega_1)$ result, on the other hand, only showed that if there is a ZFC example of a perfectly normal nonmetrizable generalized manifold, it cannot be a (locally Euclidean) manifold. Apparently this was good enough for Wilder; we do not know why—perhaps he no longer cared about his problem as originally worded.

The case of Alexandroff's problem is another matter. It remained untouched by either of Mary Ellen's results, because *Čech's definition required that the spaces be compact!* Now every compact manifold is clearly metrizable, so Alexandroff was not asking for a manifold in our sense. In fact, he plays around with a number of different meanings of "manifold," and even says in a footnote that "manifolds in the sense of Hopf" will from that point on be called "locally-Euclidean topological manifolds"! But he does not use "locally Euclidean" in connection with Čech's concept of a manifold.

Apparently, Alexandroff's problem remains unsolved to this day. Interestingly enough, we would have a complete solution to both his problem and Wilder's if we could find an affirmative answer to Problem 2, Mary Ellen's own problem of whether it is consistent that every perfectly normal, locally compact, locally connected space is metrizable.

I am saving one final bit of irony pertaining to Alexandroff's problem for the last section, where I speak of "Wilder's other problem."

6. LONG SNAKES AND COLLECTIONWISE NORMALITY

A natural question about the 1978 independence result is whether both uses of $MA(\omega_1)$ are really needed. For instance, is it simply true that every locally compact Type I perfectly normal space is subparacompact? This question was answered in the negative by a perfectly normal Aronszajn tree constructed by Hanazawa [13] using $\diamondsuit^*$, but remained open for manifolds until Mary Ellen constructed an example in 1985 using $\diamondsuit^+$ [38].

Actually, she was motivated not by the foregoing questions, but by a search for a solution to the problem of whether every perfectly normal manifold is collectionwise normal (Problem 4). She decided that her best chance to produce one was to tie together uncountably many Type I perfectly normal nonmetrizable manifolds. The examples in Section 2 were separable, hence not Type I; neither were the nonseparable examples whose existence was announced in [42] and [24], but whose details are yet to be published. So her 1985 example broke new ground.

It was not even two-dimensional, but three-dimensional to take advantage of the greater freedom of movement that the third dimension allows. It might be called a "long kinky snake." It admits a projection onto the long ray with open 2-disks for fibers. [*Fibers* in such a context means point-inverses.] This keeps it from being metrizable, of course. Being perfectly normal, it also does not have a copy of ω_1 in it. However, the initial segments $\pi^\leftarrow(0, \alpha)$ are each homeomorphic to a solid cylinder and constitute the sets U_α in Definition 4.1.

This long snake is collectionwise normal. Mary Ellen's idea was to tie together uncountably many snakes, which would collectively contain a discrete collection of non-Lindelöf closed sets that cannot be put into disjoint open sets. For a while, she thought she had succeeded with $\diamond^{++}$, but eventually had to retract this claim. She even formally "withdrew" from the attempt to produce such an example at the 1987 STACY conference. There I asked her whether she could produce a normal, not collectionwise normal example by sacrificing "perfectness." She assured me she could, and I encouraged her to write it up because it would be the first consistency result on the following problem.

PROBLEM 6: Is there a normal, locally compact, locally connected space that is not collectionwise normal?

It was at the STACY conference that Balogh created a sensation by announcing and outlining a proof that the addition of supercompact many random reals gives a model in which every normal, locally compact space is collectionwise normal. It is still not known whether large cardinals are necessary for the consistency of this statement, even with local connectedness thrown in. (For manifolds one can use cMEA as shown in [23]; this requires only a weakly compact cardinal.)

Mary Ellen's promised example appears in [40]. I must confess that I have not been through this example, and even was under the impression at the Madison conference that it was three-dimensional and built like her failed example; hence I dubbed it "Medusa." Actually it is two-dimensional, produced by tying together ω_1-many "long ribbons." Each ribbon projects to the long ray L^+ in a map π with fibers homeomorphic to an open unit interval, although it is more natural to think of the αth ribbon as projecting to $L^+\backslash(0, \alpha]$. Then in each copy, the inverse image of the set of limit ordinals is closed in the whole space, and it is these closed sets that form a discrete collection witnessing failure of collectionwise normality.

The example is not hereditarily normal, leaving the following question open.

PROBLEM 7: Is every hereditarily normal manifold collectionwise normal?

This time we do not need large cardinals for consistency of a Yes answer: see [23]. But if we weaken "manifold" to "locally compact, locally connected space," there is no improvement on Balogh's result.

Being of Type I, the examples of this section need to be produced in a different manner from the Section 3 examples. The long kinky snake is built in a kind of telescoping process, whereby a tube inside $\pi^\leftarrow(0, \alpha)$ snakes up to the fiber over α and attaches to it in a carefully prescribed way; basic open neighborhoods of points in the fiber over α are contained inside this tube. Other Type I manifolds with similar constructions appear in [28].

Unlike with the Section 3 manifolds, these do not seem to have any locally countable analogues in print until now, so I thought it worthwhile to provide a family of them in the following section.

7. LOCALLY COUNTABLE PROTOTYPES

This section presents a general construction of spaces E, each of which is of Type I, and has a continuous open projections π onto ω_1 with countably infinite discrete fibers. We will even use $\omega_1 \times \mathbf{Z}$ as an underlying set, but initial segments will be put together in such a way that only in one degenerate case will the resulting space actually be $\omega_1 \times \mathbf{Z}$ as a space.

Readers versed in fiber bundles will recognize from the description below that E is always a principal $\mathbf{Z}$-bundle over ω_1. We will not need this information at any time.

CONSTRUCTION 7.1: For each limit ordinal α, pick a cofinal $S_\alpha \subset [0, \alpha)$ of order type ω. Split each S_α into countably many disjoint subsets S_α^n ($n \in \mathbf{Z}$). We can allow all but one of these to be empty, but the nontrivial examples are those where infinitely many are nonempty. We define a locally compact topology on $[0, \alpha] \times \mathbf{Z}$ with the help of these sets S_α^i.

The topology is defined so that the vertical translations τ_n taking $\langle \alpha, k \rangle$ to $\langle \alpha, k + n \rangle$ are homeomorphisms. We define, for each ordinal ξ, a set $T_\xi \subset [0, \alpha] \times \mathbf{Z}$, "the thread leading up to $\langle \xi, 0 \rangle$," that meets each fiber $\{\eta\} \times \mathbf{Z}$, $\eta \leq \xi$ in exactly one point. The other points in $\{\xi\} \times \mathbf{Z}$ will also have threads leading up to them, vertical translations of this one. Basic neighborhoods are produced by chopping off initial segments of these threads.

However T_ξ is defined, let $T_\xi \backslash \eta$ denote $T_\xi \backslash \pi^{-1}[0, \eta]$, and let $T_\xi + k$ [respectively, $(T_\xi \backslash \eta) + k$] denote the image of T_ξ [respectively, $T_\xi \backslash \eta$] under τ_k for each $k \in \mathbf{Z}$. The local base at $\langle \xi, k \rangle$ is then defined to be $\{(T_\xi \backslash \eta) + k : \eta < \xi\}$.

Let $T_0 = \{\langle 0, 0 \rangle\}$. If $\xi = \eta + 1$, we let $T_\xi = T_\eta \cup \{\langle \xi, 0 \rangle\}$. This is enough to make every point of the form $\langle \xi, n \rangle$ isolated, where ξ is 0 or a successor ordinal.

If α is a limit ordinal, we list S_α in order as $\{\alpha(i) : i \in \omega\}$. For each i, let k_i be the unique integer k such that $\alpha(i) \in S_\alpha^k$. Let $T_\alpha^0 = T_{\alpha(0)} + k_0$. If $i > 0$, let $T_\alpha^i = (T_{\alpha(i)} \backslash \alpha(i - 1)) + k_i$. Let $T_\alpha = \cup \{T_\alpha^i : i \in \omega\} \cup \{\langle \alpha, 0 \rangle\}$. Note that this makes the sequence of points $\langle \alpha(i), k_i \rangle$ converge to $\langle \alpha, 0 \rangle$, while $\langle \alpha(i), k_i \rangle$ drags the earlier points in its own basic neighborhood $(T_{\alpha(i)} \backslash \alpha(i - 1)) + k_i$ along with it into the thread leading up to $\langle \alpha, 0 \rangle$. From this it is clear that this defines a base for a topology on T_α.

Once T_α has been defined for all $\alpha < \omega_1$, we let E denote $\omega_1 \times \mathbf{Z}$ with the resulting topology.

It is easy to see, by induction, that T_ξ is homeomorphic to $\xi + 1$ with the order topology for each countable ordinal ξ. In fact, the map that takes $\langle \eta, j \rangle$ to $\langle \eta, j + k \rangle$ for the unique k such that $\langle \eta, k \rangle \in T_\xi$ is a homeomorphism. From this it easily follows that the topology on $[0, \xi] \times \mathbf{Z}$ is homeomorphic to the product topology. Hence the space E is locally compact and locally countable; and since $[0, \xi] \times \mathbf{Z}$ is clopen, the space E is pseudonormal with this topology.

It is also easy to see that if we simply let $S_\alpha^0 = S_\alpha$ for all α, then the topology on E is the product topology. However, in the examples we now give, the topology is not even homeomorphic to the product topology.

Both examples are prototypes for the subspace T^+ of positive vectors in the tangent bundle over the long ray L^+ with respect to some specific smoothing of L^+. As explained in [28], the tangent bundle is not a single space, but a family of $2^{\aleph_1}$ distinct spaces, since its structure depends upon the differentiable structure ("smoothing") one puts on L^+. But each version is a 2-manifold with a canonical projection map $\pi: TL^+ \to L^+$ making it locally like $L^+ \times \mathbf{R}$. In fact, each initial segment $\pi^{-1}(0, \alpha)$ admits a fiber-preserving homeomorphism with $(0, \alpha) \times \mathbf{R}$. But globally it is very different. The subspace L_0 of zero vectors is a copy of L^+; but every copy of L^+ in TL^+ meets it. If we remove L_0, then TL^+ falls apart into connected mirror images T^+ and T^-, the subspaces of positive and negative vectors. Thus T^+, which is locally like $L^+ \times \mathbf{R}^+$, does not contain a copy of L^+. Similarly, neither of our specific examples of spaces E contains a copy of ω_1.

EXAMPLE 7.2 (Modeled after the first example in [28] of a T^+): In the preceding construction, let $S_\alpha^i = \{\alpha(i)\}$ for all $i \in \omega$, $S_\alpha^i = \varnothing$ otherwise. Let the resulting version of E be denoted E_1. In order to "get at $\langle \alpha, 0 \rangle$ from the left," one must follow T_α, which looks as though it were shooting up to ∞. But it is more realistic, topologically, to say that the line $[0, \alpha] \times \{n\}$ is "converging to $\langle \alpha, -\infty \rangle$," and it even "plunges down to $-\infty$ in front of each limit ordinal." Hence $\omega_1 \times \{n\}$ is a closed discrete subspace of E_1. So E_1 is σ-discrete, and is a Moore space by Theorem 8.3 below. In particular, it does not contain a copy of ω_1.

But E_1 is not metrizable; if it were, it would be the direct sum of Lindelöf, hence countable, clopen subsets C_α ($\alpha < \omega_1$). (Every locally compact, paracompact space is the direct sum of clopen σ-compact subspaces.) Then we could give each point a neighborhood that is inside the unique C_α containing it. But if we let $S \subset \omega_1$ be stationary, then the chosen neighborhoods of uncountably many $\langle \alpha, 0 \rangle$ such that $\alpha \in S$ would press down into some countable $[0, \xi] \times \mathbf{Z}$, hence uncountably many of them would have to meet, a contradiction.

A somewhat more complicated argument, also using the pressing-down lemma, shows that if two open sets meet the fibers over a stationary set, they must meet. (The analogous fact for T^+ is shown in [28], and the proof for E_1 is similar.) From this it follows that a zero set that meets the fibers over a stationary set must actually contain $\pi^\leftarrow C$ for some club set C. Hence the set of all such zero sets forms a free $\mathbf{Z}$-ultrafilter with the countable intersection property, and E_1 is not even realcompact. On the other hand, it is α-realcompact, as is any σ-discrete space of cardinality $\aleph_1$; whereas the T^+ on which it is modeled is not α-realcompact. Here is our first indication of pitfalls in the transition from locally countable spaces to manifolds. We will see others in the following section.

Our second example is modeled after the first $\diamondsuit$ example in [28]. It was meant to be included in [25] as example 3.10.

EXAMPLE 7.3 [♣]: For each limit ordinal α, we pick a cofinal $S_\alpha \subset [0, \alpha)$ in accordance with ♣. That is, for each uncountable $S \subset \omega_1$, there exists α such that $S_\alpha \subset S$. We may assume S_α is of order type ω. This time we let each S_α^n ($n \in \mathbf{Z}$) be infinite. Let this version of E be denoted E_2.

CLAIM: Let π be the projection from E_2 to ω_1. Every uncountable subset of E_2 has a set of the form $\pi^{-1}C$ in its closure, for some closed unbounded (club) subset C of ω_1.

It is enough to verify "unbounded" since each neighborhood of $\langle \alpha, n \rangle$ meets the fibers over some final segment of $[0, \alpha)$ when α is a limit ordinal. In fact, it is enough to show C is nonempty, since if S is an uncountable subset of E_2, then so is $S \backslash \pi^{-1}[0, \alpha]$ for any $\alpha < \omega_1$, and we can repeat the argument. Now if S is an uncountable subset of E_2, then so is $S(n) = S \cap (\omega_1 \times \{n\})$ for some $n \in \mathbf{Z}$. Let $S_\alpha \subset \pi''[S(n)]$; then $\langle \alpha, 0 \rangle$ is in the closure of $S \cap S_\alpha^{n-k} \times \{n - k\}$, hence $\langle \alpha, k \rangle$ is in the closure of $S \cap \pi^{-1}S_\alpha^{n-k}$ for each $k \in \mathbf{Z}$. This proves the claim.

From the claim it follows that the sets of the form $\pi^{-1}C$, where C is a club subset of ω_1, form a base for a free "closed ultrafilter" with the c.i.p. Hence E_2 is not α-realcompact.

From the claim it is immediate that if A and B are subsets of E_2 such that $A \cap \overline{B} = \varnothing = \overline{A} \cap B$, then one of them must miss the fibers over a club set; moreover, if one of them hits the fibers over a stationary set, then the other must be countable. It is easy to see from this that E_2 is hereditarily collectionwise normal. It is also totally normal.

DEFINITION 7.4: A space X is *total* if every open subset U of X has a cover by open F_σ subsets of X that is locally finite in U. A space is *totally normal* if it is both total and normal.

Normality and many related properties become hereditary in the presence of totality; see [20].

To see that E_2 is total, note that disjoint open sets are special cases of A and B as earlier. So if U is open and hits the fibers over a stationary set, then its complement, being closed, is countable. Let α be such that $(\alpha, \omega_1) \times \mathbf{Z} \subset U$; then both $(\alpha, \omega_1) \times \mathbf{Z}$ and its complement are clopen, and the complement is metrizable, so its intersection with U is an open F_σ in E_2. On the other hand, if U misses the fibers over a club C, then it is the direct sum of countable, hence metrizable, relatively clopen subsets, and each is an F_σ in E_2.

It is routine to modify the above proof to weaken ♣ to the axiom (t) of Juhász [15], which holds in any model resulting from the addition of a Cohen real. However, some axioms beyond ZFC are necessary to produce something like E_2. This is because of the following theorem of Balogh.

THEOREM 7.5 [2]: If MA(ω_1), and X is a locally compact, first countable space of cardinality $\leq \aleph_1$, exactly one of the following is true:

(1) X contains a perfect preimage of ω_1;
(2) X is a Moore space (and hence the countable union of closed discrete subspaces).

Recall that a map $f: Z \to Y$ is called *perfect* (also *proper*) if it is continuous and closed, and $f^{-1}\{y\}$ is compact for each $y \in Y$. If $Y = \omega_1$, Z is countably compact and, more strongly, has the property that every countable subset has compact closure; yet Z is not compact.

From this theorem of Balogh, it follows that, under MA(ω_1), every version of E will either contain a perfect preimage of ω_1 or else it will be the countable union of closed discrete subspaces. In neither case can the Claim for E_2 hold. (In the former case, we use the fact that every countably compact subspace of a first countable space is closed, yet the fibers of E are closed in E and not countably compact.)

Back when [51] was written, the transition between locally compact, locally countable spaces seemed strong enough that I conjectured an affirmative answer to the following question, at least for manifolds.

QUESTION 7.6: Can "first countable, of cardinality $\leq \aleph_1$" in Balogh's theorem be weakened to "locally metrizable, of weight $\leq \aleph_1$"?

In the context of manifolds, weight is $\aleph_0$ + the least cardinality of a cover by Euclidean neighborhoods. The word "weakened" is justified by the fact that locally compact, first countable spaces of cardinality $< \mathbf{c}$ are locally countable, and in a locally countable first countable space, weight equals cardinality. The "manifold" case of the Question 7.6 appeared in [24] as problem 4.20.

The conjecture was also motivated by Mary Ellen's proof that $MA(\omega_1)$ implies that perfectly normal manifolds are metrizable, which used a poset that is essentially Balogh's, but with compact subsets replacing the finite ones. There is an informal rule of thumb that "compact is almost as good as finite," which seemed tantalizingly plausible in this context; and I could not understand why perfect normality was used so heavily in Mary Ellen's proof. This bewilderment was largely dispelled in 1986 when I found a smoothing of L^+ whose T^+ is not Moore, and does not contain a perfect preimage of ω_1. This is called a Class 7 smoothing in [28]. Its associated T^+ is quasi developable, hence every countably compact subset is compact; and it is not even countably metacompact, hence not Moore.

This opened an unexpected window in the theory of nonmetrizable manifolds, which the following section looks into.

8. GENERALIZED METRIC PROPERTIES

A number of generalized metric properties have nice characterizations in the context of manifolds. Theorem 8.2 below is shown in [28].

DEFINITION 8.1: A family $\mathscr{K}$ of subsets of a space X is *relatively discrete* if it is discrete in the relative topology of $\cup \mathscr{K}$. A collection is *σ-relatively discrete* if it is the countable union of relatively discrete collections.

Here as usual, "space" means "Hausdorff space," whence local compactness implies complete regularity, as does regularity and local countability.

THEOREM 8.2: Let M be a locally compact, locally metrizable space. The following are equivalent.

  (i) M is weakly θ-refinable.
  (ii) Every subspace of M is weakly θ-refinable.
  (iii) M is quasi developable.
  (iv) M has a σ-relatively discrete cover by compact sets.

A similar result is shown for Moore spaces, and the demonstration is so short that we repeat it.

THEOREM 8.3: Let M be a locally compact, locally metrizable space. The following are equivalent.

  (i) M is θ-refinable.
  (ii) Every subspace of M is subparacompact.
  (iii) M is developable, that is, a Moore space.
  (iv) M has a σ-discrete cover by compact sets.

Proof: The equivalence of the first three conditions is shown in [48] for spaces with a base of countable order, and so is the fact that every locally metrizable space has a base of countable order. The equivalence of (ii) and (iv) is shown for locally compact spaces in the proof of theorem 4.5 of [24]. □

With the demise of the conjecture implicit in Question 7.6, the following problem, discussed in [28], becomes intriguing.

PROBLEM 8: It is consistent that every locally compact, locally metrizable space of weight $\leq \aleph_1$ either (i) contains a perfect preimage of ω_1, or (ii) is quasi developable?

This question remains open for manifolds, and $MA(\omega_1)$ has not been ruled out. Any model would have to negate Juhász's axiom (t), as Example 7.3 shows.

Most generalized metric properties destroy perfect preimages of ω_1, so we can ask many weakenings of Problem 8, such as:

PROBLEM 9: Is it consistent that every manifold of weight $\leq \aleph_1$ with a G_δ-diagonal is quasi developable?

The resemblance between Theorems 8.2 and 8.3 makes it natural to ask what is required to go from quasi developability to developability (for regular spaces, this means Moore). For instance:

PROBLEM 10: Is there a quasi-developable manifold with a G_δ-diagonal that is not developable?

Here again there may be a big difference between locally compact, locally countable spaces and manifolds. It follows from Theorem 8.2 that every locally compact, locally countable space of countable scattered (Cantor–Bendixson) height is quasi developable, and by now we know of a fair variety of such spaces that have a G_δ-diagonal but are not developable, whereas we do not even have consistency results for Problem 10, nor for the following problem, discussed in [28].

PROBLEM 11: Is every countably metacompact, quasi-developable manifold a Moore space?

For a model in which there is a locally compact, locally countable, countably metacompact, quasi-developable space that is not a Moore space, see [44]. We still do not know whether it is consistent for there to be no such space, or even:

PROBLEM 12: Is it consistent that every weakly θ-refinable, first countable, countably metacompact space is θ-refinable?

Regular spaces that are not countably metacompact were once very had to find. Now we have so many examples that it is plausible to ask questions like these, to see just how much pathology countable metacompactness can destroy.

Separating open covers of various sorts seem to tell a lot about manifolds.

DEFINITION 8.4: A collection $\mathcal{U}$ of open subsets of a space X is T_2-separating [respectively, T_1-separating] [respectively, strongly separating] if, given distinct points $x_0, x_1 \in X$, there are $U_0, U_1 \in \mathcal{U}$ such that $x_i \in U_i$ and $U_0 \cap U_1 = \varnothing$ [respectively, and $x_{1-i} \notin U_i$] [respectively, and $x_{1-i} \notin \overline{U_i}$].

Clearly, T_1-separating $\Rightarrow$ strongly separating $\Rightarrow$ T_2-separating. It is also easy to see that a space has a countable T_2-separating open cover iff it has a countable strongly separating open cover.

Mary Ellen (and, independently, Balogh and Bennett) noticed:

THEOREM 8.5 [4]: A manifold with a countable T_2-separating open cover is metrizable.

This gives an alternative proof of the Reed–Zenor theorem (Theorem 2.1): every normal Moore space of cardinality $\leq c$ has a countable T_2-separating open cover. Theorem 8.5 also meshes nicely with the following theorem.

THEOREM 8.6 [4]: Every hereditarily normal space with a countable T_1-separating open cover has a countable T_2-separating open cover.

COROLLARY 8.7: Every hereditarily normal manifold with a countable T_1-separating open cover is metrizable.

Bennett and Balogh [4] inquired whether "hereditarily" can be dropped here. Mary Ellen showed it is consistent that it cannot: in [9] she constructed a hereditarily separable Dowker manifold with a countable T_1-separating open cover using $\diamondsuit$. In the midst of a lecture series on manifolds (unfortunately, never written up for publication) at STACY, she briefly thought she had a ZFC example, but the question remains open.

PROBLEM 13: Is there a normal nonmetrizable manifold with a countable T_1-separating open cover?

Balogh and Bennett showed [4] that $MA(\omega_1)$ implies every normal manifold of weight $\leq \aleph_1$ with a countable T_1-separating open cover is metrizable. Since all manifolds are of weight $\leq c$, Mary Ellen's $\diamondsuit$ example shows $MA(\omega_1)$ cannot be dropped here.

Taking off from Corollary 8.7 in a different direction, Balogh and Bennett ask [4]:

PROBLEM 14: Is every hereditarily normal manifold with a point-countable T_1-separating open cover metrizable?

They show [4] the answer is Yes under $MA(\omega_1)$. They also show that every locally compact, locally connected space with a point-countable strongly separating open cover is metrizable. Among the other questions they pose are the following.

PROBLEM 15: Is every hereditarily normal, quasi-developable manifold metrizable?

PROBLEM 16: Is every hereditarily normal manifold with a G_δ-diagonal metrizable?

By Theorem 2.3 in [3], the answer to Problem 15 is Yes if $2^{\aleph_0} < 2^{\aleph_1}$. The answer is also affirmative if every hereditarily normal manifold is hereditarily collectionwise normal (CWN), since:

THEOREM 8.8: Every hereditarily CWN, quasi-developable, locally compact, locally metrizable space is metrizable.

Proof: In a hereditarily CWN space, a relatively discrete collection of closed sets can be expanded to a disjoint collection of open sets. In fact, this characterizes hereditarily CWN spaces: see [24] for an indication of the elementary proof. From this and Theorem 8.2 (iv) it follows that every space as in the hypothesis has a σ-disjoint cover by open relatively compact, hence second countable sets. So the space is meta-Lindelöf, hence [24, p. 637] metrizable.   □

In Problem 16, "hereditarily" cannot be dropped, as I showed back in 1982 by constructing an example [unpublished] of a ZFC construction of a separable normal

CWH (collectionwise Hausdorff) manifold with a G_δ-diagonal that is not metrizable, modifying one of van Douwen's "honest submetrizable" constructions. After I told Mary Ellen at STACY that such a construction is possible, she came up with one of her own, with a neater description than mine. It appears as the first example in [40]. It is of weight $\mathfrak{c}$, which is to be expected since MA $+ \neg$CH implies that every CWH manifold of weight $< \mathfrak{c}$ is metrizable if it does not contain a perfect preimage of ω_1 [2]. Mary Ellen [40] writes "is not countably compact" instead of this last clause, which is incorrect, as example 6.7 in [27] shows; by the way, this latter example was itself inspired by an old example of Mary Ellen's [34].

The ZFC example from [40] also answers the following question from [4] negatively.

QUESTION 8.9: Is it consistent with MA(ω_1) that every closed subspace of a manifold of weight $\geq \omega_1$ has a closed subspace of weight ω_1?

Bennett and Balogh [4] also prove a number of other results that essentially pick up where [24, sec. 4] leaves off, providing an assortment of theorems about paracompactness in locally compact, locally connected spaces under MA(ω_1). Some of the information appeared earlier, in [2], and was announced in [24]. As was said in [24, sec. 4]: it may be a long time before the last word is in on generalizations of Mary Ellen's 1978 MA(ω_1) result.

9. SOME RESEARCH TOPICS AND SOME MORE PROBLEMS

There is a great deal of room for future research on manifolds by general topologists; so much, in fact, that it seems more relevant to begin this section by writing about some research topics rather than individual problems. The first topic is one that has been overlooked until recently because it is trivial for metrizable manifolds.

9.1 Dimension Theory of Manifolds

The dimension one naturally associates with a manifold is its small inductive dimension. This is simply the unique n for which each point has a neighborhood homeomorphic to $\mathbb{R}^n$. For metrizable manifolds this is essentially the end of the story, because all the main dimension functions coincide for separable metric spaces. (Of course, there is still plenty of scope for dimension theory, which for a long time was largely confined to separable metric spaces!)

The situation is quite otherwise for nonmetrizable manifolds. Using the continuum hypothesis, V. V. Fedorchuk and V. V. Filippov [11] have constructed, for any $n \geq 3$, an example of a normal, countably compact n-manifold whose covering dimension is also n, but whose large inductive dimension is $2n - 2$. Fedorchuk [10] has also used $\diamondsuit$ to construct, for each pair of integers n, m such that $4 \leq n \leq m$, a perfectly normal, hereditarily separable, differentiable n-manifold M such that

$$n = \operatorname{ind} M < m = \dim M < \operatorname{Ind} M = m + n - 2.$$

All this, of course, raises the question of what can be done under weaker axioms, or just using ZFC. While we cannot require perfect normality, we can ask what discrepancies there can be in the dimensions of normal manifolds. Also, what kinds of peculiarities can 2-manifolds have under various set-theoretic hypotheses? Fedorchuk [9] solved both problems 3.2 and 3.3 at the end of Wilder's book [47] by constructing a ZFC example of a normal, countably compact 2-manifold M that has a nowhere dense closed subset N such that Ind N = Ind M = 2. This "dimension-preserving" behavior cannot occur in metrizable manifolds: a subset of the same dimension must have interior points [14, corollary 1 to theorem IV 3]. Are there any nice classes of manifolds (Moore? quasi developable?) to which this theorem does extend?

How much more nicely do Ind and dim behave on manifolds than on spaces in general? For instance, does the "logarithmic identity"

$$dim(X \times Y) = dim X + dim Y$$

hold if $X \times Y$ is normal? What about the logarithmic inequality $dim(X \times Y) \leq dim X + dim Y$? What if Ind is put in place of dim?

9.2 Differential Topology on Nonmetrizable Manifolds

The long paper [28] on smoothings of the long lines and their tangent bundles has finally appeared. For other nonmetrizable manifolds, the study of their smoothings has barely begun. What kinds of smoothings are possible on the examples in [24], such as the Prüfer manifold and its modifications by R. L. Moore mentioned in Section 2? Are there smoothings of the perfectly normal manifold of Kozlowski and Zenor that are different from the one given? What sorts of properties does their smoothing have? (For instance, is the tangent bundle perfectly normal?)

Are there 2-manifolds and 3-manifolds that do not admit smoothings? Is there a nonmetrizable manifold on which all smoothings are diffeomorphic? A nonmetrizable manifold with only finitely many smoothings, up to diffeomorphism?

In general, what topological properties carry over to every tangent bundle? To some tangent bundle? To a tangent bundle minus the zero vectors? For instance, how well are developability and quasi developability preserved? Are there properties that are preserved only if the manifold is metrizable? Is hereditary normality one of these, perhaps under additional axioms?

When does a product admit smoothings that are not a product of the smoothings of the factor spaces?

Does the long line have a smoothing that admits a C^∞ vector field with non-Lindelöf support? What manifolds do have this property?

9.3 Complex Analytic Manifolds

Complex-analytic manifolds are untouched by modern set theory. While all smooth manifolds can be given real-analytic atlases [17], complex-analytic atlases are very hard to come by. In fact, Radó showed [30] that every complex-analytic manifold that is locally like $\mathbb{C}^1$ is metrizable, so this is a theme involving several complex variables.

In [45, pp. 161–162], Wilhelm Stoll gave a general construction of the known nonmetrizable complex-analytic manifolds of complex dimension 2. In a private communication to Mary Ellen in 1991, he gave a geometric construction that produces these examples and others of higher dimensions, but all are Moore and have dense open separable metrizable subspaces. So our situation is similar to that of ZFC constructions of Dowker spaces: one can come up with no end of problems by taking one construction and asking about properties it does not satisfy. Two specific problems that Stoll expressed curiosity about are as follows.

PROBLEM 17: Is there a countably compact noncompact complex-analytic manifold?

PROBLEM 18: Is there a nonseparable complex-analytic manifold?

9.4 Generalized Manifolds

In a curious role reversal, algebraic and geometric topologists study a variety of concepts loosely lumped under the heading of "generalized manifolds," while general topologists' activity in this area has been virtually confined to Euclidean manifolds.

There is no one fixed definition of "generalized manifold," and over the decades it may have been used as broadly as "generalized metric space" is among us. Right in Alexandroff's 1935 article he gives his own definition, then writes of "manifolds in the sense of Lefschetz," "manifolds in the sense of Hopf," and "manifolds in the sense of Čech," all of which are compact! He also mentions the possibility of noncompact classes. Wilder's definition does not call for compactness.

We could, of course, modify almost any specific problem, except those calling for a smoothing, to generalized manifolds: does a generalized manifold with such-and-such properties exist? Some questions in dimension theory may already have answers, but on the whole we have little idea of what the added generality does for us.

A good starting point might be Wilder's problem itself: Can we extend Mary Ellen's 1978 result to show that perfectly normal generalized manifolds in Wilder's sense are metrizable under $MA(\omega_1)$, or at least under the proper forcing axiom (PFA)? Perhaps Wilder's two cohomological conditions provide enough structure, over and above local compactness and local connectedness, to push things through. And what about Alexandroff's "manifolds in the sense of Čech"? Are they the same as perfectly normal, compact generalized manifolds in the sense of Wilder?

Regardless of how these questions turn out, we ought to take some time out to learn what uses Wilder makes of perfect normality in his book. With all the advances in algebraic and geometric topology since 1949, it may be interesting to see what else can be preserved in going from metrizability to perfect normality.

9.5 Wilder's "Other Problem"

Just after posing the problem Mary Ellen did so much on, Wilder raises the question of how necessary perfect normality is to his theory of generalized manifolds.

Specifically, he wishes to know whether the book's results on Alexander and Poincaré duality can be obtained without assuming perfect normality. If it cannot be dropped altogether, we could also ask whether there are weakenings of perfect normality that allow this duality to go through.

In 1936, E. Čech announced [8] that the duality of Betti groups does extend to a definition that omits perfect normality. In the announcement, he gives six axioms, with all but "compactness" and "dimension n" involving homological concepts, and compares them to the list from which Alexandroff was working. Three axioms from that list are omitted, including perfect normality: "*Firstly,* I had supposed that R has the property that every closed subset is a G_δ, which was a strong restriction." One wonders how differently our perception of perfect normality in manifolds might have evolved, had the omission been made in his original list! Would Wilder have included it in his theory? I also wonder: What accounts for Wilder' use of it in spite of Čech's 1936 announcement?

9.6 Normality in Products

In [41] Mary Ellen gives "a plea . . . for a cottage industry into our understanding of normality in products." In the theory of manifolds, the only hint of such an industry so far has been the construction, under strong set-theoretic hypotheses, of Dowker manifolds. Ideally, we would like to have necessary and sufficient internal conditions for a pair of normal manifolds to have a normal product, but I do not even know whether there is a ZFC example of a pair of normal manifolds whose product is not normal.

9.7 Somemore Problems

Mary Ellen's own favorite unsolved problem about manifolds is the following.

PROBLEM 19: Is every normal manifold collectionwise Hausdorff?

The answer is affirmative if either $V = L$ or $\mathfrak{c}$MEA. Its status under $MA(\omega_1)$ is still unknown. Also, assuming hereditary normality does not seem to help much. On the other hand, every perfectly normal manifold is collectionwise Hausdorff; also, as remarked after Problem 7, we do not know whether every hereditarily normal, locally compact, locally connected space is (hereditarily) collectionwise normal.

A tantalizing problem discussed in [27] is as follows.

PROBLEM 20: Is there a ZFC example of a pseudocompact manifold that is not countably compact?

Some progress has been made on this problem since [27] where an example was constructed assuming $\mathfrak{b} = \omega_1$. Tree and Watson have constructed a 2-manifold example assuming the existence of a Cook set, and Steprāns [50] showed that a three-dimensional version of a Cook set is possible assuming Martin's axiom for σ-linked posets. These examples consistently answer a question reminiscent of the Reed–Zenor Theorem 2.1:

PROBLEM 21: Is there a locally compact, locally connected, pseudocompact nonmetrizable Moore space?

These last two problems, and also the following one, are beset with difficulties involving wild arcs, discussed in [27].

PROBLEM 22: Is there a separable, countably compact, noncompact manifold?

We do not have any consistency results for the following question.

PROBLEM 23: Is there a hereditarily normal Dowker manifold?

At times I have claimed to have an example assuming $\diamondsuit^*$, but this is incorrect. Even if "manifold" is weakened to "locally compact, locally connected space," we have no ideas for examples under any axioms. Moreover, the only known examples of hereditarily normal Dowker spaces require large cardinal axioms, and are not locally compact.

PROBLEM 24: Is there a ZFC example of a Dowker manifold?

We do not know of any Dowker manifolds compatible with $MA(\omega_1)$, and this axiom seems like a promising avenue to a negative answer.

Concerning anti-Dowker (countably paracompact, nonnormal) spaces, we are better off, with an easy ZFC example of a separable anti-Dowker manifold (a routine modification of example 6.7 in [27]) and a locally compact, locally countable, hereditarily countably paracompact nonnormal Moore space from a Q-set in [43, p. 57] obtained by splitting the topmost points of an ω_1-Cantor tree to kill normality just as it is killed in the wide Cantor road spaces (Section 2). However, the wide road counterpart of Shelah's construction has too many points at the tops of the branches, and countable paracompactness fails for it too. In fact, the following problem remains completely open.

PROBLEM 25 ("anti Problem 23"): Is every hereditarily countably paracompact manifold (hereditarily) normal?

REFERENCES

1. ALEXANDROFF, P. S. 1935. On local properties of closed sets. Ann. Math. **26:** 1–35.
2. BALOGH, Z. 1983. Locally nice spaces under Martin's Axiom. Comment. Math. Univ. Carolinae **24:** 63–87.
3. ———. 1986. Paracompactness in locally Lindelöf spaces. Can. J. Math. **38:** 719–727.
4. BALOGH, Z. & R. BENNETT. 1989. Conditions which imply metrizability in manifolds. Houston J. Math. **15:** 153.
5. BALOGH, Z. & G. GRUENHAGE. Preprint.
6. BING, R. H. 1951. Metrization of topological spaces. Can. J. Math. **3:** 175–186.
7. BROWN, M. 1969. The monotone union of open n-cells is an open n-cell. Proc. Am. Math. Soc. **12:** 812–814.
8. ČECH, E. 1936. On general manifolds. Proc. Natl. Acad. Sci. U.S.A. **22:** 110–111.
9. FEDORCHUK, V. V. 1992. On dimension of nowhere dense subsets of manifolds. *In* Vestnik Moscow Univ. Series I.
10. ———. A differentiable manifold with noncoinciding dimensions. In press.
11. FEDORCHUK, V. V. & V. V. FILIPPOV. 1992. Manifolds with non-coinciding inductive dimensions. Mat. Sb.

12. GRUENHAGE, G. 1980. Paracompactness and subparacompactness in perfectly normal locally compact spaces. Russ. Math. Surv. **35**(3): 49–55.
13. HANAZAWA, S. 1980. Tsukuba Math. J. **4**: 257–268.
14. HUREWICZ, & WALLMAN. 1948. Dimension Theory, rev. ed. Princeton Univ. Press. Princeton, N.J.
15. JONES, F. B. 1937. Concerning normal and completely normal spaces. Bull. Am. Math. Soc. **43**: 671–677.
16. JUHÁSZ, I., K. KUNEN & M. E. RUDIN. 1976. Two more hereditarily separable, non-Lindelöf spaces. Can. J. Math. **28**: 998–1005.
17. KOCH, W. & D. PUPPE. 1968. Differenzierbare Strukturen auf Mannigfaltigkeiten ohne abzählbare Basis. Arch. Math. **19**: 95–102.
18. KOZLOWSKI, G. & P. ZENOR. 1979. A differentiable, perfectly normal, non-metrizable manifold. Topol. Proc. **4**: 453–461.
19. LANE, D. J. 1980. Paracompactness in perfectly normal, locally compact, locally connected spaces. Proc. Am. Math. Soc. **80**: 693–696.
20. MACK, J. & M. RAYBURN. 1990. Hereditary properties and the Hodel sum theorem. *In* General Topology and Applications (Middletown, Conn., 1988), Lecture Notes in Pure and Applied Mathematics **123**: 165–181. Dekker. New York.
21. MOORE, R. L. 1942. Concerning separability. Proc. Natl. Acad. Sci. U.S.A. **28**: 56–58.
22. ———. 1962. Foundations of Point Set Theory. Colloquium Publications **13**. American Mathematical Society. Providence, R.I.
23. NYIKOS, P. 1983. Set-theoretic topology of manifolds. *In* General Topology and Its Relations to Modern Analysis and Algebra V. Proceedings of the Fifth Prague Topological Symposium 1981, J. Novak, Ed.: 513–526. Heldermann Verlag. Berlin.
24. ———. 1984. The theory of nonmetrizable manifolds. *In* Handbook of Set-Theoretic Topology, K. Kunen and J. Vaughan, Eds.: 633–684. North-Holland. Amsterdam, the Netherlands.
25. ———. 1988. Classes of compact sequential spaces. *In* Set Theory and Its Applications. Lecture Notes in Mathematics 1401: J. Steprāns and S. Watson, Eds.: 135–159. Springer-Verlag. Berlin/New York.
26. ———. 1989. The Cantor tree and the Fréchet-Urysohn property. Ann. N.Y. Acad. Sci. **552**: 109–123.
27. ———. 1990. On first countable, countably compact spaces III: The problem of obtaining separable noncompact examples. *In* Open Problems in Topology, G. M. Reed and J. van Mill, Eds.: 127–161. North-Holland. Amsterdam, the Netherlands.
28. ———. 1992. Various smoothings of the long line and their tangent bundles. Adv. Math. **93**: 129–213.
29. OSTASZEWSKI, A. 1976. On countably compact, perfectly normal spaces. J. London Math. Soc. **14**: 505–516.
30. RADÓ, T. 1925. Über den Begriff der Riemannschen Fläche. Acta Litt. Sci. Szeged **2**: 101–121.
31. REED, G. M. & P. L. ZENOR. 1974. A metrization theorem for normal Moore spaces. *In* Studies in Topology, N. M. Stavrakas and K. R. Allen, Eds.: 485–488. Academic Press. New York.
32. ———. 1976. Metrization of Moore spaces and generalized manifolds. Fundam. Math. **91**: 203–210.
33. ROTHBERGER, F. 1948. On some problems of Hausdorff and Sierpiński. Fundam. Math. **35**: 29–46.
34. RUDIN, M. E. 1965. A technique for constructing examples. Proc. Am. Math. Soc. **16**: 1320–1323.
35. ———. 1974. The metrizability of normal Moore spaces. *In* Studies in Topology, N. M. Stavrakas and K. R. Allen, Eds.: 507–516. Academic Press. New York.
36. ———. 1977. Lectures on Set Theoretic Topology. CBMS Regional Conference Series No. 23. American Mathematical Society. Providence, R.I.
37. ———. 1979. The undecidability of the existence of a perfectly normal nonmetrizable manifold. Houston J. Math. **5**: 249–252.
38. ———. 1988. A nonmetrizable manifold from $\Diamond^+$. Topol. Appl. **28**: 105–112.

39. ———. 1989. Countable point separating open covers for manifolds. Houston J. Math. **15:** 255–266.

40. ———. 1990. Two nonmetrizable manifolds. Topol. Appl. **35:** 137–152.

41. ———. 1990. Some conjectures. *In* Open Problems in Topology, G. M. Reed and J. van Mill, Eds.: 183–193. North-Holland. Amsterdam, the Netherlands.

42. RUDIN, M. E. & P. L. ZENOR. 1976. A perfectly normal nonmetrizable manifold. Houston J. Math. **2:** 129–134.

43. SHELAH, S. 1982. Proper forcing. *In* Lecture Notes in Mathematics 940. Springer-Verlag. Berlin/New York.

44. ———. 1989. A consistent counterexample in the theory of collectionwise Hausdorff spaces. Israel J. Math. **65:** 219–224.

45. STOLL, W. 1955. Über meromorphe Modifikationen IV. Math. Ann. **130:** 147–182.

46. WATSON, W. S. 1985. Separation in countably paracompact non-normal spaces. Trans. Am. Math. Soc. **290:** 831–842.

47. WILDER, R. 1949. Topology of Manifolds. Colloquium Publications **32.** American Mathematical Society. Providence, R.I.

48. WORRELL, J. M., JR. & H. H. WICKE. 1965. Characterizations of developable topological spaces. Can. J. Math. **17:** 820–830.

49. BOOTH, D. 1969. Countably indexed ultrafilters, Ph.D. thesis, University of Wisconsin, Madison.

50. STEPRĀNS, J. 1991. Almost disjoint families of paths in lattice grids. Topol. Proc. **16:** 185–200.

Mary Ellen Rudin as Advisor and Geometer

MICHAEL STARBIRD

Office of the Dean
College of Natural Sciences
University of Texas
Austin, Texas 78712

INTRODUCTION

Mary Ellen Rudin has a distinctive way of thinking. She is best known for her complicated examples in general topology, but on at least two occasions she applied her brand of cerebral convolution to problems in the real world of geometric topology in the plane and 3-space. In 1952 her "A Primitive Dispersion Set for the Plane" appeared [2], and in 1958 she published her example of "An Unshellable Triangulation of a Tetrahedron" [3]. These examples display her trademark perspective on thought.

I have had the great good fortune to learn from Mary Ellen Rudin since 1971. She taught me in the first course I took as a graduate student at the University of Wisconsin, namely, undergraduate topology. She directed my dissertation in 1974 and has continued to give me guidance and inspiration since then.

In classes, in seminars, at conferences, and in ordinary conversation she exhibits a way of approaching problems that sets her apart from others. If you know her mathematics by way of her lectures, you might think her a bit hard to follow. She has been known to say X while writing Y and intending Z—usually with ordinal subscripts. When students make these errors, they are often confused. In Mary Ellen Rudin's case, the problem is that the sequential rendering of mathematics in a lecture is a vastly different representation of the ideas than the way they arise in her mind. To her, a single notion may well require an hour or much more of lecture to unwind, and the sequential nature of the lecture seems an unnatural expression of her real idea. Working on mathematics with Mary Ellen Rudin often includes a struggle through details followed by an awakening to a sort of gestalt that makes the pieces gel.

From the perspective of a graduate student and collaborator, her most remarkable feature is the flood of ideas that is constantly bursting from her. When I was her graduate student, not a week would go by without her suggesting several ideas for me to sort and explore. The process of trying to construct the global thought from the pieces is what allowed me, anyway, to appreciate her deeper understanding of mathematics and to learn a great deal of topology. There is a difficulty with having Mary Ellen Rudin as advisor. Although you can see her work her wonders, you cannot really expect to imitate anything approaching her method of doing mathematics.

Of course, we all learn much more than mathematics from our important teachers. There is a human side to mathematics. Leaders of a field set a tone for followers. Mary Ellen Rudin sets an example of exuberance, generosity, and an unquenchable zest for living. She would not have to be a world leader in a field to be an inspirational person. She inspires us to find the best in each other, to share ideas absolutely openly and generously, and to concentrate on our mathematics. I have been greatly privileged to receive Mary Ellen Rudin's direction as a student, collaborator, and friend. She was such a sterling advisor for me, that I thought it would be useful to tell you how to follow her example with your own students.

It is easy to use the Mary Ellen Rudin model to become a great advisor. The first step is to have an endless number of great ideas. Then merely give them totally generously to your students to develop and learn from. It is really quite simple. For Mary Ellen Rudin.

THE GEOMETRIC EXAMPLES

Primitive Dispersion Set for the Plane [1]

If M is a connected set, then a subset D of M is a *dispersion set* if and only if $M - D$ is totally disconnected. If no proper subset of D is a dispersion set of M, then D is a *primitive dispersion set of M*.

Mary Ellen Rudin's construction of a primitive dispersion set for the plane is a neat iterative process based on a clear geometric insight. I will not go through her construction in detail here. Instead, I will describe two examples of unusual dispersion sets. Mary Ellen Rudin's primitive dispersion set for the plane is more complicated than these two examples; however, these examples contain several ideas that are central to an understanding of her example. If you have the following examples in mind, you will be poised to see her example as a more complicated object whose properties of being a primitive dispersion set are proved in a similar spirit.

EXAMPLE 1: Consider the Cantor set C in $\mathbb{R}^1$. For each $z \in C$, let I_z be the straight-line segment from z to the point $(\frac{1}{2}, 1)$. Let $X = \cup_{z \in C} I_z$. Consider the subset of $Y = X$ consisting of the point $(\frac{1}{2}, 1)$ together with all points (x, y) in X $(y < 1)$ such that either $(x, y) \in I_z$, where z is an inaccessible point of C and y is irrational, or $(x, y) \in I_z$, where z is an accessible point of C and y is rational. (The accessible points of the Cantor set are those that are endpoints of an interval during the defining process, so the accessible points of the standard Cantor set are those of the form $k/3^m$ for some integers k and m. The other points of the Cantor set are called inaccessible.

Using the Baire Category Theorem, this well-known example Y can be shown to be connected. Yet $Y - \{(\frac{1}{2}, 1)\}$ is totally disconnected. Thus the point $(\frac{1}{2}, 1)$ is a dispersion set.

EXAMPLE 2: We can modify this example by the following means: Instead of defining X to consist of straight-line segments between points of C and the point $(\frac{1}{2}, 1)$, we could have each line I'_z rise monotonically from a point z in the Cantor set C, but as it rises it could wiggle back and forth in a manner reminiscent of a $\sin(1/x)$

curve so that each such line approaches the whole segment $[0, 1] \times \{1\}$. See FIGURE 1. Let X' be the set of these disjoint lines I'_z together with the segment $[0, 1] \times \{1\}$.

Similar to Example 1, we can let Y' be a subset of X' consisting of the segment $[0, 1] \times \{1\}$ together with the points (x, y) in X' ($y < 1$) such that either $(x, y) \in I'_z$, where z is an inaccessible point of C and y is irrational, or $(x, y) \in I'_z$, where z is an accessible point of C and y is rational.

The same proof as for Example 1 shows that Y' is connected and that $D = ([0, 1] \times \{1\})$ is a dispersion set. The same proof also shows that no proper subset of D is a dispersion set. In fact, any point of D together with $Y' - D$ is still connected. Hence D is a primitive dispersion set of Y'.

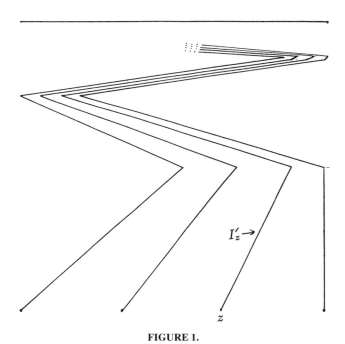

FIGURE 1.

Mary Ellen Rudin's example of a primitive dispersion set for $\mathbb{R}^2$ employs a concept somewhat similar to the construction of Y'. In her construction, she must be far more clever, however, since she produces a primitive dispersion set D for all of $\mathbb{R}^2$. Her dispersion set D divides $\mathbb{R}^2$ into arbitrarily small pieces, and during her construction of D, she associates with every point p of D a nontrivial subset Z_p in $\mathbb{R}^2 - D$ such that $Z_p \cup \{p\}$ is connected as $(Y' - D) \cup \{p\}$ is connected in Example 2. Keeping Example 2 in mind during your reading about her primitive dispersion set for the plane may well add to your understanding.

An Unshellable Triangulation of a Tetrahedron [2]

Mary Ellen Rudin is famous for examples of abstract spaces, but her example of an unshellable triangulation of a tetrahedron is as concrete as it can get. It is literally constructible out of wire or paper. In this example she shows how a real, 4-faced, rectilinear tetrahedron can be constructed from 41 real, smaller tetrahedra in such a way that the triangulation cannot be shelled. In fact, it is impossible to take the first step toward a shelling. A complete and rigorous proof of this construction would be to physically make it, and see that none of the simplices in the triangulation can be the first step in a shelling. More precisely, for each 3-simplex σ in her triangulation of the tetrahedron T, $Cl(T - \sigma)$ is not homeomorphic to a 3-cell. Her example of an unshellable triangulation of a tetrahedron is especially surprising in light of the following theorems.

THEOREM 1: Every triangulated disk in $\mathbb{R}^2$ is shellable.

This theorem shows that no example analogous to hers can exist in the plane.

THEOREM 2: Every triangulation of a convex 3-cell collapses [1].

The process of collapsing a triangulation is similar to shelling it. In fact, the next theorem shows a close relationship.

THEOREM 3: If a triangulation of a 3-cell collapses, its second barycentric subdivision shells.

Theorem 2 implies that Mary Ellen Rudin's example of an unshellable triangulation of a tetrahedron does collapse, and Theorem 3 shows how close that is to being shellable.

It is easy to check that no simplex of her triangulation can be shelled because each 3-simplex with a free face has its fourth vertex in a position to make the 3-simplex clearly nonshellable. The subtle part of her example is to see that it is geometrically possible to fit tetrahedra together in the manner she describes.

Every vertex of the triangulation lies on the boundary of the large tetrahedron T. The construction begins with thin rods around each edge of T. These rods are subdivided into several thin tetrahedra in a way that ensures that none are shellable. Then the main area of each face of T lies on a rather flat tetrahedron whose fourth vertex lies on an opposite face near a vertex. These are the type (8) tetrahedra from her paper. Finally, the main bulk of T lies in one large tetrahedron of the triangulation ($Z_1 Z_2 Z_3 Z_4$ in her paper) each of whose four vertices lies on a different face of T near a corresponding vertex of T. The large tetrahedron $Z_1 Z_2 Z_3 Z_4$ intersects the boundary only at its vertices and it is inscribed slightly skew to T.

Her example shows a symmetry and clear following of the picture of a large central tetrahedra, plates on the faces, and two types of rod arrangements about the edges of T. She aptly says in the paper that "the best way to check the construction is to draw a big picture and label the vertices." Better yet is a large model. The model is the proof. This example demonstrates a thought process of conceiving the basic form of the example and then fitting the details to make it work.

Her geometric examples show the same global insight rendered in accurate complex details as do her many wonderful examples and theorems in general topology. In learning about Mary Ellen Rudin's work, I recommend that you not omit her geometric papers from your attention.

REFERENCES

1. CHILLINGWORTH, D. R. J. 1967. Collapsing three-dimensional polyhedra, Proc. Cambridge Phil. Soc. **63:** 353–357.
2. ESTILL, M. E. (later Rudin). 1957. A primitive dispersion set of the plane. Duke Math. J. **19:** 323–238.
3. RUDIN, M. E. 1958. An unshellable triangulation of a tetrahedron. Bull. Am. Math. Soc. **64:** 90–91.

Dowker Spaces

PAUL J. SZEPTYCKI[a] AND WILLIAM A. R. WEISS

Department of Mathematics
University of Toronto
Toronto, ONT M5S 1A1, Canada

1. INTRODUCTION

Volume 3, 1951, was the best issue ever of the *Canadian Math Journal*. Thirty-two pages separate R. H. Bing's "Metrization of topological spaces" and C. H. Dowker's "On countably paracompact spaces," papers that gave rise to two of the richest subjects in the history of General and Set Theoretic Topology: the theory of normal Moore spaces and the theory of Dowker spaces. While the former is well understood, its most difficult and fundamental problems settled, the same cannot be said of the latter.

A Dowker space is a T_1, normal space whose product with the closed unit interval is not normal. Whether Dowker spaces exist was settled in 1970 by M. E. Rudin, who constructed the only known example. Her space is large; it is a P-space (G_δ are open) and it is of size and weight $\aleph_\omega^\omega$. Instead of putting the subject to rest, Rudin's construction gave rise to a flurry of activity in search of a "small" Dowker space, that is, one whose important cardinal functions (density, character, size) are small. Separable, first-countable Dowker spaces of size continuum and ω_1 have been constructed under a variety of extra set-theoretic assumptions including continuum hypothesis (CH), MA, the existence of a Souslin tree, and variations of $\diamondsuit$. Rudin's article from the *Handbook of Set Theoretic Topology* is an excellent survey of the subject up to 1984. However, a "real" small Dowker space, that is, a small Dowker space constructed with no further assumptions beyond ZFC, has eluded us. In fact, there have been no significant advances on the two central questions raised in Rudin's article: Are there "real" Dowker spaces with small cardinal functions? And are there Dowker spaces with strong global countable structures under any assumptions?

We give a survey of what has been done in the last eight years, revisit both Dowker's theorem and Rudin's space, and give a list of open problems.

2. DOWKER'S THEOREM

Dowker formulated the notion of countable paracompactness, proved that for normal spaces X, $X \times [0, 1]$ is normal if and only if it is also countably paracompact,

[a] Current address: Department of Mathematics, Ohio University, Athens, Ohio 45701.
Mathematics Subject Classification: Primary 03E35, 5402, 54A35, 54D15, 54D20, 54G20.
Key words and phrases. Normal, countable paracompact, Dowker spaces.

and he also characterized countable paracompactness in a way as to give fundamental insight into the structure of Dowker spaces.

DEFINITION 2.1: X is said to be countably paracompact (metacompact) if every countable open cover has a locally finite (point finite) open subcover.

THEOREM 2.2 [10]: The following are equivalent for X a normal space.

(a) X is countably paracompact.
(b) For every decreasing sequence of closed sets $\{D_n : n < \omega\}$ such that $\cap_{n<\omega} D_n = \varnothing$, there are open sets $\{U_n : n < \omega\}$ such that for each $n < \omega$, $U_n \supset D_n$ and

$$\cap_{n<\omega} \overline{U_n} = \varnothing.$$

(c) For every decreasing sequence of closed sets $\{D_n : n < \omega\}$ such that $\cap_{n<\omega} D_n = \varnothing$, there are open sets $\{U_n : n < \omega\}$ such that for each $n < \omega$, $U_n \supset D_n$ and

$$\cap_{n<\omega} U_n = \varnothing.$$

(d) $X \times [0, 1]$ is normal.
(e) $X \times (\omega + 1)$ is normal.

In the absence of normality, clause (b) is equivalent to countable paracompactness and clause (c) is equivalent to countable metacompactness [18]. The important equivalence (c) $\Leftrightarrow$ (d) gives us a fairly canonical description of what a Dowker space must look like.

Typically, Dowker spaces have been constructed by first fixing the point set of the space along with a countable partition, $X = \cup_{n<\omega} X_n$ (usually a set of reals or ω copies of ω_1). The topology on X is defined so that the sequence of closed sets to witness the failure of countable paracompactness is given by

$$D_0 = X, D_1 = \cup_{i \geq 1} X_i, \ldots, D_n = \cup_{i \geq n} X_i, \ldots.$$

So a basic open neighborhood of a point $x \in X_n$ is contained in the previous X_i.

For a good systematic and in-depth discussion of the basic examples, see [32].

Dowker's Theorem is also very useful to prove other theorems. To illustrate this we present a proof of Michael's hard theorem that regular σ-paracompact spaces are paracompact (see [22]) as a relatively easy corollary to Dowker's Theorem. Some would say that Michael's proof contains Dowker's proof. Recall the following.

DEFINITION 2.3: A space is σ-paracompact if every open cover has an open refinement that is the union of countably many locally finite subcollections.

THEOREM 2.4: Regular σ-paracompact spaces are paracompact.

Proof: It is easy to see that regular σ-paracompact spaces are normal. Just combine the two proofs that in the class of regular spaces

(Lindelöf $\rightarrow$ normal)   and   (paracompact $\rightarrow$ normal).

Therefore, X is regular and σ-paracompact
 $\rightarrow X \times [0, 1]$ is regular and σ-paracompact
 $\rightarrow X \times [0, 1]$ is normal
 $\rightarrow X$ is countably paracompact.
And clearly σ-paracompactness and countable paracompactness together imply paracompactness.

3. M. E. RUDIN'S CONTRIBUTION

M. E. Rudin is responsible for the most important and fundamental Dowker space constructions. Although she is most famous for constructing what is still essentially the only known ZFC example, she also constructed the first consistent example (from a Souslin tree), the first hereditarily separable example (again from a Souslin tree), the first CH example (with Juhász and Kunen), and the only example with strong global separation properties (a screenable example constructed from $\diamond^{++}$). It is fair to say that by reading only her papers, including her survey article [32], one would not miss very much. In this section we sketch a few of her more important constructions, omitting most of the details and referring the interested reader to the original papers. As well, we totally omit Rudin's generalization of Dowker spaces to higher cardinals (see [33]).

Soon after C. H. Dowker published his famous paper, M. E. Rudin constructed a Dowker space from a Souslin tree. Although at the time it was not known if the existence of a Souslin tree was consistent, it is the first construction of a Dowker space under any assumption and involves techniques employed in almost every other consistent Dowker space construction.

EXAMPLE 3.1 [27]: If there exists a Souslin tree, there is a Dowker space of size $\aleph_1$.

Let T be a Souslin tree. A topology is constructed on $X = T \times \omega$ so that letting $D_n = T \times (\omega \backslash n)$, then $\{D_n : n < \omega\}$ is the decreasing sequence of closed sets witnessing that the space is not countably metacompact. For each $\alpha < \omega_1$ and each $x \in \mathrm{Lev}_\alpha(T)$ we choose a sequence $\{y_i : i < \omega\}$ so that for each $i, y_i <_T x$ and so that $\sup\{\mathrm{lev}(y_i : i < \omega\} = \alpha$. Then we choose for each $i, x_i \in \mathrm{Lev}_\alpha(T)$ such that $x_i >_T y_i$. Then a topology is defined on X by declaring $U \supseteq X$ open if and only if for all $(x, n) \in U$

   (i) $\exists m < \omega, \{(x_i, n - 1) : i > m\} \subseteq U$,
   (ii) $\exists y <_T x, \{(z, n) : y <_T z <_T x\} \subseteq U$.

So a typical open neighborhood of (x, n) must look down in the nth copy of T and must also pick up a tail of a sequence from the same level as x in the $(n - 1)$st copy of T. Rudin proves that the space is not countably metacompact by proving that if T is Souslin and if for every n if $U_n \supseteq D_n$ is open, then $\{x \in T : (x, 1) \notin U_n\}$ is nowhere dense in T. Since nowhere dense sets are countable in Souslin trees, X is not countably metacompact. To see that X is normal, suppose H and $K \subseteq X$ are closed disjoint. Let $H_i = \{x \in T : (x, i) \in H\}$, and similarly define K_i. Then normality follows from the following.

LEMMA 3.2: There is an $\alpha < \omega_1$ such that for all $x \in \mathrm{Lev}_\alpha(T)$ for all $i, j < \omega$ if H_i contains elements above x, then K_j does not.

Therefore above the αth level of T, H and K can be trivially separated, while below the αth level of tree we need to separate the countable closed sets $H \cap (T \upharpoonright \alpha) \times \omega$ and $K \cap (T \upharpoonright \alpha) \times \omega$ in the clopen set $(T \upharpoonright \alpha) \times \omega$.

This example can be easily modified to be first countable and locally compact (the topology must be defined inductively). As well it generalizes up to yield the following.

EXAMPLE 3.3 [29]: If there exists a κ-Souslin tree, then there exists a Dowker space of size κ.

And later Rudin modified the example again [30] to construct a hereditarily separable example.

M. E. Rudin is certainly most famous for her ZFC Dowker space. The example was inspired by an example of Miščenko [23]. He constructed a space that was finally compact in the sense of complete accumulation points but not Lindelöf and remarked that if it were normal it would be a Dowker space. Rudin proved that it wasn't normal and after a number of modifications produced her example.

Let X and X' be subspaces of the box product $\square_{i<\omega}(\omega_i + 1)$ defined by

$$X' = \{ f \in \square(\omega_i + 1) : \forall i \; \mathrm{cof}\,(f(i)) > \omega \}$$

and

$$X = \{ f \in X' : \exists n \forall i \; \mathrm{cof}\,(f(i)) \leq \omega_n \}.$$

X is Rudin's Dowker space. The closed sets that are to witness that the space is not countably metacompact are given by

$$D_n = \{ f \in X : \forall i \leq n \, f(i) = \omega_i \}.$$

Clearly $\cap_{n<\omega} D_n = \varnothing$, and it is not hard to prove that for any open sets $U_n \supseteq D_n$ $\cap_{n<\omega} U_n \neq \varnothing$. The proof that X is normal, in fact collectionwise normal, is more involved. One approach, taken by K. P. Hart (see [14], [15], and [17]) is outlined by the following two lemmas. Both of them are essentially proved in Rudin's original paper, though not explicitly stated.

LEMMA 3.4: X' is paracompact.

LEMMA 3.5: All disjoint closed A and $B \subseteq X$ have disjoint closures in X'.

Lemmas 3.4 and 3.5 easily imply that X is collectionwise normal. In addition, X is a P-space. This apparent defect has been exploited to get a Dowker topological group [17] and an extremally disconnected Dowker space [9] by modifying Rudin's example.

EXAMPLE 3.6 [21]: CH implies that there is a hereditarily separable first-countable Dowker space of size $\aleph_1$.

The example is constructed by fixing ω disjoint, mutually $\aleph_1$-dense Lusin subsets $\{ L_i : i < \omega \}$ of 2^ω. The usual topology ρ is refined to τ so that:

(i) $\forall n \; \cup_{i<n} L_i$ is τ-open.
(ii) $\forall$ uncountable $X \subseteq 2^\omega$, the ρ-closure of X and the τ-closure of X differ by at most a countable set.

Letting $D_n = \cup_{i \geq n} L_i$, then (i) implies that each D_n is closed. Clearly $\cap_{n < \omega} D_n = \varnothing$ and (ii) implies that if $U_n \supseteq D_n$ is open, then $U_n \cap L_0$ is cocountable (otherwise $\mathrm{cl}_\tau(L_0 \backslash U_n) \cap L_n \neq \varnothing$). Therefore $(2^\omega, \tau)$ is not countably metacompact. As well (ii) implies that if A and B are disjoint τ-closed sets, then $\mathrm{cl}_\tau A \cap \mathrm{cl}_\tau B$ is countable. A slightly stronger version of (ii) not only implies that A and B can be separated, but that $(2^\omega, \tau)$ is hereditarily separable.

One of the more important questions raised by M. E. Rudin is whether there is a Dowker space with a σ-disjoint base. Such an example seems very difficult to construct; however, there are no partial negative results precluding a ZFC example. The best result we have is Rudin's screenable example. A space is screenable if every open cover has a σ-disjoint refinement. An old question of Bing asked whether normal, Hausdorff, screenable spaces are paracompact. Nagami later proved that any counterexample must be a Dowker space. In [31] Rudin constructs a screenable Dowker space from the combinatorial principle $\diamondsuit^{++}$. Any partial answers to the following question would be quite enlightening.

QUESTION 1: Is there under any (consistent) assumptions a Dowker space with a σ-disjoint base?

A result showing it is consistent that there is no first-countable Dowker space with a σ-disjoint base would be equally interesting.

4. RECENT RESULTS

The results since Rudin's survey [32] can be grouped into two classes:

(1) Negative results telling us that our techniques for constructing examples won't yield a ZFC Dowker space.
(2) Modifications of existing constructions to obtain new Dowker spaces from relatively weaker assumptions or with relatively stronger properties.

There are classes of spaces in which normality implies countable paracompactness. The classic results include that there are no perfect, hence, no Moore Dowker spaces. As well, there are no monotonically normal [32], hence, no linearly ordered or GO Dowker spaces. In addition linear spaces [37] and pseudocompact spaces are never Dowker. No Dowker space can be of the form $\Pi_{n < \omega} X_n$; that is, if a space X is the product of countably many other spaces, then if X is normal, then it is also countably paracompact [42]. As well, any space X endowed with the order topology induced by a tree ordering is not Dowker [11].

There are precious few such theorems, even when one ventures beyond ZFC to consistency results. It is remarkable how hard we must work to prove seemingly very weak results. All the theorems we mention are unsatisfying in so far as they give us little insight into whether there are real small Dowker spaces. They do, however, point out the inadequacy of our techniques to yield ZFC examples.

The earliest and most typical theorem is Fleissner's.

THEOREM 4.1 [12]: MA $+ \neg$CH implies that de Caux's space is countably metacompact.

Later Balogh proved a better but similar result:

THEOREM 4.2 [1]: The proper forcing axiom (PFA) implies that there are no first countable, locally compact, locally countable, σ-discrete Dowker spaces of size ω_1.

And recently C. Good has announced that Fleissner's result can be improved to:

THEOREM 4.3: MA + $\neg$CH implies that there are no locally countable Dowker spaces of size $< 2^\omega$ of countable scattered height.

We have no idea whether or not MA_{ω_1} or even PFA imply that there are no Dowker spaces of size ω_1, even in the class of first-countable spaces. It is worth noting, however, that modifying Rudin's construction [29] of a Dowker space from an $\aleph_2$-Souslin tree together with a Q-set of size $\aleph_2$ (obtained from $MA_{\aleph_3}$) one can get a separable Dowker space of size $\aleph_2$ consistent with $MA_{\aleph_3}$. Note that indeed it is true that $MA_{\aleph_3}$ is consistent with the existence of an $\aleph_2$-Souslin tree. For if T is an $\aleph_2$-Souslin tree, then T is $\aleph_2$-Baire. So forcing with T doesn't add any new sets of size ω_1. Therefore if P is a countable chain condition (ccc) poset, then $\Vdash_T \check{P}$ is ccc, which implies that $T \times P$ has the $\aleph_2$-cc (chain condition). Therefore $\Vdash_P \check{T}$ is $\aleph_2$-cc, which implies that $\Vdash_P \check{T}$ is Souslin. Therefore forcing MA over a model of $V = L$ leaves all $\aleph_2$-Souslin trees Souslin.

Nyikos has also recently proved that PFA implies that there are no hereditarily normal, locally compact, separable, first-countable Dowker spaces (in fact, all such spaces are Moore; see [25] and [26]).

There is one natural class of Dowker spaces for which we have both consistency and independence results. In [30] Rudin constructs a hereditarily separable Dowker space, and as a corollary to Todorčević's theorem that PFA implies no S-spaces, we have the following theorem.

THEOREM 4.4 [38]: PFA implies there are no hereditarily separable Dowker spaces.

The best evidence we have to date that there may be real small Dowker spaces is Good's recent modification of de Caux's construction. It shows that the consistency, for example, of no first-countable Dowker spaces entails the consistency of large cardinals.

EXAMPLE 4.5 [13]: If no inner model contains a measurable cardinal, then there is a first-countable, locally compact, locally countable σ-discrete, θ-refinable Dowker space.

Therefore the consistency of no first-countable Dowker spaces requires at least a measurable cardinal. Hidden in the hypothesis is the important theorem of Jensen and Dodd [8]:

THEOREM 4.6: If no inner model contains a measurable cardinal, then there exist regular uncountable cardinals κ such that:

(a) $\exists E \subset \kappa$, E is a nonreflecting stationary subset of ω-cofinal ordinals (we call such sets E-sets);
(b) $\diamondsuit(E)$ holds.

Good uses the E-set along with $\diamondsuit(E)$ to push de Caux's construction through on κ. The E-set is essential. De Caux's inductive construction works because initial segments of his space are countable and therefore can be kept metrizable. The nonstationarity of initial segments of the E-set enabled Good to keep initial segments of his space metrizable. Otherwise the construction is the same as the first-countable, locally compact modifications of de Caux's original example (see [32]).

Recently, the first author with Todorčević has constructed a Dowker space from a Lusin set.

EXAMPLE 4.7 [36]: If there exists a Lusin set of reals, then there exists a first-countable, locally countable, Dowker space of size ω_1.

The backbone of the local base structure for the space is determined by Todorčević's set function H [39]:

LEMMA 4.8: Let $A = \{x_\alpha : \alpha < \omega_1\} \subseteq {}^\omega\omega$ be such that each uncountable subset is unbounded under the order of eventual dominance (i.e., assume $\mathfrak{b} = \omega_1$). Then there is a function $H : \omega_1 \to [A]^{\leq \omega}$ such that:

(a) For each α, $H(\alpha)$ is either finite or a sequence convergent to x_α.
(b) For each infinite $B \subset A$ and for each uncountable $C \subset \overline{B}$

$$\{x_\alpha : H(\alpha) \cap B \text{ is infinite}\}$$

is uncontrollable.

The space is constructed by fixing ω copies of A, $\{A_i : i < \omega\}$ and inductively defining a countable local neighborhood base at each point in $X(A) = \cup_{i < \omega} A_i$, so that for each i and each α, the copy of $H(\alpha)$ in A_i converges to the copy of x_α in A_{i+1}. Then $X(A)$ is not countable metacompact. If, in addition, A is a Lusin set, then the space is normal. The Lusin set is essential. If there is a Sierpiński set of size $\aleph_2$, then regardless of the choice of A, $X(A)$ is not normal [36].

Nonetheless we wonder whether $\mathfrak{b} = \omega_1$ or even $\mathfrak{p} = \omega_1$ implies that there is a first-countable Dowker space. In light of Bell's [2] and Weiss's [41] examples of Dowker spaces assuming $\mathfrak{p} = c$ we ask the following.

QUESTION 2: Does $\mathfrak{c} = \omega_2$ imply that there exists a first-countable Dowker space of size $\leq \mathfrak{c}$?

Also it would be curious to establish a measure theoretic analogue to Example 4.6:

QUESTION 3: Does the existence of a Sierpiński set of reals imply that there exists a Dowker space of size ω_1?

It is possible that the existence of a Lusin set is equivalent to the existence of some class of small Dowker spaces:

QUESTION 4: Is the existence of a Lusin set of reals equivalent to the existence of a first-countable, locally countable Dowker space of size ω_1?

Z. Balogh remarked to us that almost all the constructions of first-countable Dowker spaces from versions of $\diamond$ can be made locally compact, while none of the examples using CH or weaker axioms are locally compact. This raises the following question:

QUESTION 5: Does CH imply the existence of a first-countable, locally compact Dowker space?

As a partial answer to this question we can prove, analogously to Theorem 4.1:

THEOREM 4.9: It is consistent with CH that the locally compact versions of de Caux's space are never normal.

Sketch of the Proof: Recall that de Caux's space has as its point set $\omega_1 \times \omega$ and its topology is built using a sequence $\{\xi_\delta : \delta \in \mathrm{Lim}\,(\omega_1)\}$ such that each ξ_δ is of order type ω and is unbounded in δ. We call any such sequence a *ladder system*. The important facts that are relevant to us are:

(a) For each $\alpha < \omega_1$, $(\alpha + 1) \times \omega$ is clopen.
(b) For each $\alpha < \omega_1$ and each $n < \omega$, there is an $\eta_\alpha(n) \subset \alpha$ of order type $\leq \omega^{n-1}$ such that for each basic open neighborhood U of (α, n), $U \cap \omega_1 \times \{1\} = $ a tail of $\eta_\alpha(n)$.

TABLE 1. A List of Dowker Spaces

Reference	Hypothesis	Description
[27]	Souslin tree	First countable, size $\aleph_1$
[28]	ZFC	P-space of size and weight $\aleph_\omega^\omega$
[29]	κ Souslin tree	Size κ, P_κ-space
[30]	Souslin tree	Hereditarily separable, first countable
[6]	$\clubsuit, \diamondsuit$	Locally countable, σ-discrete, size $\aleph_1$
[21]	CH	Hereditarily separable, first countable, size $\aleph_1$
[20]	$\diamondsuit$	Hereditarily separable, first countable, locally compact
[41]	$MA_{\omega_1} + \diamondsuit_{\omega_2}$	First countable, locally compact, size 2^ω
[2]	$\mathfrak{p} = \mathfrak{c}$	First countable, size 2^ω
[9]	ZFC	Extremally disconnected
[31]	$\diamondsuit^{++}$	Screenable
[17]	ZFC	Topological group
[3]	$\diamondsuit$	Dowker product
[40]	$\exists$strongly compact	σ-discrete, hereditarily normal
[4]	CH	Dowker product
[13]	$\exists$ an E-set $+ \diamondsuit(E)$	First countable, locally compact
[36]	$\exists$ a Lusin set	First countable, size $\aleph_1$

(c) As a subspace of X, $\omega_1 \times 2$ is the ladder system space based on the sequence $\{\eta_\delta : \delta \in \mathrm{Lim}(\omega_1)\}$.

For a very general treatment of these spaces see [32] or [24].

The heart of the proof is contained in the following two lemmas. The first is due to Shelah, and its proof can be gleaned from [35, chap. 8].

LEMMA 4.10: It is consistent with CH that for any sequence $\{\eta_\delta : \delta \in \mathrm{Lim}\,(\omega_1)\}$ such that for each $\delta \in \mathrm{Lim}\,(\omega_1)$:

(i) The order type of $\eta_\delta < \omega^\omega$;
(ii) $\sup(\eta_\delta) = \alpha$,

there is a club $C \subset \omega_1$ such that for each $\eta_\delta \in \mathrm{Lim}\,(\omega_1)$, $C \cap \eta_\delta$ is bounded below δ.

So in particular, if $\{\eta_\delta : \delta \in \text{Lim } (\omega_1)\}$ is a ladder system, then there is a club C such that for each δ, $C \cap \eta_\delta$ is finite. Therefore ♣ fails very badly.

LEMMA [7]: (CH) If $X = \omega_1 \times 2$ is a ladder system space, then X is not normal. Furthermore for any club $C \subseteq \omega_1$ there are disjoint H and $K \subseteq C$ such that $H \times \{1\}$ and $K \times \{1\}$ can't be separated in $\omega_1 \times \{0\} \cap (C \times \{1\})$.

While this lemma does not follow directly from the fact that CH implies no ladder system is uniformizable on a club, it is an easy corollary to the proof (see [7, theorem 5.1]).

Now the proof of nonnormality of de Caux's spaces follows easily from 4.10 and 4.11. For each $n < \omega$ fix a club C_n so that for each δ, $C_n \cap \eta_\delta(n)$ is bounded below δ. Therefore for each $n < \omega$, $\overline{C_n \times \{1\}} \cap (\omega_1 \times \{n\}) = \varnothing$. Letting $C = \cap_{n<\omega} C_n$, then $C \times \{1\}$ is a closed subset of X. Then 4.11 implies that the X is not normal.   □

EXAMPLE 4.12 [40]: If there exists a compact cardinal κ, then there exists a hereditarily normal Dowker space of size κ.

Watson's example is far from small. It is of size κ and of character 2^κ. The space is built by iterating a version of Bing's G [5] up to height ω: the space is built on $\kappa \times \omega$, where for each $n < \omega$, $\kappa \times \{n, n + 1\}$ is homeomorphic to a version of Bing's G. The hereditary normality of Bing's space implies that X is also hereditarily normal, regardless of how else the space is put together. The trick is to put them together in such a way to ensure that the canonical decreasing sequence of closed sets cannot be fattened up to open sets. Watson uses the compact cardinal to build a countably complete ultrafilter and glues the spaces together to satisfy the hypothesis of the following lemma.

LEMMA 4.13: Let $\mathscr{F}$ be a countably complete filter on κ. Suppose $Y = \kappa \times 2$ is a hereditarily normal space such that $\kappa \times \{0\}$ is a set of isolated points, $\kappa \times \{1\}$ is closed discrete, and such that whenever U is an open subset of X, $U^{-1}(1) \in \mathscr{F} \rightarrow U^{-1}(0) \in \mathscr{F}$. Then there is a hereditarily normal Dowker space.

Watson conjectures that the countably complete ultrafilter is overkill and that the construction can be carried out to yield a ZFC hereditarily normal Dowker space of size ω_1. Unlike the de Caux type examples, there are no negative consistency results indicating otherwise. Currently, this may be the best hope for an out and out ZFC Dowker space of size ω_1.

QUESTION 6: Does there exist a hereditarily normal Dowker space of size ω_1?

The next set of results provide Dowker spaces with richer or more esoteric global properties.

In [34] Rudin and Starbird show that if X is countably compact and M is a metric space, then $X \times M$ is countably paracompact, and they raised the question whether the product of two countably paracompact spaces can be Dowker. Bešlagić in [3] and [4] has examples of these Dowker products assuming $\diamond$ and CH, respectively. The best result is the following.

EXAMPLE 4.14 [4]: (CH) For each $n < \omega$ there is a countably paracompact space X such that X^n is countably paracompact and X^{n+1} is Dowker.

In the case $n = 1$ Bešlagić builds two topologies τ_1, τ_2 on ω_1 so that the diagonal of $(\omega_1, \tau_1) \times (\omega_1, \tau_1)$ is homeomorphic to a first-countable Dowker space. And X is then taken to be the disjoint sum of the two spaces. Of course, care must be taken to ensure that X is countably paracompact and that X^2 is normal.

EXAMPLE 4.15 [17]: There is a Dowker topological group.

In [17], Hart, Junilla, and van Mill construct a Dowker topological group. Consider the following modification of Rudin's space.

For any increasing sequence of regular cardinals $\{\kappa_i : i < \omega\}$, let

$$X = \{f \in \square_{i \in \omega}(\kappa_i + 1) : \forall i \in \omega, \mathrm{cof}\,(f(i)) \geq \kappa_1\},$$

and let

$$X' = \{f \in X : \exists n \in \omega, \forall i \in N, \mathrm{cof}\,(f(i)) < \kappa_n\}.$$

It's easy to see that X is a P_{κ_0}-space, and it is a corollary to Rudin's proof of normality that the Lindelöf degree of X, $L(X) = 2^\omega$. If for each i, $\kappa_i = \aleph_{i+1}$, then X' is just Rudin's original example. Hart $et\ al.$ use one of these modifications to build a Dowker group in ZFC [17].

For a set X, the free Abelian group containing X is

$$B(X) = \{f \in 2^X : |f^{-1}(1)| < \omega\}.$$

For any topological space X they define a natural topology on $B(X)$ and prove the following lemmas.

LEMMA 4.16: For every space z and $Y \subset Z$ if $B(Z)$ is a topological group, then $B(Y)$ is as well. Furthermore, if there exists a κ such that Z is a P_κ-space such that $L(Z) < \kappa$, then $B(Z)$ is a topological group.

LEMMA 4.17: If X' is the modification of Rudin's space based on a sequence $\{\kappa_i : i \in \omega\}$ of regular cardinals, then $B(X')$ is also a Dowker space.

Letting X' be the Dowker subspace of $\square_{i \in \omega}(\kappa_i + 1)$ where

$$\kappa_1 = (2^\omega)^+, \ldots, \kappa_{i+1} = (\kappa_i)^+, \ldots,$$

then $B(X')$ is a $P_{(2^\omega)^+}$-space of Lindelöf degree 2^ω. Therefore $B(X')$ is a Dowker group.

For the convenience of the reader we list in TABLE 1 all Dowker spaces that have appeared in the literature. We omit those examples merely announced (e.g., [19]), including only those explicitly constructed. The reader is invited to increase the size of the list by constructing a new Dowker space.

REFERENCES

1. BALOGH, Z. 1983. Locally nice spaces under MA. Comment. Univ. Carolinae **24**: 63–87.
2. BELL, M. G. 1981. On the combinatorial principle $P(\mathfrak{c})$. Fundam. Math. **114**: 149–157.
3. BEŠLAGIĆ, A. A Dowker product. Trans. Am. Math. Soc. **292**: 519–530.
4. ———. 1991. Another Dowker product. Topol. Appl. **36**: 253–264.
5. BING, R. H. 1951. Metrization of topological spaces. Can. J. Math. **3**: 175–186.
6. DE CAUX, P. 1976. A collectionwise normal weakly θ refinable Dowker space which is neither irreducible nor realcompact. Topol. Proc. **1**: 66–77.
7. DEVLIN, K. J. & S. SHELAH. 1978. A version of $\Diamond$ that follows from $2^{\aleph_0} > 2^{\aleph_1}$ Isr. J. Math. **29**: 239–247.

8. DODD, A. J. & R. JENSEN. 1982. The covering lemma for K. Ann. Math. Logic **22:** 1–30.
9. DOW, A. & J. VAN MILL. 1982. An extremally disconnected Dowker space. Proc. Am. Math. Soc. **86:** 669–672.
10. DOWKER, C. H. 1951. On countably paracompact spaces. Can. J. Math. **3:** 219–224.
11. FLEISSNER, W. 1980. Remarks on Souslin properties and tree topologies. Proc. Am. Math. Soc. **80:** 320–326.
12. ———. 1980. Martin's Axiom implies that de Caux's space is countably metacompact. Proc. Am. Math. Soc. **80:** 495–498.
13. GOOD, C. Large cardinals and small Dowker spaces. Proc. Am. Math. Soc. In press.
14. HART, K. P. 1981. Strong collectionwise normality and M. E. Rudin's Dowker space. Proc. Am. Math. Soc. **83:** 802–806.
15. ———. 1982. More on M. E. Rudin's Dowker space. Proc. Am. Math. Soc. **86:** 508–510.
16. HARDY, K. & I. JUHÁSZ. 1976. Normality and the weak cb property. Pac. J. Math. **64:** 167–172.
17. HART, K. P., H. JUNILLA & J. VAN MILL. 1985. A Dowker group. Comment. Math. Univ. Carolina. **26**(4): 799–810.
18. ISHIKAWA, F. 1963. On countably paracompact spaces. Proc. Japan Acad. **39:** 95–97.
19. JAKOVLEV, N. N. 1976. On the theory of o-metrizable spaces. Sov. Math. Dokl. **17:** 1217–1219.
20. JUHÁSZ, I. 1977. Consistency results in topology. *In* The Handbook of Mathematical Logic, J. Barwise, Ed. North-Holland. Amsterdam, the Netherlands.
21. JUHÁSZ, I., K. KUNEN & M. E. RUDIN. 1976. Two more hereditarily separable non-Lindelof spaces. Can. J. Math. **28:** 998–1005.
22. MICHAEL, E. 1953. A note on paracompact spaces. Proc. Am. Math. Soc. **4:** 831–838.
23. MIŠČENKO, 1962. Finally compact spaces. Sov. Math. Dokl. **3:** 1199–1202.
24. NYIKOS, P. 1983. A topological test space for many axioms of set theory. Usp. Mat. Nauk. **38:** 97–103.
25. ———. Hereditary normality versus countable tightness in countably compact spaces. Preprint.
26. NYIKOS, P. & B. VELIČKOVIČ. Complete normality and countable compactness. Preprint.
27. RUDIN, M. E. 1955. Countable paracompactness and Souslin's problem. Can. J. Math. **7:**543–547.
28. ———. 1972. A normal space X for which $X \times I$ is not normal. Fundam. Math. **73:** 179–186.
29. ———. 1972. Souslin trees and Dowker spaces. Colloq. Math. Soc. János Bolyai. **8:** 557–562. Keszthely.
30. ———. 1973. A separable Dowker space. Symposia Mathematica: 125–132. Instituto Nazionale di Alta Mathematica.
31. ———. 1983. A normal screenable nonparacompact space. Topol. Appl. **15:** 313–322.
32. ———. 1984. Dowker spaces. *In* The Handbook of Set Theoretic Topology, K. Kunen and J. Vaughan, Eds.: 761–780. North-Holland. Amsterdam, the Netherlands.
33. ———. 1985. κ-Dowker spaces. *In* Aspects of Topology In Memory of Hugh Dowker; I. M. James and E. H. Kronheimer, Eds.: 175–193. Cambridge Univ. Press. New York.
34. RUDIN, M. E. & M. STARBIRD. 1975. Products with a metric factor. Gen. Topl. Appl. 5: 235–248.
35. SHELAH, S. Proper Forcing, Revised version.
36. SZEPTYCKI, P. A Dowker space from a Lusin set. Topl. Appl. In press.
37. TKACHUK, V. V. 1991. Methods of the theory of cardinal invariants and the theory of mappings as applied to spaces of functions. Siberian Math. J. (Transl. from Russian), **32:** 93–107.
38. TODORČEVIĆ, S. 1983. Forcing positive partition relations. Trans. Am. Math. Soc. **280:** 703–720.
39. ———. Partition problems in topology. Contemporary Mathematics **84.** American Mathematical Society. Providence, R.I.
40. WATSON, S. 1990. A construction of a Dowker space. Proc. Am. Math. Soc. **109:** 835–841.
41. WEISS, W. 1981. Small Dowker spaces. Pac. J. Math. **94:** 485–492.
42. ZENOR, 1971. Countable paracompactness in product spaces. Proc. Am. Math. Soc. **30:** 199–201.

Some Applications of S and L Combinatorics[a]

STEVO TODORČEVIĆ

Department of Mathematics
University of Toronto
Toronto, ONT M5S 1A1, Canada
and
Matematički Institut
Kneza Mihaila 35
11001 Beograd, p.p. 367
Yugoslavia

1. INTRODUCTION

In our Madison lecture we gave an exposition of the Ramsey-theoretic point of view of the well-known problem that asks whether hereditary separability and hereditary Lindelöfness are equivalent restrictions on a regular toplogical space. Except for one or two new results, the exposition followed closely that of [8], so we see no reason to repeat all this again. Instead, we present some applications of the point of view to the theory of topological groups as well as to the theory of the function spaces of the form $C_p(X)$ in the course of solving several well-publicized problems of these two subjects. The vast literature devoted to the S and L problem (see, e.g., [8]) contains many other surveys and many other points of view reflecting an activity that became especially strong after the problem was connected with the Souslin Hypothesis by D. Kurepa and M.E. Rudin and after A. Hajnal and I. Juhász started the use of methods of Combinatorial Set Theory and Forcing in the area. The early results were usually about counterexamples to the two implications called *S-spaces* and *L-spaces*. One group of researchers (M. E. Rudin, A. Ostaszewski, V. Fedorchuk, K. Kunen, E. van Douwen, F. Tall, W. Weiss, and others) would concentrate their efforts on S-spaces and L-spaces with additional topological properties like normality or compactness. Other researchers (A. Hajnal, I. Juhász, J. Roitman, and others) would get their S-spaces and L-spaces in a simple way from partitions $c : [\omega_1]^2 \to 2$ with properties very reminiscent of the classic partitions of Sierpiński, Kurepa, Erdös, Hajnal, Rado, Milner, and others from the subject of Partition Calculus of Set Theory. The common drawback of these beautiful examples was that nobody knew how to construct any of them without additional set-theoretical assumptions. (In some cases, though, it was known that additional assumptions were necessary, but most of them were left open.) In 1984 we showed ([7]) that the difficulties lie with the word "countable" (or the parameter "ω_1") in the statement of the basic problem, that is, that, without any additional set-theoretical assumptions there exist many cardinals θ for which one can construct an "S_θ-space"

[a] Research supported by the Science Fund of Serbia Grant 0401A and by the National Sciences and Engineering Research Council of Canada.

or an "L_θ-space." For example, one such θ is the minimal cardinal of unbounded subset of ω^ω under the ordering of eventual dominance. The combinatorics developed in the production of such examples will also find its application in this article. In fact, one of our basic applications essentially uses only the result, established in [7], that there exists θ such that 3^{θ^+} contains a subspace X with the property that the Lindelöf degree of X is equal to θ^+, but every subspace of X^ω has a dense subset of size $\leq \theta$. In the literature such a space X is usually called a "strong S_{θ^+}-space" and it is well-known that its existence is equivalent to the existence of the corresponding "strong L_{θ^+}-space."

Some of the specific contributions of this paper are as follows. In Section 2 we construct a topological group G whose square has bigger cellularity than G. It is well known that asking a similar result for groups satisfying the countable chain condition (ccc) is a more delicate matter that cannot be resolved without additional axioms. In Section 2 we also construct such a ccc group assuming that the Lebesgue measure on the reals can be extended to a countably additive measure defined on all sets of reals. In Section 4 we prove similar results about the cardinal invariants density, tightness, and the Lindelöf degree. In Section 4 we construct a completely regular space X containing a subspace Y such that the density of Y is bigger than the supremum of densities of closed subspaces of X. It is well known that this cannot happen if X is a compact space. It is well known that every compact topological group supports a strictly positive measure and has every uncountable regular cardinal as precaliber. In Section 5 we give an application of the combinatorics of section 2 of [8] to the problems whether every compactly generated group supports a strictly positive measure and whether it has $\aleph_1$ as a precaliber. We show that the free topological group of the one-point compactification of the discrete space of size continuum does not support a strictly positive measure. In particular, we show that answers to both questions are negative assuming that ω^ω contains a subset of size $\aleph_1$ unbounded under the ordering of eventual dominance. In Sections 6–8 we consider the problem of productiveness of tightness and the Lindelöf degree in some special classes of spaces. A well-known result of Malykhin says that tightness of the product of two spaces, one of which is compact, is not bigger than the tightness of the factors. In Section 6 we show that replacing compactness by σ-compactness is not sufficient to get the same conclusion. In fact, we construct two σ-compact Frechet–Urysohn groups G and H such that $G \times H$ is not countably tight. They are constructed using the combinatorics of the special Aronszajn tree of [25], which itself is closely related to the S and L combinatorics where it finds many applications (see, e.g., [8, secs. 2 and 6]). Most of the examples about the nonpreservation of tightness in products involve spaces with single nonisolated point. For example, it is well-known that the square of the Frechet–Urysohn fan with ω_1 edges is not countably tight. What if we require that the space X with unique nonisolated point is Lindelöf? (Obviously the Frechet–Urysohn fan with ω_1 edges is not Lindelöf.) In Section 8 we show that this is the right question by showing that, under the proper forcing axiom (PFA), the product of any two Lindelöf countably tight spaces X and Y with single nonisolated points is countably tight. In fact, we show that for such X and Y the function spaces $C_p(X)$ and $C_p(Y)$ are Lindelöf and that their product $C_p(X) \times C_p(Y)$ is also Lindelöf. The use of the additional axiom is necessary since again pathological counterexamples can be constructed assuming that Lebesgue measure can be extended to all sets of reals (Section 7). In Section 9, we present a criterion for compactness of subsets of $C_p(X)$. It can be considered as a direction of generalizing the classical criteria of A. Grothendieck ([40], [41]) for compactness of countably compact subspaces of $C_p(X)$. In Section 10, we consider Velicko's problem ([6]) of whether the hereditary Lindelöf degree of space $C_p(X)$ is preserved by taking finite powers. For example, we show (modulo $MA_{\aleph_1}$) that the basic implications S and L hold for subspaces of any space X for which $C_p(X)$ contains no uncountable free sequence. This is an analogue of [8, 7.10] where finite powers of X play the role of $C_p(X)$. In Section 11 we show that the product of every two countably tight initially ω_1-compact spaces is countably tight. The proof of this result does not use any additional axiom of set theory, but our uses of PFA in the previous sections were quite instrumental in finding the result. So this is another instance of the phenomenon (which unfortunately is not so well-known) that a PFA analysis of a problem at hand quite frequently also leads to an absolute result. The last four sections are meant for those topologically inclined readers interested in further study of the S and L phenomena occurring in their subject, as we have included there only some hints to a vast range of subjects where the phenomena might occur and to a number of problems left open by this study.

We finish the introduction with a short review of Mary Ellen Rudin's work in this area. The S and L problems have always been in the center of M. E. Rudin's mathematical interests, not only for their own sake but also because the patterns often occur in other problems. For example, she devoted a whole chapter to them in her lectures on set-theoretic topology [10]. The terms "S-space" and "L-space," which are predominant in most of the literature on this subject, are also first found in the lectures [10]. This shows a great influence not only of [10] but also of M. E. Rudin's personality on the generation of mathematicians working in this area, since it is rather unusual in mathematics to talk about certain statements in terms of their counterexamples. She is also the first to construct an S-space using a Souslin tree ([11]). The space turned out to be normal, so in [12] she produced a machine for turning certain S-spaces into nonnormal ones. (Today we know (see [17]) that every S-space can be turned into a nonnormal one.) In one of the most often cited papers in this area [15], she showed how to modify the Kunen Line (a certain CH-example of an S-space) to get a Dowker space, that is, a normal space that is not countably paracompact. She has been also very successful in using the methods of the S and L theory in treating many other seemingly unrelated problems. For example, in her proof ([13]) from $MA_{\aleph_1}$ that perfectly normal manifolds are metrizable, she makes essential use of the fact that there are no compact S-spaces under that hypothesis ([18]). That some assumption is necessary follows from her paper with Zenor [14], which contains two constructions [using diamond and the Continuum Hypothesis (CH)] of perfectly normal nonmetrizable manifolds that are S-spaces. Her survey article [16] is still one of the best introductions to the subject. It starts with a list of five problems, two basic ones and three derived. As expected, problems (1) and (2) are stated as "Does there exist an S space?" and "Does there exist an L space?" (i.e., in terms of counterexamples to the two basic implications). Problem (3) asks "Are problems (1) and (2) the same?" Problem (4) reads "If a T_3 space has no uncountable discrete subspace, must it be the union of a hereditarily separable subspace and a hereditarily Lindelöf subspace?" Problem (5) asks "Are height [hereditary Lin-delöf number] and width [hereditary density] ever different?" All these problems except (2) have been solved since then (see [8]). Under an additional set-theoretic assumption (see Section 8), hereditary separability implies hereditary Lindelöfness in regular spaces giving us answers to (1) and (4); (1) and (2) are not the same problems ([8, sec. 9]); without any additional set-theoretical assumption, there is a regular space having different height and width ([8, sec. 0]).

2. CELLULARITY OF TOPOLOGICAL GROUPS

Eight years ago we proved [7] that cellularity is not a productive property in the class of compact topological spaces. In [8, p. 12] we announced that similar methods can be used to show that the same is true about the class of topological groups. This announcement was apparently overlooked, so the question was asked again during the 7th Prague Topological Symposium. Motivated by the renewed interest, in this section, we present the proof.

THEOREM 0: There is a topological group G such that the cellularity of G^2 is bigger than the cellularity of G.

Proof: The proof will also use some ideas of [4] that prove the consistency of this statement. We start with a cardinal $\theta \geq \omega$ and $p : [\theta^+]^2 \to 3$ such that:

(1) For every disjoint family $\{a_\alpha : \alpha < \theta^+\}$ of finite subsets of θ^+ all of the same size m and for every $h : m \to 3$ there exist $\alpha < \beta$ such that $p(a_\alpha(i), a_\beta(j)) = h(i)$ for all $i, j < m$.[b]

Removing a small subset of θ^+ we may assume that:

(2) For every $\alpha < \theta^+$ and $\epsilon < 3$ there exist cofinally many β's such that $p(\alpha, \beta) = \epsilon$.

Let $X = \theta^+ \times 2$. An element (α, i) of X will also be denoted by α^i. Let G be the set of all finite subsets of X and let $+_2$ denote the operation of symmetric difference between the elements of G. The group $\langle G, +_2 \rangle$ is made topological as follows: A typical subbasic open set containing $\varnothing$ is the set $U_\beta (\beta < \theta^+)$ of all a in G such that β^0, $\beta^1 \notin a$ and for all $\alpha < \beta$:

(3) $p(\alpha, \beta) = 0$ implies $\alpha^1 \notin a$;
(4) $p(\alpha, \beta) = 1$ implies $\alpha^0 \notin a$;
(5) $p(\alpha, \beta) = 2$ implies that either both α^0 and α^1 are in a or both are out of a.

By (2) the intersection of $U_\beta (\beta < \theta^+)$ is equal to $\{\varnothing\}$. Note also that U_β is a subgroup of G for all β. It follows that G with this topology is a completely regular topological group. For finite subset b of θ^+ let U_b be the intersection of $U_\beta (\beta \in b)$. Then $U_b (b \in [\theta^+]^{<\omega})$ is the neighborhood basis of $\varnothing$ in G.

CLAIM 1: $c(G) \leq \theta$.

---

[b] Here $a_\xi(i)$ denotes the ith element of the set a_ξ in the increasing order.

Proof: Let $a_\xi +_2 U_{b_\xi}(\xi < \theta^+)$ be a given sequence of basic open sets of G. We need to find two of them with nonempty intersection. First of all, we may assume a_ξ and b_ξ form Δ-systems with roots a and b, respectively. Let $\bar{a}_\xi = a_\xi \backslash a$ and $\bar{b}_\xi = b_\xi \backslash a (\xi < \theta^+)$. Clearly, we may assume $\bar{a}_\xi$ and $\bar{b}_\xi$ are nonempty. For $\xi < \theta^+$ let d_ξ be the set of $\alpha < \theta^+$ such that $\alpha^\epsilon \in \bar{a}_\xi$ for some $\epsilon < 2$. So we may assume that all d_ξ have the same size m. For $\xi < \theta^+$, let $h_\xi : m \to 3$ be defined by $h_\xi(i) = 0$ if $d_\xi(i)^0 \in \bar{a}_\xi$ and $d_\xi(i)^1 \notin \bar{a}_\xi$; $h_\xi(i) = 1$ if $d_\xi(i)^1 \in \bar{a}_\xi$ and $d_\xi(i)^0 \notin \bar{a}_\xi$; $h_\xi(i) = 2$ if both $d_\xi(i)^0$ and $d_\xi(i)^1$ are elements of $\bar{a}_\xi$. Refining further, we may assume that h_ξ are all equal to some fixed h. By (1) there exist $\xi < \eta$ such that every ordinal of $\bar{b}_\xi \cup d_\xi$ is bigger than every ordinal of b but smaller than every ordinal of $\bar{b}_\eta \cup d_\eta$ and such that:

(6) $p(d_\xi(i), \beta) = h(i)$ for $i < m$ and β in $\bar{b}_\eta$.

Tracing back the definitions one easily checks that $a_\xi +_2 a_\eta (= \bar{a}_\xi \cup \bar{a}_\eta)$ is an element of the intersection of $a_\xi +_2 U_{b_\xi}$ and $a_\eta +_2 U_{b_\eta}$. This finishes the proof of Claim 1. □

CLAIM 2: $c(G^2) = \theta^+$.

Proof: It suffices to show that

$$(\{\beta^0\} +_2 U_\beta) \times (\{\beta^1\} +_2 U_\beta) \qquad (\beta < \theta^+)$$

is a disjoint family of open subset of G^2. Let $\beta < \gamma < \theta^+$ be given. Suppose first that $p(\beta, \gamma) \neq 0$. In this case we show that $\{\beta^0\} +_2 U_\beta$ and $\{\gamma^0\} +_2 U_\gamma$ are disjoint. For suppose a is in the intersection. Since no element of U_β contains β^0 nor β^1, we must have $\beta^0 \in a$ and $\beta^1 \notin a$. For the same reason we also must have that $a = \bar{a} \cup \{\gamma^0\}$ for some $\bar{a}$ in U_γ. So $\bar{a}$ is an element of U_γ with the property $\beta^0 \in \bar{a}$ and $\beta^1 \notin \bar{a}$, contradicting conditions (4) and (5). If $p(\beta, \gamma) \neq 1$, then a similar argument shows that in this case $\{\beta^1\} +_2 U_\beta$ and $\{\gamma^1\} +_2 U_\gamma$ must be disjoint. This proves Claim 2. □

It remains to show that there is a cardinal θ and a partition $p : [\theta^+]^2 \to 3$ with the property (1). This has been essentially proved in [7] (see also [8, sec. 0]), but for the convenience of the reader we sketch the argument. To avoid a complete repetition we actually make a slight twist of the old argument. First of all note that the oscillation mapping of [8, sec. 1] satisfies a rather strong form of (1). This is the content of theorem 1.2 of [8]. We can therefore assume that the continuum is at least as big as the first weakly inaccessible cardinal. (This assumption is actually not needed for the rest of this argument, but we want to point out the group associated with Theorem 1.2 of [8] because it might have same further properties.) By [1, sec. 4] there is a cardinal $\theta \leq c$ of cofinality ω (e.g., $\theta = \aleph_\omega$) and an increasing sequence $\theta_i(i < \omega)$ cofinal with θ such that $\Pi_i\theta_i$ contains a subset A of size θ^+ cofinal in the ordering of eventual dominance. For f and g in A let

$$\Gamma(f, g) = \min \{i < \omega : f(i) < g(i)\}.$$

The argument of lemma 0.12 of [8] shows that for every disjoint family $a_\xi(\xi < \theta^+)$ of finite subsets of A and for almost all $i < \omega$ we can find $\xi < \eta$ such that $\Gamma(f, g) = i$ for f in a_ξ and g in a_η. Composing Γ with a mapping from ω onto ω gives us a $q : [\theta^+]^2 \to \omega$ such that:

(7) For every disjoint family $a_\xi(\xi < \theta^+)$ of finite subsets of θ^+ and every $l < \omega$ then exist $\xi < \eta$ such that $q(\alpha, \beta) = l$ for all α in a_ξ and β in a_η.

Let $h_i(i < \omega)$ be an enumeration of all mappings

$$h : [0, 1]^n \times [0, 1]^n \to \omega,$$

where $n < \omega$. Let n_i be the n that corresponds to h_i. Choose also a 1–1 sequence $\{r_\alpha : \alpha < \theta^+\}$ of elements of $[0, 1]^\omega$. Define $p : [\theta^+]^2 \to \omega$ by $p(\alpha, \beta) = h_i(r_\alpha \upharpoonright n_i, r_\beta \upharpoonright n_i)$ where $i = q(\alpha, \beta)$. Tracing back the definitions one easily shows that p has the following property that is more than needed for (1).

(8) For every disjoint family $a_\xi(\xi < \theta^+)$ of finite subsets of θ^+ all of the same size m and for every $h : m \times m \to \omega$ then exist $\xi < \eta$ such that $p(a_\xi(i), a_\eta(j)) = h(i, j)$ for all $i, j < m$. $\square$

As explained in [7, sec. 2] almost every example of nonproductiveness of cellularity must produce a cardinal θ and a partition $p : [\theta^+]^2 \to 3$ with property (1). Hence in every such instance we can have a topological group witnessing the nonproductiveness. Clearly every such θ must be uncountable or else Martin's axiom would be equivalent to CH. This means that examples of ccc groups with non-ccc squares can only be obtained axiomatically or in certain forcing extensions. For example, in [4] such a group was constructed in the one Cohen-real extension. To see an axiomatic solution, let RVM denote the well-known assertion (considered by S. Banach, H. Lebesgue, and S. Ulam early in this century) that there is a countably additive measure extending the Lebesgue measure and defined on all sets of reals (see [3]).

THEOREM 0^c (RVM): There is a ccc group whose square is not ccc.

Proof: Start with an $e : [\omega_1]^2 \to \omega$ such that for all $\alpha < \omega_1, e_\alpha = e(\cdot, \alpha) : \alpha \to \omega$ is 1–1 and $\{\xi < \alpha : e(\xi, \alpha) \neq e(\xi, \beta)\}$ is finite for all $\alpha < \beta$. For $r \in \omega^\omega$, define $e_r : [\omega_1]^2 \to \omega$ by

$$e_r(\alpha, \beta) = r(e(\alpha, \beta)).$$

The theorem now follows from the following Claim and the fact (previously proved) that the existence of a partition with property (7) implies the existence of one with properties (8) or (1). $\square$

CLAIM 1 (RVM): There is an $r \in \omega^\omega$ for which e_r satisfies (7).

Proof: We may assume to have countably additive $\mu : \mathscr{P}(\omega^\omega) \to [0, 1]$ extending the standard product measure of ω^ω. Let $\mathscr{I}^+ = \{X \subseteq \omega^\omega : \mu(X) > 0\}$ considered as a poset ordered by $\subseteq$. By standard arguments (see [3]), in $V^{\mathscr{I}^+}$, there is a generic elementary embedding $j : V \to M$ such that $\omega^\omega \subseteq M$ (and more). Thus if there is an r in $V^{\mathscr{I}^+}$ with e_r satisfying (7), then there is one in M and, therefore, one in V by the elementarity of j. By Maharam's theorem, forcing with $\mathscr{I}^+$ (at least below one of its conditions) is equivalent to the standard forcing for adding a number of random reals. Since (7) is preserved by any Property K poset (so, in particular, by a measure algebra), Claim 1 follows from the following fact whose proof uses some arguments of [5]. $\square$

CLAIM 2: If r is a random (or a Cohen) real, then e_r satisfies (7).

Proof: Suppose we have $m \geq 1$, disjoint $\{a_\xi : \xi < \omega_1\} \subseteq [\omega_1]^m$, and a sequence B_ξ $(\xi < \omega_1)$ of compact subsets of ω^ω of positive measure. Then we can find $\xi_0 <$

$\xi_1 < \cdots < \xi_\omega$ such that $B = \cap_{i \leq \omega} B_{\xi_i}$ has positive measure and such that $a_{\xi_0} < a_{\xi_1} < \cdots < a_{\xi_\omega}$. Let $\gamma = \sup(\cup_{i < \omega} a_{\xi_i})$ and let $F = \{\alpha < \gamma : e_\beta(\alpha) \neq e_\gamma(\alpha)$ for some $\beta \in a_{\xi_\omega}\}$. Removing some of the a_{ξ_i} for $i < \omega$, we can assume $a_{\xi_i} \cap F = \varnothing$ for all $i < \omega$. For $i < \omega$, let $b_i = \{e_\gamma(\alpha) : \alpha \in a_{\xi_i}\}$ and $A_i = \{x \in \omega^\omega : x(n) = l$ for all $n \in b_i\}$. Then A_i are independent and all have the same measure $(\frac{1}{2}^{l+1})^m$, so their union has measure 1. So, there is an $i < \omega$ such that $A_i \cap B$ has positive measure. It is clear that this set forces $\dot{r}(e(\alpha, \gamma)) = l$ for all $\alpha \in a_{\xi_i}$. This finishes the proof of the random-real case. The Cohen-real case follows from the fact that the union of A_i is dense open in ω^ω. $\square$

It should be noted that the arguments of Kunen and Roitman in [5] use a different stepping-up tool $e : [\omega_1]^2 \to \omega$ from ours, but that they are sufficient to prove the existence of a $p : [\omega_1]^2 \to 2$ with property (1) in one Cohen or random-real extension. Papers [5] and [6] have some applications of properties like (1) to the problem of separability versus Lindelöfness in regular topological spaces as well as groups.

3. DENSITY, TIGHTNESS, AND THE LINDELÖF DEGREE OF TOPOLOGICAL GROUPS

Why can the results of Section 2 be considered as "applications of S and L combinatorics"? Well, the basic tool was definitely the partition $p : [\theta^+] \to 3$ with the property (1) and its existence is only a minor strengthening of the following statement.

(1)* There is a subspace $X \subseteq 3^{\theta^+}$ with Lindelöf degree equal to θ^+ but that has the property that the hereditary density of X^ω is $\leq \theta$.

The two properties are so closely related that it is still unknown whether they are, in fact, equivalent. On the other hand, almost every construction of a space satisfying (1)* gives the partition p with property (1). Partition property (1) is also sufficient to show the following.

THEOREM 1: Taking square of a topological group does not necessarily preserve its Lindelöf degree, its density, or its tightness.

Proof: For $\beta < \theta^+$ let $x_\beta^0, x_\beta^1 \in \{0, 1\}^{\theta^+}$ be defined by $x_\beta^0(\alpha) = x_\beta^1(\alpha) = 0$ for $\alpha \leq \beta$; $x_\beta^0(\alpha) = 0$ and $x_\beta^1(\alpha) = 1$ if $p(\beta, \alpha) = 1$; $x_\beta^0(\alpha) = 1$ and $x_\beta^1(\alpha) = 0$ if $p(\beta, \alpha) = 0$; $x_\beta^0(\alpha) = x_\beta^1(\alpha) = 1$ if $p(\beta, \alpha) = 2$. Let G_s^θ be the subgroup of $\{0, 1\}^{\theta^+}$ generated by $\{x_\beta^0, x_\beta^1 : \beta < \theta^+\}$. By property (1), G_s^θ has hereditary density $\leq \theta$, while $\{\langle x_\beta^0, x_\beta^1 \rangle : \beta < \theta^+\}$ is a subset of its square that accumulates to $\langle 0, 0 \rangle$ but no smaller subset of it accumulates to $\langle 0, 0 \rangle$, that is, the square of G_s^θ has tightness $> \theta$. Replacing $<$ by $>$ in the preceding definitions of x_β^0 and x_β^1 one gets a subgroup G_l^θ of $\{0, 1\}^{\theta^+}$ with hereditary Lindelöf degree $\leq \theta$ while the Lindelöf degree of its square is $> \theta$. $\square$

As we have already noted (without any additional axiomatic assumptions) there are many cardinals θ with partitions $p : [\theta^+]^2 \to \omega$ satisfying (8) and therefore many groups like G_s^θ and G_l^θ given earlier. For $\theta = \omega$ such a partition was just produced under the additional axiomatic assumption RVM, so we have the following theorem.

THEOREM 1^c (RVM): There is a hereditarily Lindelöf group whose square is not Lindelöf and also there is a hereditarily separable group whose square is not countably tight.

It is still open, however, whether there is a Lindelöf group whose square is not Lindelöf or a countably tight group whose square is not countably tight if one is not willing to go beyond the usual axioms of set theory (see [2]). In Section 6 we show (see [9]) that there exist *two* Lindelöf countably tight topological groups whose *product* is not countably tight. Some other examples of similar pairs of topological groups can be found in [2] and [4].

4. DENSITIES OF CLOSED SUBSPACES

The partition property (1) is the combinatorial essence of our proof that hL and hd are different cardinal functions in the class of all regular spaces, that is, that the higher cardinal analogues of the S and L implications are false. In the first two sections of this paper we have seen that this combinatorial essence can also be used in attacking some other problems. In this section, however, we return to the original construction of [7] because the constructed spaces have some other interesting properties. For example, the inequalities $hL < hd$ and $hd < hL$ can be witnessed by first-countable 0-dimensional spaces. Also, the diference between these two cardinal functions can be made maximally large, that is, $hd = 2^{hL}$ and $hL = 2^{hd}$. Note that this is strikingly different from the behaviour of the class of all compact spaces where we always had $hd \leq hL^+$ by a result of Shapirovskii [55]. In this section we present another such difference. Namely, again by a result of Shapirovskii [55], if X is a compact space, then for every subspace Y of X there is a closed $F \subset X$ such that $d(Y) = d(F)$. Here we show that this is no longer true in the larger class of completely regular spaces. This answers a question asked in [19, p. 226].

THEOREM 2: There is a completely regular space X such that $hd(X)$ is bigger than the supremum of densities of closed subspaces of X.

Proof: As in Section 2 choose an increasing sequence $\theta_i(i < \omega)$ of regular cardinals with supremum θ such that the product $\Pi_i \theta_i$ contains a subset A of size θ^+ cofinal in $\Pi_i \theta_i$ in the ordering $\leq$ of everywhere dominance and that is well ordered in type θ^+ by the ordering $<^*$ of eventual dominance. [For f and g in $\Pi_i \theta_i$ and $n < \omega$, set

$$f \leq_n g \qquad \text{iff} \qquad f(i) \leq g(i) \quad \text{for} \quad i \geq n.$$

Thus $\leq \ = \ \leq_0$ is the ordering of everywhere dominance. The ordering $<^*$ of eventual dominance is defined by letting $f <^* g$ if there is n such that $f <_n g$, that is, if there is an n such that $g(i) < g(i)$ for all $i \geq n$. In [7], we have taken θ_i above the continuum, but according to a more recent work of Shelah ([1]) every cardinal θ of countable cofinality is the supremum of an increasing sequence $\theta_i(i < \omega)$ with the property that the product $\Pi_i \theta_i$ contains a set A as before.] Let D be the set of all $d \in \Pi_i(\theta_i + 1)$ such that there is an $n < \omega$ such that

$$d \upharpoonright n \in \Pi_{i<n} \theta_i \qquad \text{and} \qquad d(i) = \theta_i \quad \text{for all} \quad i \geq n.$$

Let $X = A \cup D$ considered as a subspace of the product space $\Pi_i(\theta_i + 1)$ where each

$\theta_i + 1$ is topologized as follows: the points of θ_i are all isolated while the complements of neighborhoods of the point θ_i must have size $< \theta_i$. Let $X[A, \geq]$ be the refinement of the topology of X by making each

$$\{g \in X : f \leq g\} \qquad (f \in A)$$

as a new open set (see [8, sec. 0]). Clearly $hd(X[A, \geq]) = \theta^+$ since A is a left-separated subspace of type θ^+.

CLAIM: $d(F) = \theta$ for every closed subspace F of $X[A, \geq]$.

Proof: Let $Y = F\backslash(\overline{F \cap D})$. It suffices to show that Y has size $\leq \theta$. Otherwise, suppose Y has size θ^+. Shrinking Y, if necessary, assume that there is $m < \omega$ such that:

(a) $f \upharpoonright m = g \upharpoonright m$ for all f and g in Y, but
(b) $f \upharpoonright m \neq g \upharpoonright m$ for all $f \in Y$ and $g \in F \cap D$.

Let $\Pi^* \theta_i$ be the union of all finite subproducts $\Pi_{i<n} \theta_i (n < \omega)$. For t in $\Pi^*\theta_i$ let Y_t be the set of all elements of Y that extend t. Shrinking Y further, assume that each Y_t is either empty or it has size θ^+. Then by our assumption, for every t in $\Pi^*\theta_i$ of length $\geq m$, if d is the member of D extending t such that $d(i) = \theta$ for $i \geq |t|$, then d is not in the closure of $F \cap A$ and, in particular, not in the closure of Y itself. This reduces to the following fact:

(c) For every t in $\Pi^*\theta_i$ of length $n \geq m$ there is $f_t \in A$ such that $f_t \nleq_n g$ for all $g \in Y_t$.

Let f^* be the element of $\Pi_i\theta_i$ defined by:

$$f^*(n) = 0 \quad \text{for} \quad n < m,$$

$$f^*(n) = \sup\left\{ f_t(n) : t \in \bigcup_{j=m}^{n} \prod_{i=0}^{j-1}\theta_i \right\} \quad \text{for} \quad n \geq m.$$

Since Y is also $<^*$-cofinal in the product, there is $g \in Y$ such that $f^* <^* g$. Pick a $n \geq m$ such that $f^*(i) < g(i)$ for all $i \geq n$ and let $t = g \upharpoonright n$. Then by the definition of f^*, we have

$$f_t \leq_n f^* \leq_n g$$

Since $g \in Y_t$ this is in contradiction with (c). This finishes the proof of the Claim and also the proof of Theorem 2. $\square$

We also have the following countable version of Theorem 2, which can serve two purposes. It gives us another example witnessing Theorem 2 when the cardinal $\mathfrak{b}$ is not weakly inaccessible. But it also gives us a completely regular space X that is not hereditarily separable, but every closed subspace of X is separable under the (consistent) assumption $\mathfrak{b} = \omega_1$. [Here, $\mathfrak{b}$ is the minimal cardinality of an $<^*$-bounded subset of ω^ω.]

THEOREM 3: There is a completely regular space X such that $hd(X) = \mathfrak{b}$, but $d(F) < \mathfrak{b}$ for every closed subspace F of X.

Proof: Choose a subset A of ω^ω consisting entirely of increasing functions such that A is unbounded in ω^ω, $<^*$ and such that $<^*$ well orders A in order type $\mathfrak{b}$. The construction will be similar to the previous proof, but since in this case we have only unbounded rather than a cofinal subset we give some details. Of course, the combinatorics itself has again its roots in [7] or [8, sec. 0]. Let D be the set of all increasing elements d of $(\omega + 1)^\omega$ such that for some $n < \omega$, $d \upharpoonright n \in \omega^n$ and $d(i) = \omega$ for all $i \geq n$. Let $X = A \cup D$ with the topology induced from the product $(\omega + 1)^\omega$, where $\omega + 1$ is taken with the order topology. Again, the space will be $X[A, \geq]$, that is, obtained from X by making each $\{g \in X : f \leq g\}$ ($f \in A$) open. So, again the main part of the proof is in the following fact.

CLAIM 2: $d(F) < \mathfrak{b}$ for every closed subspace F of $X[A, \geq]$.

Proof: Let $Y = F \backslash (\overline{F \cap D})$, so it suffices to get a contradiction from the fact that Y has size $\mathfrak{b}$. As before, shrinking Y we assume to have an $m < \omega$ such that (a) and (b) given earlier are satisfied. Choose a countable elementary submodel M of some large enough structure of the form H_θ such that M contains all these objects. Let $h \in A$ be such that $f <^* h$ for every f in $A \cap M$. Then $h <^* g$ for almost all elements g of Y so there must be an integer p such that

$$Y_0 = \{g \in Y : h \leq_p g\}$$

has cardinality $\mathfrak{b}$, so, in particular, it is unbounded in ω^ω. Let n ($\geq m$) be a minimal integer such that $g(n)(g \in Y_0)$ is unbounded in ω. Then we can find a t in ω^n and $g_i(i < \omega)$ in Y_0 such that

(d) $t \subset g_i$ for all $i < \omega$,
(e) $g_i(n) < g_j(n)$ for $i < j < \omega$.

Pick $f \in Y \cap M$ extending t and let d be the element of D such that

$$d \upharpoonright n = t \quad \text{and} \quad d(i) = \omega \quad \text{for} \quad i \geq n.$$

Note that d is also an element of M and by (b) d is not in F so, in particular, it is not in the closure of $F \cap A$. So by the elementarity of M there is a $f_t \in A \cap M$ such that

(f) $f_t \nleq_n f$ for every $f \in Y$ extending t.

Since $f_t <^* h$ we can fix a $k \geq p$ such that $f_t \leq_k h$. By (e), the sequence $g_i(n)$ ($i < \omega$) is unbounded in ω, so there is an i such that

(g) $f_t(k) \leq h(k) \leq g_i(n)$.

Since all the functions are monotonic (and since $h \leq_p g_i$ and, therefore, $h \leq_k g_i$) this easily gives that $f_t \leq_n g_i$. But we also know that g_i extends t [see (d)], so this is in contradiction with the condition (f). This finishes the proof. □

REMARKS: The results of this section were announced in [9]. Note that any space X witnessing Theorem 2 must contain a subspace Y such that $hL(Y) < hd(Y)$. Namely, the Y can be any left-separated subspace of X of size bigger than the supremum of $d(F)$'s for F a closed subspace of X. In the proof of Theorem 3 we could have also taken (in the notation of [8, sec. 0]) the space

$$(\omega + 1)^\omega[A, \geq]$$

to witness this result. We believe that this space is of independent interest and that it might have some other interesting properties.

5. STRICTLY POSITIVE MEASURES IN COMPACTLY GENERATED GROUPS

A space X carries a *strictly positive measure* ([46]) if there is a probability measure μ defined on a σ-field generated by a π-basis of X such that $\mu(B) > 0$ for every nonempty open set B from the σ-field of μ. It is well known (see [50]) that every compact group supports a strictly positive measure, and until recently that was the only way to prove that compact groups satisfy the countable chain condition. Recently M. G. Tkačenko ([47]) found a direct argument using Ramsey's Theorem, which moreover, shows that every compactly generated group is ccc. In fact, [47] shows that every compactly generated group has the property $K_{\theta n}$ (see [46]) for every finite n and every cardinal θ of uncountable cofinality. On the other hand, it is well known (see [46]) that every carrier of a strictly positive measure also has property $K_{\theta n}$ for every finite n and every cardinal θ of uncountable cofinality. So it is natural to ask whether every compactly generated group supports a strictly positive measure. It is also natural to ask (see [47]) whether every uncountable cardinal is a precaliber of every compactly generated group, since this is so for compact topological groups. In [48], Shakhmatov showed that in a special model of set theory obtained by adding uncountably many Cohen reals, the free topological group of the one-point compactification of the discrete space of size $\aleph_1$, $F(\omega_1)$, does not have precaliber $\aleph_1$. Note that the Cohen reals form a second category set of reals of size $\aleph_1$, so in the model of [48] we also have that $\aleph_1$ is not a precaliber of any nontrivial measure algebra. Hence this rules out the obvious route to the first question. In this section we show that unlike the case of compact groups there is no natural way to define strictly positive measures on compactly generated groups. This will be done by applying the combinatorics of [8, sec. 2], that is, the combinatorics of the assumption that ω^ω contains a subset of size $\aleph_1$ unbounded in the ordering of eventual dominance. This assumption, henceforth denoted by $\mathfrak{b} = \omega_1$, will also give us that the free topological group $F(\omega_1)$ [in fact, its Abelian counterpart $A(\omega_1)$] does not have precaliber $\aleph_1$, so we shall also have the result of [48] as a corollary.

Let $A(\mathbb{R}^*)$ be the free Abelian group with generators equal to the set $\mathbb{R}^* = \mathbb{R} \cup \{*\}$. Thus $A(\mathbb{R}^*)$ can be viewed as the set of all formal sums of the form $\Sigma_{i=0}^m n_i x_i$, where $n_i \in \mathbb{Z}$ and $x_i \in \mathbb{R}^*$ ($i \leq m < \omega$). For an (not necessarily strictly) $\subseteq$-increasing sequence $\vec{F} = \langle F_i : i < \omega \rangle$ of finite subsets of $\mathbb{R}_1$ let $U(\vec{F})$ be the set of all sums of the form $\Sigma_{i=0}^m (x_i - y_i)$, where $m < \omega$ and where $x_i, y_i \in \mathbb{R}^*$ and $x_i, y_i \notin F_i$ for all $i \leq m$. Then if $\mathscr{F}$ denotes the set of all increasing ω-sequences of finite subsets of $\mathbb{R}$, the family $U(\vec{F})(\vec{F} \in \mathscr{F})$ satisfies all the standard conditions ([50, chap. II, sec. 4]) sufficient for generating a neighborhood basis of 0 in $A(\mathbb{R}^*)$. For example, it is clear that $-U(\vec{F}) = U(\vec{F})$ for every $\vec{F}$. Note also that

$$U(\vec{E}) + U(\vec{E}) \subseteq U(\vec{F})$$

where $\vec{E}$ is determined from $\vec{F}$ by $E_{2i} = E_{2i+1} = F_i$ for $i < \omega$. Also,

$$U(\vec{E} \cup \vec{F}) \subseteq U(\vec{E}) \cap U(\vec{F})$$

where $\vec{E} \cup \vec{F}$ denote the sequence whose ith element is equal to $E_i \cup F_i$. It is also easily checked that the group topology generated by this basis when restricted to the set $\mathbb{R}^*$ of generators gives $\mathbb{R}^*$ a compact topology with $*$ as the only nonisolated point. Thus $A(\mathbb{R}^*)$ is a compactly generated group homeomorphic to the free topological group of the one-point compactification of a discrete space of size continuum. (The proof of the second fact can be found in [47], although this will not be used here.)

THEOREM 4: The compactly generated group $A(\mathbb{R}^*)$ does not carry a strictly positive measure and if $\mathfrak{b} = \omega_1$ it does not have precaliber $\aleph_1$.

We first give a version of the combinatorics of [8, sec. 2] that is going to be used in the proof of Theorem 4. Choose a sequence $\{f_\alpha : \alpha < \omega_1\}$ of elements of ω^ω, every uncountable subsequence of which is unbounded in the ordering of eventual dominance. Fix also an $e : [\omega_1]^2 \to \omega$ as in the proof of Theorem 0^c. Then to every $\beta < \omega_1$ we can associate an increasing sequence $F_i(\beta)(i < \omega)$ of finite subsets of $\beta + 1$ as follows:

$$F_i(\beta) = \{\alpha < \beta : \Delta(f_\alpha, f_\beta) \leq i \text{ and } e(\alpha, \beta) < f_\beta(\Delta(f_\alpha, f_\beta))\} \cup \{\beta\}.$$

We need the following version of lemma 2.0 of [8].

LEMMA 0: For every uncountable $A \subseteq \omega_1$ there exist $k < \omega$ and an increasing sequence $\beta_i(i < \omega)$ of elements of A such that for all $i < j < \omega$

(a) $\beta_i \in F_k(\beta_j)$
(b) $\Delta(f_{\beta_i}, f_{\beta_j}) = k$.

Proof: Choose a countable elementary submodel M of some large enough structure of the form H_θ that contains all these objects and let $\delta = M \cap \omega_1$. Then we can find an $e : \delta \to \omega$ and uncountable $B \subseteq A \setminus \delta$ such that $e \subseteq e_\beta$ for all $\beta \in B$. Since $\{f_\beta : \beta \in B\}$ is unbounded in ω^ω, we can find a $k < \omega$ and $t \in \omega^k$ such that for every $n < \omega$ there exist $\beta \in B$ such that $f_\beta \upharpoonright k = t$ and $f_\beta(k) > n$. Let β_0 be any element of $A \cap M$ such that $t \subseteq f_{\beta_0}$, which exists by the elementarity of M. Choose β in B such that $f_\beta(k)$ is bigger than $e(\beta_0)$ and $f_{\beta_0}(k)$. Thus $\beta_0 \in F_k(\beta)$. By the elementarity of M we can choose β_1 in $A \cap M$ such that $t \subseteq f_{\beta_1}, f_{\beta_1}(k) > f_{\beta_0}(k)$ and $\beta_0 \in F_k(\beta_1)$, and so on. Proceeding in this way we get $\{\beta_i : i < \omega\} \subseteq A \cap M$ satisfying (a) and (b) just given. This finishes the proof. $\square$

We shall also need the following concept (see [46]). Let $\vec{U} = \langle U_0, \ldots, U_{n-1} \rangle$ be a finite sequence of (not necessarily distinct) nonempty open subsets of a space X. By cal $\vec{U}$ we denote the largest integer k such that there is $I \subseteq n$ of size k such that $U_i(i \in I)$ has nonempty intersection. For a family $\mathscr{U}$ of nonempty open subsets of X we let

$$\text{int } \mathscr{U} = \inf \{\text{cal } \vec{U}/n : 1 \leq n < \omega, \vec{U} \in \mathscr{U}^n\}$$

be the intersection number of $\mathscr{U}$. A classic result of Kelley [49] says that X supports a strictly positive measure iff the family of nonempty open subsets of X can be split into countably many subfamilies with positive intersection numbers.

For our purpose here, we let ω^ω play the role of the set $\mathbb{R}$ of real numbers. For every f in ω^ω we associate in a uniform way a sequence $\{g_i\}$ converging to f as follows.

Let $n_0 < n_1 < \cdots$ be the list of all n such that $f(2n + 1) \neq 0$. For a given $i < \omega$, let $(f)_i$ be the real determined by the following conditions

(c) $(f)_i \upharpoonright n_k = f \upharpoonright n_k$,
(d) $(f)_i(n_k + j) = f(2^{i+1}(2n_k + 2j + 1))$ for $j < \omega$, where $k = k(i)$ is minimal with the property.
(e) $f(2n_0 + 1) + \cdots + f(2n_k + 1) > i$; if such a k does not exist, let $(f)_i = f$. For $j < \omega$, let
(f) $F_j(f) = \{(f)_i : i < \omega \text{ and } \Delta((f)_i, f) \leq j\} \cup \{f\}$.

CLAIM 1: For every decomposition $H_n(n < \omega)$ of ω^ω there exist $k, m < \omega$ and distinct $f_i(i < \omega)$ in H_m such that $f_i \in F_k(f_j)$ for all $i < j < \omega$.

Proof: We attempt to build a member of f of ω^ω as follows. For $i < \omega$ let

$$N_i = \{2^i(2j + 1) : j < \omega\}.$$

We define $f \upharpoonright N_0$ and $f \upharpoonright N_2$ by finite initial approximations, while for $i > 1, f$ on most of the N_i may be defined in a single step of the recursion. Suppose we have an integer $k = 2(2m + 1)$ in N_1 such that, in particular, we have defined $f \upharpoonright k$, but $f(k)$ is undefined. Let D_k be the set of all integers n such that $f(n)$ is defined. If there is $h \in H_m$ extending $f \upharpoonright D_k$ we take it and call it h_0. Take $p_1 \neq h_0(k)$ and ask if there is $h \in H_m$ extending $(f \upharpoonright D_k) \cup \{\langle k, p_1 \rangle\}$ such that $h_0 \in F_k(h)$. If there is such h we take it and call it h_1. Now, taking $p_2 \neq p_1, h_0(k)$, we ask for $h \in H_m$ such that $h_0, h_1 \in F_k(h), \ldots$ etc. If this process never stops, we are done. Otherwise we have $h_0, \ldots, h_n \in H_m$ extending $f \upharpoonright D_k$ such that $h_i(k) \neq h_j(k)$ for $i \neq j \leq n$ and such that for some $p \notin \{h_i(k) : i \leq n\}$ there is no h in H_m extending $(f \upharpoonright D_k) \cup \{\langle k, p \rangle\}$ such that $h_i \in F_k(h)$ for all $i \leq n$. In this case let $f(2k + 1) = n + 1, f(k) = p$ and for

$$f(1) + f(3) + \cdots + f(2k - 1) \leq i < f(1) + f(3) + \cdots + f(2k + 1)$$

and all $j < \omega$, let

$$f(2^{i+1}(2k + 2j + l)) = h_l(k + j),$$

where

$$l = i - f(1) - f(3) - \cdots - f(2k - 1).$$

If there is no $h \in H_m$ extending $f \upharpoonright D_k$ at all, we let $f(2k + 1) = f(k) = 0$. If k is not in N_1 but still undefined, let $f(k) = 0$.

Suppose at no stage we were able to satisfy the conclusion of Claim 1 and look at the constructed function f. Then it follows rather easily from the construction that f cannot be a member of any of the H_m. This completes the proof. $\square$

CLAIM 2 ($\mathfrak{b} = \omega_1$): There is a sequence $f_\alpha(\alpha < \omega_1)$ of elements of ω^ω every uncountable subsequence of which is unbounded such that for every $\alpha < \beta$, if $e(\alpha, \beta) < f_\beta(\Delta(f_\alpha, f_\beta))$, then there is $i < \omega$ such that $f_\alpha = (f_\beta)_i$.

Proof: Choose an $<^*$-increasing unbounded sequence $g_\alpha(\alpha < \omega_1)$ in ω^ω such that g_α is an increasing function for every α. The sequence $f_\alpha(\alpha < \omega_1)$ is constructed recursively on α as follows. Suppose, we have $f_\alpha(\alpha < \beta)$. We define f_β by partial approximations similarly as before. Only $f_\beta \upharpoonright N_0$ is now defined by finite initial approximations, while for $i > 0, f_\beta$ on most of the N_i may be defined in a single step of the recursion. Let $f_\beta(0) = 0, f_\beta(1) = 1$, and

$$f_\beta(2(2j + 1)) = g_\beta(2(2j + 1)) \quad \text{for} \quad j < \omega.$$

This will provide our sequence $f_\alpha(\alpha < \omega_1)$ with the property that all of its uncountable subsequences are unbounded in ω^ω. Supposing we are at some stage k. If $f_\beta(k)$ is not defined let it be equal to 0. Suppose $f_\beta(k)$ has already been determined at previous stages. Let $\alpha_0, \ldots, \alpha_n$ list all $\alpha < \beta$ such that

(g) $f_\alpha \upharpoonright k = f_\beta \upharpoonright k, f_\alpha \upharpoonright k \neq f_\beta \upharpoonright k$, and $e(\alpha, \beta) < f_\beta(k)$.

In this case we let $f_\beta(2k + 1) = n + 1$, and for

$$f_\beta(1) + f_\beta(3) + \cdots + f_\beta(2k - 1) \leq i < f_\beta(1) + f_\beta(3) + \cdots + f_\beta(2k + 1)$$

and all $j < \omega$, let

$$f_\beta(2^{i+1}(2k + 2j + 1)) = f_{\alpha_i}(k + j),$$

where

$$l = i - f_\beta(1) - f_\beta(3) - \cdots - f_\beta(2k - 1).$$

If there is no α satisfying (g), set $f_\beta(2k + 1) = 0$. It is clear that a so-constructed sequence $f_\alpha(\alpha < \omega_1)$ is as required. □

For x in $\mathbb{R} = \omega^\omega$, let $\vec{F}(x)$ denote the sequence defined in (f) just given and let $U(\vec{F}(x))$ be the neighborhood of 0 of $A(\mathbb{R}^*)$ determined by $\vec{F}(x)$ as previously. Let

$$V(x) = x + U(\vec{F}(x)).$$

CLAIM 3: The family $V(x)$ $(x \in \mathbb{R})$ of open subsets of $A(\mathbb{R}^*)$ is not the union of countably many subfamilies of positive intersection numbers.

Proof: Suppose $H_n(n < \omega)$ is a decomposition of $\mathbb{R}$ such that

$$\text{int } \{V(x) : x \in H_n\} > 0$$

for all n. Choose $k, m < \omega$ and distinct $x_i(i < \omega)$ in H_m such that $x_i \in F_k(x_j)$ for all $i < j < \omega$ as in the Claim 1. It is clear that we shall get a contradiction if we show that

$$\bigcap_{i \in I} V(x_i) = \emptyset$$

for every $I \subseteq \omega$ of size $> 2k + 1$. So, fix such finite I, let j be its maximum, and suppose the intersection is nonempty. Fix an element w in the intersection and write it in the form of a typical element of $V(x_j)$, that is, as

$$w = x_j + \sum_{l=0}^{m} (u_l - v_l),$$

where $u_l, v_l \in \mathbb{R}^*$ and $u_l, v_l \notin F_l(x_j)$ for all $l \leq m$. Then there is $i \in I \cap j$ such that $x_i \neq u_l$ and $x_i \neq v_l$ for all $l < k$. Since $x_i \in F_l(x_j)$ for all $l \geq k$, it follows that the generator x_i does not appear in w. This contradicts the obvious fact that every element of $V(x_i)$ contains x_i with coefficient 1. □

To complete the proof of the second part of Theorem 4, assuming $\mathfrak{b} = \omega_1$, fix a sequence $x_\alpha(\alpha < \omega_1)$ of elements of ω^ω satisfying the conclusion of Claim 2, that is, every subsequence of $x_\alpha(\alpha < \omega_1)$ is unbounded in ω^ω while $e(\alpha, \beta) < x_\beta(\Delta(x_\alpha, x_\beta))$ implies that $x_\alpha = (x_\beta)_i$ for some i. It should now be clear that Lemma 0 and the proof of Claim 3 gives us the following fact, which finally finishes the proof of Theorem 4.

CLAIM 4: No uncountable subfamily of $V(x_\alpha)(\alpha < \omega_1)$ has the finite intersection property.

REMARK: The similarity of the proofs of Claims 1 and 2 is not an accident. In fact, Claim 1 can formally be deduced from Claim 2 using Lemma 0 and going to a forcing extension that doesn't add reals but forces $\mathfrak{b} = \omega_1$.

6. A PRODUCT OF TWO COMPACTLY GENERATED GROUPS

There are many examples of spaces showing that tightness is not a productive cardinal function ([19]). Similarly, there are many examples showing that the Lindelöf degree is not productive ([19]). On the other hand, there are not many results asserting that the two functions are productive in certain classes of spaces. In fact, the only such class of spaces we know of is the class of all compact spaces ([18]). So, it is natural to ask whether a similar result can be proved in a slightly wider class of spaces such as σ-compact or Lindelöf spaces (see [20, problems 18 and 19]). In this section we shall give a negative answer to this question (announced in [9]) even in the class of topological groups.

THEOREM 5: There exist two σ-compact (or compactly generated) Frechet–Urysohn topological groups whose product is not countably tight.

Proof: Let T be a rooted binary special Aronszajn tree with the topology generated by sets of the form

$$B_s(t) = \{u \in T : s \le u \quad \text{and} \quad t \not< u\},$$

where $s < t$ if t is a limit node and $s = t$ if t is a root or a successor mode of T. This is the *path topology* on T that has been studied in some detail in [21] from where we take some of the arguments below. For a set S of countable limit ordinals, let $T(S)$ be the set of all t in T whose height is an element of $\{0\} \cup S \cup \{\alpha + 1 : \alpha < \omega_1\}$. We consider $T(S)$ as a space with the topology induced from T. Let T^0 be the set of all elements of T of successor height including the root of T. Let $G(S)$ be the subalgebra of the clopen algebra of $T(S)$ generated by the sets of the form

$$T'(S) = \{u \in T(S) : t \le u\} \qquad (t \in T^0).$$

We consider $G(S)$ as a topological group with symmetrical difference as the group operation and with the topology generated by the sets of the form

$$\{X \in G(S) : E \subset X \text{ and } F \cap X = \varnothing\} \qquad (E, F \in [T(S)]^{<\omega}).$$

We need the following representations of elements of the Boolean algebra $G(S)$, which can be found in [22, sec. 16]: An X in $G(S)\setminus\{\varnothing\}$ is uniquely determined by a sequence of the form $B_t(t \in A)$ such that:

(a) A is a finite subset of T^0 and for $t \in A$, B_t is a finite antichain of successors of t in T^0,

(b) For every $s \neq t$ in A, either $s \| t$ or there is u in B_s such that $u \leq t$, or conversely there is u in B_t such that $u \leq s$,

(c) X is the set of all $v \in T(S)$ such that for some $t \in A$, $t \leq v$, and if t is the maximal such element, then $u \nleq v$ for all $u \in B_t$.

To show that $G(S)$ is countably tight if S is stationary it suffices to prove that for every subset H of $G(S)$ accumulating to the element $T(S)$ of $G(S)$ there is a countable subset of H that also accumulates to $T(S)$. To see this, choose a countable elementary submodel M of some large enough structure of the form $\langle H_\theta, \in \rangle$ containing all these objects such that $\delta = M \cap \omega_1$ is an element of S. We shall show that for every finite $F \subseteq T(S)$ there is an X in $H \cap M$ such that $F \subseteq X$. This will finish the proof. Let

$$F_0 = F \cap T(S) \restriction \delta,$$

$$F_1 = \{s \in T(S)_\delta : s \leq t \text{ for some } t \text{ in } F\}.$$

Increasing F, we may assume that $F_1 \subseteq F$. Since H accumulates to $T(S)$ there is an X_0 of H such that $F \subseteq X_0$. Let X_0 be determined by some sequence B_t^0 $(t \in A^0)$ in the sense of (a), (b), and (c). Let

$$T_{X_0} = A^0 \cup \bigcup_{t \in A^0} B_t^0$$

and let $W = T_{X_0} \cap M$. By the elementarity of M, for each $\gamma > 0$ there is an X_γ in H such that if $B_t^\gamma(t \in A^\gamma)$ is the sequence that generates X_γ, then

$$W = T_{X_\gamma} \cap (T(S) \restriction \gamma),$$

and $B_t^\gamma(t \in A^\gamma)$ and $B_t^0(t \in A^0)$ are isomorphic over W, that is, the subtrees T_{X_0} and T_{X_γ} are isomorphic via an isomorphism that fixes W and sends A^0 to A^γ and B_t^0 to B_t^γ. We may assume that the sequences of X_γ and B_t^γ $(t \in A^\gamma)$'s are elements of M. For $\gamma < \omega_1$, let $W_\gamma = T_{X_\gamma} \setminus W$. A standard lemma about Aronszajn trees (see, e.g., [42, lemma 5.9]) applied to the sequence $W_\gamma(\gamma < \omega_1)$ shows that there exists an arbitrarily large level β of $T(S)$ such that for every finite $D \subseteq T(S)_\beta$ there is some $\gamma > \beta$ such that the projection of W_γ to $T(S)_\beta$ is disjoint from D. By the elementarity of M there is such a β in M above the height of elements of W. Let D be the projection of F_1 to $T(S)_\beta$. Then there is a $\gamma > \beta$ in M such that the projection of W_γ to $T(S)_\beta$ is disjoint from D. Then it is easily checked [using (a), (b), and (c)] that X_γ is an element of $M \cap H$ containing F.

We first prove that $G(S)$ is Lindelöf since the argument will be needed in the proof of σ-compactness. It suffices to show that every sequence $X_\xi(\xi < \omega_1)$ in $G(S)$ has a complete accumulation point. Let X_ξ be determined by $B_t^\xi(t \in A^\xi)$ in the preceding sense. Refining the sequence we may assume that the A^ξ form a Δ-system with root A, and moreover that for every t in A, $B_t^\xi(\xi < \omega_1)$ form a Δ-system with root B_t. Let X be the element of $G(S)$ determined by $B_t(t \in A)$. Then it is easily checked that X is a complete accumulation point of (X_ξ) in $G(S)$.

To prove the stronger property that $G(S)$ is σ-compact we use the fact that T is a special tree, that is, that there exists an $a : T \to \omega$ such that $a^{-1}(n)$ is an antichain of T for all $n < \omega$. Let X be an element of $G(S)$ determined by the sequence $B_t(t \in A)$ in the preceding sense. By T_X we denote the finite subtree

$$A \cup \bigcup_{t \in A} B_t$$

of $T(S)$. For $m < \omega$, set

$$G_m(S) = \{X \in G(S) : |T_X| \le m \text{ and } a(t) \le m \text{ for all } t \in T_X\}.$$

Since $G_m(S)$'s cover $G(S)$ it suffices to show that each $G_m(S)$ is compact. Note that in the preceding proof of Lindelöfness of $G(S)$, if each X_ξ is taken from $G_m(S)$, the produced complete accumulation point X has the property $T_X \subseteq T_{X_\xi}$ for uncountably many ξ. It follows that X is also an element of $G_m(S)$. This shows that each $G_m(S)$ is Lindelöf so we are left with the proof that each $G_m(S)$ is countably compact. So let X_i $(i < \omega)$ be a sequence of elements of $G_m(S)$. Let X_i be generated by $B_t^i(t \in A^i)$ in the preceding sense. Since A^i have sizes bounded by m, a standard application of Ramsey's theorem shows that going to a subsequence we can assume that A^i form a Δ-system with root A (and that they have all some fixed size $\le m$). Similarly, we can assume that for every $t \in A$, the sequence B_t^i $(i < \omega)$ forms a Δ-system with root B_t (and have all members of some fixed size $\le m$). We claim that $B_t(t \in A)$ satisfies the conditions (a) and (b) given earlier. Since (a) is trivially satisfied, we assume (b) fails and work for a contradiction. So suppose there is $s < t$ in A such that $u \not\le t$ for all $u \in B_s$. Since (b) is true for every $B_t^i(t \in A^i)$, this means that for every i there is $u_i \in B_t^i\backslash B$ such that $u_i \le t$. Then $u_i(i < \omega)$ forms an infinite chain of elements of $T(S)$ such that $a(u_i) \le m$ for all i. This clearly contradicts the fact that a is a specializing map. It follows that $B_t(t \in A)$ is a legitimate sequence for generating an element X of $G_m(S)$ in the sense of (c) given earlier. We prove that X is the limit of $X_i(i < \omega)$. For this it suffices to show that every $v \in X$ is an element of X_i for all but finitely many i's and that every $w \notin X$ is not an element of X_i for all but finitely many i's. For suppose there is a $v \in X$ and infinite $I \subseteq \omega$ such that $v \notin X_i$ for all $i \in I$. Let t be the maximal element of A below v. The preceding argument applied to the Δ-system $A_i(i \in I)$ and the specializing map a shows that we can assume (by removing finitely many elements of I) that for every $i \in I$, t is also the maximal element of A^i below v. So by (c) applied to the facts $v \in X$ and $v \notin X_i(i \in I)$, we get that for every $i \in I$ there is $u_i \in B_t^i\backslash B_t$ such that $u_i \le v$. It follows that $u_i(i \in I)$ forms an infinite chain. This and the fact that $a(u_i) \le m$ for all $i \in I$ gives us a contradiction. Suppose now that there exist a $w \notin X$ and infinite $J \subseteq \omega$ such that $w \in X_i$ for all $i \in J$. For $i \in J$ let t_i be the maximal element of A_i such that $t_i \le w$. It follows that t_i for $i \in I$ are comparable, and since $a(t_i) \le m$ for all $i \in I$, there must be infinite $K \subseteq J$ such that for some t, $t_i = t$ for all $i \in K$. It follows that $t \in A$ and, moreover, that t is the maximal element of A such that $t \le w$. The condition (c) applied to $w \notin X$ implies that there must be $u \in B_t$ such that $u \le w$. Applying (c) to the fact $w \in X_i$ for some $i \in K$, we know that $u \not\le w$ for all $u \in B_{t_i}^i$. This clearly contradicts the fact that B_t is a subset of B_t^i for all i and finishes the proof that $G(S)$ is σ-compact.

We have shown that $G(S)$ is countably tight if S is stationary. To show the stronger property that $G(S)$ is Frechet–Urysohn, it suffices to prove that if a countable set $H \subseteq G(S)$ accumulates to the element $T(S)$ of $G(S)$, then there is a sequence $X_i(i < \omega)$ of elements of H such that

$$\lim_i X_i = T(S).$$

Let δ be a large enough (limit) ordinal from S such that if an X of H is determined by $B_t(t \in A)$ in the sense of (a), (b), and (c), then the heights of all elements of

$$A \cup \bigcup_{t \in A} B_t$$

are $< \delta$. For a finite set $F \subseteq T(S)$ let $\pi_\delta(F)$ be the set of all elements u of $T(S) \restriction (\delta + 1)$ such that either $u \in F$ or else $u \notin F$, height $(u) = \delta$ and there exists a t in F such that $u < t$. Then a simple analysis of (a), (b), and (c) shows that for every finite $F \subset T(S)$ and $X \in H$, $F \in X$ iff $\pi_\delta(F) \in X$. Let $F_i(i < \omega)$ be an increasing sequence of finite subsets of $T(S) \restriction (\delta + 1)$, which covers this set. Then the assumption that $T(S)$ is in the closure of H implies that for every i there exists an $X_i \in H$ such that $F_i \subseteq X_i$. Now using the fact about the retraction π_δ it follows easily that $X_i(i < \omega)$ converges to $T(S)$.

Note that the argument for the Frechet–Urysohn property proves that $G(S)$ is a *monolithic space,* that is, that the closure of every countable subset of $G(S)$ is second countable. To see this, for δ as above, let

$$G^\delta(S) = \{X \in G(S) : T_X \subseteq T(S) \restriction (\delta + 1)\}.$$

Then the argument using the retraction π_δ shows that $G^\delta(S)$ is a closed second-countable subgroup of $G(S)$.

From now on we fix two disjoint stationary sets S and S' of countable limit ordinals. Recall that T^0 is the set of all elements of T of successor height, including the root of T. For t in T let t^0 and t^1 be the immediate successors of t in T. Now to every pair $p = (E_p, F_p)$ of finite subsets of $T(S)$ and $T(S')$, respectively, we associate an element (X_p, Y_p) of $G(S) \times G(S')$ as follows. For t in $(E_p \cup F_p) \backslash T^0$, let $\hat{t}$ be the minimal element $u < t$ in T^0 having the property that if an element of $E_p \cup F_p$ is a successor of u, then it is also a successor of t, and such that the height of u is bigger than the heights of all elements of $E_p \cup F_p$ lying in levels lower than the level of t. Let

$$\hat{E}_p = (E_p \cap T^0) \cup \{\hat{t} : t \in E_p \backslash T^0\}$$

$$\hat{F}_p = (F_p \cap T^0) \cup \{\hat{t} : t \in F_p \backslash T^0\}.$$

Let X_p be the element of $G(S)$ determined by the sequence $b_t(t \in \hat{E}_p)$, where:

(d) $B_t = \{t^0, t^1\}$ for $t \in E_p \cap T^0$
(e) $B_{\hat{t}} = \{t^0, t^1\}$ for $t \in E_p \backslash T^0$.

Similarly, one defines Y_p in $G(S')$ interchanging the roles of E_p and F_p in the preceding definition. The following two facts follow immediately from the definitions and the fact that S and S' are disjoint:

(f) $E_p \subseteq X_p$ and $F_p \subseteq Y_p$,
(g) If $t \in T^0$ has height above the heights of all elements of $E_p \cup F_p$, then $t \notin X_p$ or $t \notin Y_p$.

Then (X_p, Y_p) $(p \in [T(S)]^{<\omega} \times [T(S')]^{<\omega})$ is a subset of $G(S) \times G(S')$ that accumulates to the point $(T(S), T(S'))$, but for every countable $P \subset [T(S)]^{<\omega} \times [(T(S'))]^{<\omega}$

$$(T(S), T(S')) \notin \overline{\{(X_p, Y_p) : p \in P\}}.$$

This shows that $G(S) \times G(S')$ has uncountable tightness and finishes the proof of Theorem 5 for the σ-compact case. To get a compactly generated example we use the idea of Pestov [35] of embedding σ-compact groups into compactly generated ones. To this end, let T^* be the (disjoint and incomparable) sum of the sequence $T \times \{n\}$ $(n \in \mathbb{Z})$ of copies of the tree T. Clearly, T^* is still a special Aronszajn tree and the previous notions and proofs work equally well for T^* in place of T. Thus, by $G^*(S)$ we denote the corresponding group associated to T^* and a set S of countable limit ordinals. The point of $G^*(S)$ is that for every X in $G^*(S)$ and integer n in $\mathbb{Z}$ we can define the shift $n \cdot X$ obtained moving X by $|n|$ places forward or backward depending on whether n is positive or negative. Let $a : T^0 \to \mathbb{Z}$ be an antichain-decomposition of the tree T and let

$$D = \{T'(S) \times \{a(t)\} : t \in T^0\}.$$

Note that $D \cup \{\varnothing\}$ is a compact subspace of $G^*(S)$ with $\varnothing$ as the only nonisolated point, that is, $D \cup \{\varnothing\}$ is homeomorphic to the one-point compactification of a discrete space of size $\aleph_1$. Unfortunately, $D \cup \{\varnothing\}$ does not generate $G^*(S)$, but if we can somehow embed $G^*(S)$ into a bigger group $G^*_{\mathbb{Z}}(S)$ containing also the group $\mathbb{Z}$ such that the product of an n in $\mathbb{Z}$ and an X in $G^*(S)$ is the shift $n \cdot X$, the set

$$D \cup \{\varnothing\} \cup \{1\}$$

would generate $G^*(S)$ and it would still be homeomorphic to the one-point compactification of the discrete space of size $\aleph_1$. This indeed can be done by letting $G^*_{\mathbb{Z}}(S)$, as topological space, be equal to the product $G^*(S) \times \mathbb{Z}$, while the group operation is defined as follows:

$$(X, m) \cdot (Y, n) = (X \, \Delta \, m \cdot Y, m + n).$$

It is easily checked that this fulfills all our needs when $G^*(S)$ is identified with $G^*(S) \times \{0\}$ and when $\mathbb{Z}$ is identified with $\{\varnothing\} \times \mathbb{Z}$. This finishes the proof of Theorem 5. $\square$

REMARK: Let G be the product of the two groups $G(S)$ and $G(S')$ from the previous result and, in the notation of its proof, let G_m be the product

$$G_m(S) \times G_m(S'),$$

for $m < \omega$. Then G is not a countably tight space, but it can be written as an increasing union of the sequence $G_m(m < \omega)$ of compact Frechet–Urysohn subspaces. (The proof that $G_m(S)$ is Frechet–Urysohn also shows that G_m is Frechet–Urysohn.) It follows that G contains no uncountable free sequences. This answers a question asked in [52; chap III, sec. 2] in connection with the well-known result of Shapirovskii ([54], [55]) and Arhangel'skii [53] that characterizes the tightness of compact spaces in terms of free sequences and its corollary, the result of Rancin [56], which says that if a compact space G can be written as a countable union of compact countably tight subspaces, then it must be countably tight. Our result shows that neither of these two results can be extended to σ-compact spaces or to the wider class of Lindelöf Σ-spaces as asked in [52]. Note also that if we started with $G_{\mathbb{Z}}^{*}(S)$ and $G_{\mathbb{Z}}^{*}(S')$ in place of $G(S)$ and $G(S')$, the resulting space $G_{\mathbb{Z}}^{*}$ will be, moreover, a topological group compactly generated by the one-point compactification of a discrete space of size $\aleph_1$. This is the second place in this paper (besides that of Section 5) where it is shown that groups compactly generated by the one-point compactification of a discrete space of size $\aleph_1$ can be quite pathological indeed.

The spaces $T(S)$ are worth investigating even from the point of view of the C_p-theory.

THEOREM 6: If S is a stationary subset of ω_1, then $C_p(T(S))$ is a Frechet–Urysohn monolithic space.

Since $T(S)$ is clearly an uncountable first-countable space, this answers question 18 of [37]. To see this one also needs to know the result of Gerlits [38], which says that in this context the Frechet–Urysohn property is equivalent to the property of being a k-space.

Proof of Theorem 6: One first observes that $C_p(T(S))$ is countably tight, which follows (see [51] and [39]) from the fact proved in [21] that every finite power of $T(S)$ is Lindelöf. So it remains to prove only that the closure of every countable subset of $C_p(T(S))$ is second countable. This is an easy consequence and, in fact, equivalent (see [51, chap. II, sec. 7]) to the following property of the space $T(S)$:

CLAIM: Every continuous function f from $T(S)$ into a second-countable space M has countable range.

Proof: Let $\langle U_i, V_i \rangle$ $(i < \omega)$ be a list of all pairs $\langle U, V \rangle$ of elements of a fixed countable basis of M such that $\overline{U} \cap \overline{V} = \varnothing$. For $i < \omega$, set

$$T_i = \{t \in T(S) : \text{there exist } u, v \geq t \quad \text{such that} \quad u \in U_i \quad \text{and} \quad v \in V_i\}.$$

Then no T_i can have any accumulation points in $T(S)$, so Lindelöfness of $T(S)$ gives us that each T_i is countable. Let δ be the index of a level of $T(S)$ that has no points in common with any of the T_i. Then

$$f''T(S) = f''T(S) \upharpoonright (\delta + 1),$$

and this finishes the proof. $\square$

Note that if S and S' are two disjoint stationary subsets of ω_1, then the product

$$C_p(T(S)) \times C_p(T(S'))$$

is not countably tight. This follows from the fact (see [20] and [39]) that countable tightness of $C_p(X)$ implies the Lindelöfness of X in all finite powers and the fact (see [21]) that the product $T(S) \times T(S')$ is not Lindelöf. Thus, we have proved the following, which solves problem II.3.10 of [51].

THEOREM 7: There exist two Frechet–Urysohn function spaces $C_p(X)$ and $C_p(Y)$ whose product $C_p(X) \times C_p(Y)$ is not countably tight.

REMARKS: More than ten years ago we constructed a pair of space $C_p(X)$ and $C_p(Y)$ with similar properties using Martin's axiom. Some of the ideas of that proof were presented in [36]. This result in the literature is usually quoted with the assumption of the diamond principle rather than MA because in the only place where the result was indicated ([36]) one finds our diamond-construction of a set of reals with properties much stronger than needed for production of such a pair of function spaces. Finally, we note that similar pairs of function spaces have been recently announced by K. Alster and R. Pol. We finish this section with the remark that $G(S)$ can also be viewed as a (dense) subgroup of $C_p(T(S))$. This gives us another way for proving that $G(S)$ is a countably tight group, that is, by simply deducing it from the fact that every finite power of $T(S)$ is Lindelöf.

7. COHERENT SEQUENCES

Known instances of nonproductiveness of tightness usually involve space X with a single nonisolated point $*$. So it is natural to ask whether the question of [20] can be answered negatively also in the class of spaces with a unique nonisolated point. In the next section we show that this cannot be done. While examples of spaces obtained under additional set-theoretical assumptions frequently turn out to be not very informative about a problem in hand, we now review some of the "dishonest" (see [23]) examples relevant to the problem under consideration.

A sequence $A_\alpha (\alpha < \omega_1)$ of subsets of ω_1 is *coherent* if $A_\alpha =^* A_\beta \cap \alpha$ (mod. finite) for all $\alpha < \beta$. A coherent sequence $\{A_\alpha\}$ is nontrivial if there is no $A \subset \omega_1$ such that $A_\alpha =^* A \cap \alpha$ for all α. The first nontrivial coherent sequence constructed without additional set-theoretical assumptions was given in [24] (see also [25]). Warren's example $\{W_\alpha\}$ has the additional property that ω_1 can be covered by countably many sets $B_n (n < \omega)$ such that $W_\alpha \cap B_n$ is finite for all α and n. In particular, there are no uncountable $\{W_\alpha\}$-*homogeneous* sets, that is, sets $C \subseteq \omega_1$ such that $C \cap \alpha \subseteq^* W_\alpha$ for all α. Today we know (see [8, pp. 81–82]) that for every coherent sequence $\{A_\alpha\}$ with no uncountable homogeneous set must have the property of Warren's sequence. A coherent sequence $\{A_\alpha\}$ naturally determines a space $X\{A_\alpha\}$ on $\omega_1 \cup \{*\}$ where the points of ω_1 are isolated, while for $B \subseteq \omega_1$, $* \notin \overline{B}$ iff B is $\{A_\alpha\}$-homogeneous. Then $X\{A_\alpha\}$ is a Frechet–Urysohn space that is Lindelöf iff $\{A_\alpha\}$ does not have uncountable homogeneous sets. Let $A_\alpha^c = \alpha \backslash A_\alpha$. Then $\{A_\alpha^c\}$ is also a coherent sequence. Note also that if $\{A_\alpha\}$ is nontrivial, then the product $X\{A_\alpha\} \times X\{A_\alpha^c\}$ has uncountable tightness. To have all this relevant to the problem under consideration we need coherent sequence $\{A_\alpha\}$ such that neither $\{A_\alpha\}$ nor $\{A_\alpha^c\}$ has uncountable homogeneous sets. The result of [8, pp. 81–82] just quoted shows that such a sequence cannot be constructed without additional set-theoretical assumptions, so we offer the following.

THEOREM 8 (RVM): There is a coherent sequence $\{A_\alpha\}$ such that both $X\{A_\alpha\}$ and $X\{A_\alpha^c\}$ are Lindelöf (and Frechet–Urysohn), but their product is not countably tight.

Proof: The proof will again require the stepping-up tool of Section 4. Thus, we start with an $e : [\omega_1]^2 \to \omega$ such that e_α are 1–1 and such that $e_\beta \upharpoonright \alpha =^* e_\alpha$ for all $\alpha < \beta$. For a real $r \in \{0, 1\}^\omega$, define $\{A_{r\alpha}\}$ by

$$\xi \in A_{r\alpha} \quad \text{iff} \quad \xi < \alpha \quad \text{and} \quad r(e(\xi, \alpha)) = 1.$$

Clearly $\{A_{r\alpha}\}$ is coherent for all r, but we need an r such that neither $\{A_{r\alpha}\}$ nor $\{A_{r\alpha}^c\}$ has uncountable homogeneous sets. By RVM it suffices to show that this is true when r is a random real. This is contained in the argument from Claim 2 of Section 4. □

REMARKS: Note that $X\{W_\alpha\}$, where $\{W_\alpha\}$ is the coherent sequence of Warren, is an interesting space. For example, it can be shown that the corresponding function space $C_p(X\{W_\alpha\})$ is both Lindelöf and countably tight.

The first examples of countably tight Lindelöf spaces X and Y with single nonisolated points whose product is not countably tight were constructed in [31] using a consequence of Jensen's combinatorial principle diamond. Other examples are given in [26] in one Cohen or random real extensions. The latter examples are not of the form $X\{A_\alpha\}$ where $\{A_\alpha\}$ is a coherent sequence on ω_1.

The spaces $X\{A_\alpha\}$ and $X\{A_\alpha^c\}$ of Theorem 8 also have the property that the corresponding function spaces $C_p(X\{A_\alpha\})$ and $C_p(X\{A_\alpha^c\})$ are Lindelöf and countably tight (compare this with the results of [26] and [30]).

8. TIGHTNESS OF SPACES WITH STRONG CONDENSATION PROPERTIES

Results giving sufficient condition for productiveness of countable tightness are potentially very useful. Even more useful can be results about the productiveness of the closely related property of not having an uncountable free sequence, that is, a sequence x_ξ ($\xi < \theta$) with the property $\overline{\{x_\xi : \xi \leq \alpha\}} \cap \overline{\{x_\xi : \alpha < \xi < \theta\}} = \varnothing$ for all $\alpha < \theta$ (see [8, sec. 7]). In this section we give a few results in that direction using the PFA. The examples of the previous section indicate that this might be a reasonable assumption for treating problems of this kind. Our use of this axiom is essentially contained in the following lemma, which is just a reformulation of the statement (*) of [28], which in turn is a slight strengthening of the partition property (P) of [27] (see also [8, secs. 8 and 9]) very closely related to the problem S. In fact the statement (P) is all that we need below, but the stronger version is used in order to indicate the possibilities for further study.

LEMMA 1 (PFA): If $\mathscr{I}$ is a nonprincipal $\aleph_1$-generated ideal on a set S, then either:

(a) There is uncountable $A \subseteq S$ such that $A \cap B$ is finite for every B in $\mathscr{I}$, or
(b) S can be decomposed into countably many sets S_n ($n < \omega$) such that $[S_n]^\omega \subseteq \mathscr{I}$ for all n.

[In the partition property (P) the second alternative is weakened to the statement that there is an uncountable $A \subseteq S$ such that $[A]^\omega \subseteq \mathscr{I}$. The proof of Lemma 1 (see [8, sec. 7]) shows that this statement can be strengthened in various directions by adding to it "side conditions" related to certain structures that can be put on S. In this paper, however, we choose the simplicity over the strength for we shall have no use of it. We should, however, point out at least one such strengthening just to illustrate what one can do with that proof. For an ideal $\mathscr{I}$ on S (which is always assumed to contain all finite subsets of S), let $\hat{\mathscr{I}}$ denote the set of all $A \subseteq S$ such that $[A]^\omega \subseteq \mathscr{I}$. Clearly, $\hat{\mathscr{I}}$ is also an ideal on S. If $\mathscr{F}$ is a family of subsets of S, then by $(\mathscr{F})$, respectively, $(\mathscr{F})_\sigma$, we denote the ideal, respectively, σ-ideal, on S generated by $\mathscr{F}$. So, here is a version of Lemma 1 that should be sufficient in many applications of this lemma that require "side conditions:"

LEMMA 1* (PFA): Let $\mathscr{I}$ be an ideal on S and let $\mathscr{F}$ be a family of size $\aleph_1$ of subsets of S such that $(\hat{\mathscr{I}} \cup \mathscr{F})_\sigma$ is a proper σ-ideal on S. Then for every $f : S \times \omega_1 \to \mathscr{I}$ there is an uncountable set $A \subseteq S$ such that:

(a) $A \cap B$ is countable for every B in $\mathscr{F}$,
(b) $A \cap f(x, \alpha)$ is finite for every (x, α) in $A \times \omega_1$.

Thus, Lemma 1 is a special case of Lemma 1* by letting $\mathscr{F} = \varnothing$ and $f(x, \alpha) = B_\alpha$ for all x in S and $\alpha < \omega_1$, where B_α ($\alpha < \omega_1$) is a fixed family of subsets of S which generates the ideal $\mathscr{I}$.] To see what this Lemma is saying about a countably tight space X, suppose $x \in X$ and $S \subseteq X\backslash\{x\}$ are such that $x \in \bar{S}$. Let $\mathscr{I}$ be the ideal of subsets of S that don't accumulate to x, and suppose $\mathscr{I}$ is $\aleph_1$-generated or, equivalently, that x has character $\aleph_1$ relative to $S \cup \{x\}$. The alternative (b) of Lemma 1 reduces to the statement that x is G_δ in $S \cup \{x\}$. (It is interesting that almost every known instance of nonproductiveness of tightness involve X, x, and S with x G_δ relative to S.) The alternative (a) means that there is uncountable $A \subset S$ such that (if X is T_2), $A \cup \{x\}$ is a compact space with single nonisolated point x.

We shall say that a (completely regular) space X is *condensed* if there is a single point $*_X$ in X such that every uncountable subset of X accumulates to $*_X$. (We shall frequently suppress the index X and write $*$ in place of $*_X$ even when considering several spaces from this class. This should not cause any confusion since it will be clear from the context which condensation point is in the focus.) Note that X^ω is Lindelöf for every condensed space X. The productiveness of tightness in this class of spaces is a more delicate matter as we shall now see (see also Section 7).

THEOREM 9 (PFA): If X is a countably tight condensed space, then its countable power X^ω is also countably tight.

Proof: It suffices to show that every finite power X^n is countably tight. This is proved by induction on n. The case $n = 1$ is our hypothesis. Suppose $S \subset X^n$ accumulates to some $x = (x_0, \dots, x_{n-1})$. It is easily seen that we can assume that $x_i = *$ for all i since all other cases easily reduce to the inductive hypothesis. Since X is condensed around $*$ we can find a sequence $x^\xi = (x_0^\xi, \dots, x_{n-1}^\xi)$ ($\xi < \omega_1$) of elements of S accumulating to $x = (*)^n$. If no countable subset of $x^\xi(\xi < \omega_1)$ accumulates to $(*)^n$, for each $\gamma < \omega_1$ there exist closed sets $K_0^\gamma, \dots, K_{n-1}^\gamma$, which do not contain $*$ such that:

(c) For every $\alpha < \gamma$ there is $i < n$ such that $x_i^\alpha \in K_i^\gamma$.

Let $S_0 = \{x_0^\xi : \xi < \omega_1\}$ and let $\mathscr{I}$ be the ideal on S_0 generated by $K_0^\gamma \cap S_0$ ($\gamma < \omega_1$) and the finite sets. The alternative (b) of Lemma 1 cannot happen since every uncountable $B \subseteq S_0$ accumulates to $*$ and since $[B]^\omega \subseteq \mathscr{I}$ means that no countable subset of B accumulates to $*$ (contradicting the countable tightness of X). The alternative (a) means that there is uncountable $A_0 \subseteq S_0$ such that $A_0 \cap K_0^\gamma$ is finite for all $\gamma < \omega_1$. From this we get uncountable sets $A \subseteq A_0$ and $B \subseteq \omega_1$ such that $A \cap K_0^\gamma = \varnothing$ for every γ in B. Reindexing the subsequences $x^\xi(\xi \in A)$ and $K_0^\gamma, \dots, K_{n-1}^\gamma(\gamma \in B)$ we see that (c) holds for $n - 1$ in place of n. This contradicts our inductive hypothesis and finishes the proof. $\square$

It should be clear that the preceding proof also gives the following slightly more general result.

THEOREM 10 (PFA): The product of countably many countably tight condensed spaces is countably tight.

It follows that no countably tight condensed space X has an uncountable free sequence in any of its finite powers. Since it is clearly locally countable in all points except $*$, theorem 7.13 of [8] shows that if X has size at most $\aleph_1$ it must be σ-discrete. If moreover $\chi(*, X) = \aleph_1$ the σ-discrete decomposition

$$X \setminus \{*\} = \bigcup_{n < \omega} X_n$$

given by [8, theorem 7.13] can be taken to have the property that $X_n \cup \{*\}$ is compact for every $n < \omega$. Thus, we have the following fact that tells us something about the structure of small condensed spaces.

THEOREM 11 (PFA): Every condensed space X of size at most $\aleph_1$ is σ-discrete. If, moreover, the character of $*$ is at most $\aleph_1$, X is σ-compact.

REMARK: Choose $e : [\omega_1]^2 \to \omega$ such that e_α are 1–1 and such that $e_\alpha =* e_\beta \upharpoonright \alpha$ for $\alpha < \beta$. Let $X = (\omega_1 \times \omega) \cup \{*\}$ with the topology determined by letting every point of $\omega_1 \times \omega$ be isolated, while for $F \subseteq \omega_1 \times \omega$, the point $*$ is not in the closure of F iff there is a $\beta < \omega_1$ such that for all but finitely many (α, n) in F, $\alpha < \beta$ and $n \le e(\alpha, \beta)$. Then X is a σ-compact [each $(\omega_1 \times \{n\}) \cup \{*\}$ is compact] Frechet–Urysohn space of character $\aleph_1$ whose product with the Frechet–Urysohn fan with $\aleph_1$ edges, $F_{\omega_1\omega}$, is not countably tight. (Consider the subset $\{\langle\langle\alpha, e(\alpha, \beta)\rangle, \langle\beta, e(\alpha, \beta)\rangle\rangle : \alpha < \beta < \omega_1\}$ of the product $X \times F_{\omega_1\omega}$.) The space $F_{\omega_1\omega}$ is not Lindelöf or else this would contradict Theorem 10. So this example gives us some ideas about the sharpness of Theorems 9, 10, and 11.

The property of not having an uncountable free sequence is very closely related to countable tightness and Lindelöfness. The experience shows that studying the productiveness of this property might be, in fact, more important than studying the productiveness of tightness and the Lindelöf degree. To explain this point, let us reformulate theorem 7.10 of [8], which shows the effect of MA on the S and L problems, a result obtained after an unusually long search. (See [8, pp. 64–65] for an explanation of this point.)

THEOREM 12 ($\mathrm{MA}_{\aleph_1}$): A subspace Y of a regular space X is either both separable and Lindelöf or it contains an uncountable discrete subspace, or some finite power of X contains an uncountable free sequence.

This result might also be worth formulating in the following form.

THEOREM 13 ($\mathrm{MA}_{\aleph_1}$): Let X be a regular space that contains no uncountable free sequence in any of its finite powers. Then a subspace of X is hereditarily separable iff it is hereditarily Lindelöf.

This shows that the basic implications S and L are true for any space Y that can be embedded into a space X that contains no uncountable free sequence in any of its finite powers. More importantly this shows that pulling down the free sequence from some finite power of X into X itself is equivalent to proving S and L in general. Clearly not every uncountable free sequence of a product can be pulled down to one of its factors (consider the second diagonal of the arrow space), so one needs to give a careful analysis of the free sequence from the third alternative of Theorem 12. It turns out that the free sequence obtained under the assumption that Y is not Lindelöf but that it contains no uncountable discrete subspace is quite different from the free sequence obtained under the assumption that Y is not separable and contains no uncountable discrete subspace. We have been able (see [8]) to pull down the first kind of free sequence and therefore to prove the implication S in general using the stronger axiom PFA. Lemma 1 given earlier was discovered in the course of that proof, and in fact the pulling-down procedure was very similar to the proof of Theorem 9. In the rest of this paper we give more examples of these kinds of arguments. The combinatorial essence of our arguments will be in transformations of certain free sequences from $C_p(X)$ to some finite power of X and then to X itself, or conversely.

THEOREM 14 (PFA): Let X be a Lindelöf space with a single nonisolated point $*$. Then X is countably tight iff its function space $C_p(X)$ is Lindelöf.

Proof: The converse implication is a result of Asanov [29], so let us concentrate on the direct implication. Let $f_\xi(\xi < \omega_1)$ be a given sequence of elements of $C_p(X)$. Let M be a countable elementary submodel of some large structure H_θ such that X, $\{f_\xi : \xi < \omega_1\} \in M$. Let $g \in C_p(X \cap M)$ be a condensation point of

$$f_\xi \upharpoonright (X \cap M) \qquad (\xi < \omega_1).$$

Extend g to a function $\bar{g} : X \to \mathbb{R}$ by letting $\bar{g}(x) = g(*)$ for all $x \in X \backslash M$. Note that $\bar{g} \in C_p(X)$. We claim that $\bar{g}$ is a condensation point of the sequence f_ξ. For suppose that there is a basic open neighborhood U of $\bar{g}$ that contains only countably many f_ξ. We can assume that all intervals of U have rational endpoints, so $V = U \cap M$ is an element of M. Then

$$A = \{\xi < \omega_1 : f_\xi \in V\}$$

is also an element of M. Note that by our choice of g, A is uncountable. Note also that M thinks that for every $D \in [X]^{\aleph_0}$ there is basic open set U_D of $C_p(X)$ with domain $F_D \subseteq X \backslash D$ such that

(d) There is a single rational interval I of $g(*)$ such that $U_D(x) = I$ for all $x \in F_D$
(e) $f_\xi \notin U_D$ for all but countably many ξ in A.

Since this statement must be true in H_θ, we can build two sequences $D_\alpha(\alpha < \omega_1)$ and $U_\alpha(\alpha < \omega_1)$ of countable subsets of $X \cup \omega_1$ and basic open subsets of $C_p(X)$, respectively, such that

(f) $F_\alpha = \text{dom}(U_\alpha)$ is disjoint from D_α
(g) There is a single rational interval I containing $g(*)$ such that $U_\alpha(x) = I$ for all $\alpha < \omega_1$ and $X \in F_\alpha$
(h) $D_\alpha(\alpha < \omega_1)$ is an increasing sequence of sets and $D_\alpha \cap \omega_1$ $(\alpha < \omega_1)$ is a strictly increasing sequence of ordinals
(i) $f_\xi \notin U_\alpha$ for all $\xi \in A \backslash D_{\alpha+1}$.

To reformulate (i), for $\xi < \omega_1$, let

$$K_\xi = \{x \in X : f_\xi(x) \notin I\}.$$

Note that each K_ξ is a closed subset of X and that (i) becomes:

(j) For all $\xi \in A \backslash D_{\alpha+1}$ there is $x \in F_\alpha$ such that $x \in K_\xi$.

So we are in the situation of the proof of Theorem 9 where it is shown that conditions like (j) imply that X cannot be countably tight. Of course, a similar proof shows that any ω_2-sequence of elements of $C_p(X)$ has a complete accumulation point. This is sufficient to give us the Lindelöfness of $C_p(X)$ since we can assume that our space X has weight at most continuum (which in our context is equal to ω_2) by simply replacing X with $X \cap M^*$, where M^* is an elementary submodel of H_θ closed under ω-sequences. □

THEOREM 15 (PFA): Let X and Y be two Lindelöf spaces with at most one nonisolated point. Then the product $C_p(X) \times C_p(Y)$ is Lindelöf provided that each factor is Lindelöf.

Proof: The proof of Theorem 14 works equally well for the sum $X \oplus Y$. □

REMARKS: The results of this and the previous two sections were announced in [9] and [8, p. 88]. Note that Theorem 15 solves problem 1 of [30]. In the proof of Theorem 14 we could have used a result of [30] that says that for such X, $C_p(X)$ is Lindelöf if every finite power of X is countably tight. Since the proof of [30] is quite indirect using results from the literature not accessible to us, we have opted for the direct argument. (Note that our argument can also be used to give another proof of the result of [30].) Finally, we note that Theorem 14 cannot be extended to the more general class of condensed spaces. To see this let K be the Cantor-tree space, that is, the tree $\{0, 1\}^{\leq \omega}$ with the interval topology and let K^* be the one-point compactification of K. Then K^* is a condensed compact space, but $C_p(K^*)$ contains a closed discrete subset of size continuum (the set of characteristic functions of maximal branches of K).

9. COMPACT SETS IN $C_p(X)$

A classical result of Grothendieck ([40], [41]) says that if X is a compact space, then every countably compact subset of $C_p(X)$ is compact. In this section we give another class of spaces X that enjoys the same property. It gives us an opportunity to expose the following variation of Lemma 1, which is proved along similar lines (see [33] and [8, sec. 8])

LEMMA 2 (PFA): Let Y be a subspace of a countably tight countably compact space X and let $\mathscr{F}$ be a maximal countably complete filter of closed subsets of X such that $F \cap Y$ is nonempty for every F in $\mathscr{F}$. Suppose that for every y in Y we have associated neighborhoods $U(y)$ and $V(y)$ of y (in X) such that $U(y) \subset V(y)$ and $X \backslash V(y) \in \mathscr{F}$. Then either:

(a) There is an uncountable closed discrete subspace Z of Y, or
(b) There is $\{y_\xi : \xi < \omega_1\} \subseteq Y$ such that for all $\alpha < \omega_1$, $\{y_\xi : \xi \leq \alpha\} \subseteq U(y_\alpha)$ and $\{y_\xi : \xi > \alpha\} \cap V(y_\alpha) = \varnothing$.

Proof: Choose a large enough structure of the form $\langle H_\theta, \in, < \rangle$ containing all these objects. (Here $<$ is a well-ordering of H_θ.) For a countable elementary submodel M of $\langle H_\theta, \in, < \rangle$, let y_M be the minimal point of

$$Y \cap \bigcap \{\overline{F \cap M} : F \in \mathcal{F} \cap M\}$$

if this intersection is nonempty. Otherwise, we take y_M to be the minimal point of that intersection but with X in place of Y. Let the phrase "y_M exists" mean that the point is an element of Y rather than of $X \backslash Y$. Let z_M be the minimal point of

$$Y \cap \bigcap \{F : F \in \mathcal{F} \cap M\}.$$

Let $\mathcal{P}$ be the set of all finite *elementary chains* (see [32, p. 210]) of submodels of $(H_\theta, \in, <)$, that is, sequences of the form $\vec{N} = \langle N_\alpha : \alpha \in A \rangle$ where A is a finite subset of ω_1, where each N_α is a countable elementary submodel of $(H_\theta, \in, <)$ containing all relevant objects, $N_\alpha \in N_\beta$ for $\alpha < \beta$ in A, and there exists a *continuous* $\in$-chain $\vec{M} = \langle M_\alpha : \alpha < \omega_1 \rangle$ of countable elementary submodels of $\langle H_\theta, \in, < \rangle$ such that $M_\alpha = N_\alpha$ for all $\alpha \in A$ (i.e., $\vec{M} \restriction A = \vec{N}$). For $\vec{M} = \langle M_\alpha : \alpha \in A \rangle$ and $\vec{N} = \langle N_\beta : \beta \in B \rangle$ in $\mathcal{P}$, set $\vec{M} \leq \vec{N}$ iff

(c) $\vec{M} \supseteq \vec{N}$,
(d) For every $\beta \in B$ and for every $\alpha \in A \cap (\max (B \cap \beta), \beta)$, $y_{M_\alpha} \in U(y_{N_\gamma})$ for every $\gamma \in B \backslash \beta$ for which y_{N_γ} exists and for which $U(y_{N_\gamma})$ contains y_{N_β}.

By now standard arguments (see [8, sec. 8]) show that $\mathcal{P}$ is a proper poset, so applying PFA to an appropriately chosen family of $\aleph_1$ dense subsets of $\mathcal{P}$ gives us a filter $\mathcal{G}$ of $\mathcal{P}$ such that

$$\bigcup \mathcal{G} = \langle M_\alpha : \alpha < \omega_1 \rangle$$

is a *continuous* $\in$-chain of countable elementary submodels of $\langle H_\theta, \in, < \rangle$. It follows that for every limit $\delta < \omega_1$ for which y_{M_δ} exists there is $\gamma < \delta$ such that

$$y_{M_\alpha} \in U(y_{M_\delta})$$

for all $\alpha \in [\gamma, \delta)$ for which y_{M_α} exists. So if the set of such δ's is stationary, the pressing down lemma would give us the alternative (b). So assume there is a club $D \subseteq \omega_1$ such that

$$Y \cap \bigcap \{\overline{F \cap M_\delta} : F \in \mathcal{F} \cap M_\delta\} = \varnothing$$

for all δ in D. Let

$$Z = \{z_{M_\delta} : \delta \in D\}.$$

Then it is easily checked that Z satisfies the alternative (a) of the Lemma.

REMARK: The proof of Lemma 2 allows a number of interpretations, many of which have been already observed before (see [33], [57], [59]):

(1) The countable compactness of X can be replaced by the fact for every countable elementary submodel M of H_θ, the intersection of $\{\overline{F \cap M} : F \in F \cap M\}$ is nonempty. The countable compactness and countable tightness of X, however, are

not that essential restrictions here, since every completely regular countably tight space Y can be embedded into a countably compact countably tight space X (not necessarily regular). The real restriction in Lemma 2 is the interaction of the filter $\mathscr{F}$ and the subspace Y.

(2) If Y is countably compact, then there is no need to assume that it is countably tight, for we can increase its topology to get countable tightness (see the proof of Theorem 16 below). Also, in this case there is no need for any filter $\mathscr{F}$ of closed set, but only the requirement that no countable subfamily of $\{V(y) : y \in Y\}$ covers Y, that is, that the family is a witness that Y is not Lindelöf, the only property of some real interest and of some use in this paper. This explains why the proof of Lemma 2 turns out to be a rather simple modification of the proof of Conjecture S from PFA ([27]). Note also that in this case the conditions of $\mathscr{P}$ can simply be finite $\in$-chains rather than elementary chains.

(3) Suppose now $\mathscr{F}$ is a maximal countably complete filter of closed subsets of Y such that $Y\backslash V(y)$ is in $\mathscr{F}$ for every y in Y, but that we don't have any assumption of countable compactness of countable tightness of X or on Y. It is then natural to try to force the alternative (b) with z_{M_α} in place of y_{M_α}. In the case the proof that $\mathscr{P}$ is proper needs the assumption that for every $Z \subseteq Y$ intersecting every element of $\mathscr{F}$ (i.e., $Z \in \mathscr{F}^+$) there exist countable $Z_0 \subseteq Z$ such that $\overline{Z} \in \mathscr{F}$ (the closure taken in Y). If this fails for some Z in $\mathscr{F}^+$, then recursively we can construct a sequence $z_\alpha(\alpha < \omega_1)$ of elements of Z and a decreasing sequence $F_\alpha(\alpha < \omega_1)$ of elements of $\mathscr{F}_\alpha$ such that for all $\alpha, z_\alpha \in F_\alpha$ and $\overline{\{z_\xi : \xi < \alpha\}} \cap F_\alpha = \varnothing$. It follows that $z_\alpha(\alpha < \omega_1)$ is a free sequence in Y. Thus, starting with an arbitrary pair of spaces $Y \subseteq X$, a neighborhood assignment $U(y) \subset V(y)$ ($y \in Y$), and a maximal countably complete filter $\mathscr{F}$ of closed subsets of Y such that $Y\backslash V(y)$ is in $\mathscr{F}$ for every y in Y, we get the conclusion of Lemma 2 with the alternative (a) replaced by the existence of an uncountable free sequence in the subspace Y. Unfortunately, this restatement of Lemma 2 is not as useful as it looks since, as we shall see later, the free sequence $z_\alpha(\alpha < \omega_1)$, not being separated by any natural sequence of open sets, is quite difficult to deal with.

THEOREM 16 (PFA): Suppose X contains no uncountable free sequence. Then every countably compact subspace Y of $C_p(X)$ is compact.

Proof: Suppose Y is not compact. Then Y is not Lindelöf and so for every f in Y we can fix a basic open set $V(f)$ in $C_p(X)$ such that $f \in V(f)$ and such that

$$Y\backslash \bigcup_{f \in Y_0} V(f) \neq \varnothing$$

for every countable $Y_0 \subseteq Y$. We can assume that for every f in Y and x in dom $(V(f))$, $V(f)(x)$ is an interval with rational endpoints. Choose now for each f in Y basic open neighborhood $U(f)$ in $C_p(X)$ such that

$$\text{dom } (U(f)) = \text{dom } (V(f))$$

and for each $x \in$ dom $(U(f))$, $U(f)(x)$ is an open rational interval whose endpoints are both elements of the open interval $V(f)(x)$. We would like to apply Lemma 2 so we need to provide the other two inputs. To this end, call a subset Z of Y countably closed if $\overline{Z_0} \subseteq Z$ for every countable $Z_0 \subseteq Z$ (the closure taken in Y). Extend the topology of Y by proclaiming every countably closed subset of Y to be closed. Note Y still remains countably compact so that the outcome (a) of Lemma 2 does not happen even in the extended topology. The point of the new topology is that Y is now countably tight. The countable compactness of Y also helps us to (recursively) extend $Y \backslash V(f)$ ($f \in Y$) to a maximal countably complete filter of closed subsets of Y. By Lemma 2, there exists $\{ f_\xi : \xi < \omega_1 \} \subseteq Y$ such that for every $\alpha < \omega_1$,

(e) $\{ f_\xi : \xi \le \alpha \} \subseteq U(f_\alpha)$ and $\{ f_\xi : \xi > \alpha \} \cap V(f_\alpha) = \varnothing$.

We can assume that

$$F_\alpha = \mathrm{dom}\ (U(f_\alpha)) = \mathrm{dom}\ (V(f_\alpha))$$

have all fixed size n, and that if $x_{\alpha i}(i < n)$ is a fixed enumeration of F_α, then for some sequences $I_i(i < n), J_i(i < n)$ of rational intervals and for all $\alpha < \omega_1$,

$$U(f_\alpha)(x_{\alpha i}) = I_i \qquad \text{and} \qquad V(f_\alpha)(x_{\alpha i}) = J_i$$

for all $i < n$. For $\beta < \omega_1$ and $i < n$, set

$$U_{\beta i} = \{ x \in X : f_\beta(x) \in I_i \}$$
$$V_{\beta i} = \{ x \in X : f_\beta(x) \in J_i \}.$$

Then $U_{\beta i}$ and $V_{\beta i}$ are two open neighborhood of $x_{\beta i}$ in X such that $\overline{U}_{\beta i} \subseteq V_{\beta i}$. Note also that for all $\alpha < \beta < \omega_1$:

(f) There is $j < n$ such that $x_{\alpha j} \notin V_{\beta j}$
(g) $x_{\beta i} \in U_{\alpha i}$ for all $i < n$.

Note that (f) gives us a partition of $[\omega_1]^2$ into n pieces, so by Lemma 1, we can find $j < n$, an uncountable $A \subseteq \omega_1$ and an uncountable disjoint family B of finite subsets of ω_1 such that for every $\alpha \in A$ and $b \in B$ with $\alpha < \min\ (b)$ there is $\beta \in b$ such that $x_{\alpha j} \notin V_{\beta j}$. For b in B, set

$$U_{bj} = \bigcap_{\beta \in b} U_{\beta j} \qquad \text{and} \qquad V_{bj} = \bigcap_{\beta \in b} V_{\beta j}.$$

Then U_{bj} and V_{bj} are two open subsets of X such that $\overline{U}_{bj} \subseteq V_{bj}$. Note also that for every $\alpha \in A$ and $b \in B$ with $\alpha < \min\ (b)$:

(h) $x_{\alpha j} \notin V_{bj}$.

On the other hand, condition (g) says that for every $\alpha \in A$ and $b \in B$ with $\alpha > \max\ (b)$,

(i) $x_{\alpha j} \in U_{bj}$.

Since we clearly may assume (by shrinking A) that for every $\alpha > \beta$ in A there is b in B such that $\alpha < \min\ (b) \le \max\ (b) < \beta$, it follows that $\{ x_\alpha : \alpha \in A \}$ is an uncountable free sequence in X, a contradiction. This finishes the proof. $\square$

The countably compact subset Y of $C_p(X)$ in Grothendieck's Theorem is a Frechet–Urysohn space with a dense set of G_δ-points (see [39], [51]). It is interesting that the space Y of our Theorem 16 is also a compact sequential space with a dense set of G_δ-points. This follows from the results of Balogh and Dow (see [57], [59], and [8, sec. 8]) and the following fact.

THEOREM 17 (PFA): Suppose X contains no uncountable free sequence. Then every (countably) compact subspace of $C_p(X)$ is countably tight.

Proof: Let K be a compact subspace of $C_p(X)$ and suppose that for some $g \in K$ and $Z \subseteq K, g \in \overline{Z}$ but $g \notin \overline{Z_0}$ for every countable $Z_0 \subseteq Z$. Let

$$Y = \cup\{\overline{Z_0} : Z_0 \subseteq Z, Z_0 \text{ countable}\}.$$

Then Y is a countably compact subspace of K and therefore of $C_p(X)$. By Theorem 16, Y is compact and so, in particular, it is closed in $C_p(X)$. This contradicts the fact that $g \in \overline{Y}\backslash Y$. $\square$

REMARKS: In [34], Baturov showed that if X is compact, then a subspace of Y of $C_p(X)$ is Lindelöf iff Y contains no uncountable closed discrete subspace. We believe that the method of this section can be extended to give us the same conclusion for the class of spaces without uncountable free sequences. Let $L(\theta)$ be the one-point Lindelöfication of the discrete subspace of size $\theta > \omega$. Then $L(\theta)$ is Lindelöf and the space $C_p(L(\theta))$ is homeomorphic to the Σ-product of θ copies of $\mathbb{R}$ (see [43]). On the other hand, the space $C_p(L(\theta))$ contains the Σ-product of the unit interval as a closed countably compact noncompact subset. Note that $L(\theta)$ does contain an uncountable free sequence or else this would contradict our Theorem 16.

10. *S* AND *L* AND THE SPACE $C_P(X)$

A well-known result of Velicko [61] and Zenor [62] says that finite powers of X are hereditarily Lindelöf (hereditarily separable) iff finite powers of $C_p(X)$ are hereditarily separable (hereditarily Lindelöf). Velicko [61] goes a bit further in the direction of eliminating the assumption that all finite powers of $C_p(X)$ are hereditarily separable (hereditarily Lindelöf), but only that $C_p(X)$ itself is hereditarily separable (hereditarily Lindelöf). He was able to prove this for hereditary separability, but the following two interesting conjectures were left open in [61]:

(V_0) Hereditary Lindelöf number is preserved by taking finite powers of spaces of the form $C_p(X)$.

(V_1) The spread of every finite power of $C_p(X)$ is equal to the spread of $C_p(X)$.

In [29], Asanov remarks that Velicko's arguments give the conjectures (V_0) and (V_1) provided that the diagonal of X^2 is the intersection of at most $hl(C_p(X))$ [respectively, $s(C_p(X))$] open subsets of X^2. Note that (V_0) is equivalent to the statement that $hd(X^n) \le hL(C_p(X))$ for all $n < \omega$ and that (V_1) is equivalent to the statement $s(X^n) \le s(C_p(X))$ for all $n < \omega$. It follows that (V_1) is stronger than (V_0). Velicko [61] showed that $hd(X^2) \le hL(C_p(X))$ and that $s(X^2) \le s(C_p(X))$. Thus, the simplest open cases of (V_0) and (V_1) are the problems of similar bounds for the cube of X. Note that the result of Velicko about X^2 and the observation of Asanov about the diagonal of X^2 immediately give us that the basic implication S solves the countable cases of (V_0) and (V_1). [Thus, $s(X^2) = \aleph_0$ implies $hL(X^2) = \aleph_0$, and this implies that the diagonal of X^2 is G_δ.] It follows that PFA implies the countable cases of (V_0) and (V_1), a fact first explicitly stated in [60]. Combining this with a well-known result of Kunen [58] (a consequence of Theorem 12) gives us the following result.

THEOREM 18 (PFA): If $C_p(X)$ has no uncountable discrete subspace, then every finite power of $(X$ and) $C_p(X)$ is hereditarily Lindelöf and hereditarily separable.

[Note that it has been conjectured (see [8, sec. 3]) that, in fact, in this situation we might be able to conclude that X (and therefore $C_p(X)$) is a continuous image of a separable metric space that would be a considerable strengthening of this result.]

Free sequences of finite powers of X correspond to free sequences of finite powers of $C_p(X)$, giving us another cardinal invariant that is self-dual. The following analogue of Theorem 12 is an instance where certain free sequences of finite powers of $C_p(X)$ can be pulled down to $C_p(X)$ itself.

THEOREM 19 ($MA_{\aleph_1}$): Assume $C_p(X)$ contains no uncountable free sequence. Then a subspace of X is hereditarily Lindelöf iff it is hereditarily separable.

Proof: Suppose we have a hereditarily Lindelöf subspace Y and X that is not separable. We need to produce a free sequence in $C_p(X)$. Clearly, we can assume that there is a well-ordering $<$ of Y of type ω_1 such that every initial segment of Y, $<$ is closed in Y. Then for every y in Y we can fix two neighborhoods U_y and V_y (in X) such that for some continuous function $f_y : X \to [0, 1]$:

 (a) $\overline{U}_y = f_y^{-1}(1)$,
 (b) $X \backslash V_y = f_y^{-1}(0)$.
 (c) $V_y \cap \{x \in Y : x < y\} = \varnothing$.

Let $\mathscr{P}$ be the poset of all finite subsets p of Y such that $y \notin U_x$ for all $x < y$ in p. Since $MA_{\aleph_1}$ holds and since Y is hereditarily Lindelöf, the poset $\mathscr{P}$ cannot be ccc. By taking an uncountable antichain of $\mathscr{P}$, forming a Δ-system, and then removing the root, we can assume that there is a sequence $p_\alpha(\alpha < \omega_1)$ of elements of $\mathscr{P}$ such that for all $\alpha < \beta < \omega_1$:

 (d) $x < y$ for $x \in p_\alpha$ and $y \in p_\beta$,
 (e) There exist $x \in p_\alpha$ and $y \in p_\beta$ such that $y \in U_x$.

For $\alpha < \omega_1$, set

$$g_\alpha = \sum_{x \in p_\alpha} f_x.$$

We claim that $g_\alpha(\alpha < \omega_1)$ is a free sequence in $C_p(X)$. To see this, for every $\alpha < \omega_1$, we consider the following two open sets in $C_p(X)$:

$$W_\alpha^0 = \{f \in C_p(X) : f(x) > 0 \text{ for some } x \in p_\alpha\}$$

$$W_\alpha^1 = \{f \in C_p(X) : f(x) > \frac{1}{2} \text{ for some } x \in p_\alpha\}.$$

Clearly, $\overline{W}_\alpha^1 \subseteq W_\alpha^0$ for all α while the conditions (c), (d), and (e) reduce to the following two facts.

 (f) $f_\beta \notin W_\alpha^0$ for all $\beta > \alpha$
 (g) $f_\xi \in W_\alpha^1$ for all $\xi \leq \alpha$.

This finishes the proof that $g_\alpha(\alpha < \omega_1)$ is a free sequence in $C_p(X)$.

A similar argument gives us a free sequence in $C_p(X)$ under the assumption that X contains a hereditarily separable non-Lindelöf subspace. The free sequence $g_\alpha(\alpha < \omega_1)$ so obtained has the open sets W_α^0 and W_α^1 identically defined, while conditions (f) and (g) will have the following dual forms:

(h) $g_\xi \notin W_\alpha^0$ for all $\xi < \alpha$,
(i) $g_\beta \in W_\alpha^1$ for all $\beta \leq \alpha$.

This completes the proof of Theorem 19. $\square$

REMARK: The preceding proof indicates that the crucial step in proving the implications *S* and *L* can also be viewed as a way of transforming certain free sequences from $C_p(X)$ back to X. We have seen in previous sections that the stronger forcing axiom PFA (via Lemma 1) gives us such a way for free sequences of the form (h) and (i), but the problem of transforming the free sequence $f_\xi(\xi < \omega_1)$ with properties (f) and (g) is combinatorially quite different and is currently open.

11. COUNTABLE TIGHTNESS UNDER RESTRICTED COMPACTNESS

A space X is *initially* ω_1-*compact* if every open cover of X of size at most $\aleph_1$ has a finite subcover. This restricted form of compactness is sufficient for many purposes. For example, this is the segment of compactness needed to prove that X is countably tight iff X contains no uncountable free sequence. It is interesting, therefore, that countable tightness might be exactly the property needed to give us full compactness out of initial ω_1-compactness. This is still an open conjecture true under various additional set-theoretical assumptions (see [59]). (For example, the reader may easily check that the conjecture follows from Lemma 2.) We have already mentioned a result of Malykhin that tightness is preserved by taking products of compact spaces. So, a weaker form of the conjecture would ask whether the same is true in the class of initially ω_1-compact spaces. This appears as problem 8.27(i) of [44]. The purpose of this section is to give a positive answer to this question as another instance of the combinatorics, previously exposed, of pulling down free sequences from products to the factors. The proof now will not require Lemma's 1 and 2 nor any additional set-theoretic assumption, but these two lemmas have been quite instrumental in finding this result. We expect many more results to be found using exactly this route or reasoning.

THEOREM 20: The product of every two countably tight initially ω_1-compact spaces is countably tight.

Proof: We use the notion of the free sequence of regular pairs from [45], that is the algebraic form of this notion rather than the topological one. Thus, a *regular pair* of a space T is a pair of the form $\langle F, G \rangle$, where F is a nonempty closed subset of T, G is an open subset of T, and $F \subseteq G$. A sequence $\langle F_\xi, G_\xi \rangle$ ($\xi < \theta$) of regular pairs is *free* if

$$F_{KL} = \bigcap_{\xi \in K} F_\xi \cap \bigcap_{\eta \in L} (T \backslash G_\eta)$$

is nonempty for every two finite subsets K and L of θ such that $K < L$ (i.e., $\xi < \eta$ for every $\xi \in K$ and $\eta \in L$). Note that a sufficient amount of compactness of X would transform $\langle F_\xi, G_\xi \rangle$ $(\xi < \theta)$ into a free sequence $x_\xi(\xi < \theta)$ of points of T (satisfying the usual definition).

Let X and Y be two countably tight initially ω_1-compact spaces. Suppose there exist p in $X \times Y$ and $A \subseteq X \times Y$ such that $p \in \overline{A}$ but $p \notin \overline{A_0}$ for every countable $A_0 \subseteq A$ and work for a contradiction. Clearly, we assume $X = Y$, and we shall do this for notational simplicity. We would like to build an uncountable free sequence in X^2 and then try to pull it down to one of the coordinates. Unfortunately, the known methods for building such a sequence use a bit of compactness that is not available to us. (We don't know whether X^2 is, for example, countably compact.) Instead, we try to construct a free sequence $\langle F_\xi, G_\xi \rangle$ $(\xi < \omega_1)$ of regular pairs of the space $T = X^2$ that are "rectangles" of X^2, that is, such that for all ξ:

(a) $F_\xi = F_\xi^0 \times F_\xi^1$ and $G_\xi = G_\xi^0 \times G_\xi^1$,
(b) $p \in \text{int } F_\xi$.

Suppose we have constructed $\langle F_\xi, G_\xi \rangle$ $(\xi < \alpha)$ for some $\alpha < \omega_1$. Let

$$\mathcal{U}_\alpha = \{\overline{A} \cap \text{int } F_{KL} : K, L \subseteq \alpha \text{ finite with } K < L\}.$$

The family $\mathcal{U}_\alpha$ is countable and since p has uncountable π-character in the subspace $\overline{A}$ of X^2, there is an open rectangle $G_\alpha = G_\alpha^0 \times G_\alpha^1$ containing p such that $U \not\subseteq \overline{G}_\alpha$ for all nonempty U in $\mathcal{U}_\alpha$. Let $F = F_\alpha^0 \times F_\alpha^1$ be any closed rectangle contained in G_α such that p is an element of its interior.

CLAIM 1: $\langle F_\xi, G_\xi \rangle$ $(\xi < \omega_1)$ is a free sequence of regular pairs in X^2.

Proof: Fix finite $K, L \subseteq \omega_1$ such that $K < L$. We shall prove by induction on the size of L that

$$\overline{A} \cap \text{int } F_{KL} \neq \varnothing.$$

This will clearly finish the proof of Claim 1. Note that this is true when $L = \varnothing$ by the property (b). So, assume L is nonempty and let α be the maximal element of L. Let $M = L\setminus\{\alpha\}$. By the induction hypothesis

$$U = \overline{A} \cap \text{int } F_{KM}$$

is a nonempty member of $\mathcal{U}_\alpha$. By our choice of G_α, the set $U\setminus\overline{G}_\alpha$ is a nonempty set, every point of which belongs to $\overline{A} \cap \text{int } F_{KL}$. This completes the inductive step and also the proof of Claim 1. $\square$

We would like now to pull the free-sequence $\langle F_\xi, G_\xi \rangle$ $(\xi < \omega_1)$ of regular pairs down to one of the coordinates. For this purpose for $\epsilon = 0$ or 1 and finite $K, L \subset \omega_1$ with $K < L$ we extend the previous notation and write

$$F_{KL}^\epsilon = \bigcap_{\xi \in K} F_\xi^\epsilon \cap \bigcap_{\eta \in L} (X\setminus G_\eta^\epsilon).$$

Case 1: For every $\alpha < \omega_1$ there is $\beta \geq \alpha$ such that for every finite $K \subseteq \alpha$ and $L \subseteq \omega_1\setminus\beta$, $F_{KL}^0 \neq \varnothing$. Let D be the set of all limit ordinals $\delta < \omega_1$ such that for every $\alpha < \delta$ the β that works for α is $< \delta$. It is then clear that $\langle F_\delta^0, G_\delta^0 \rangle$ $(\delta \in D)$ is a free sequence of regular pairs in X. But X has a sufficient amount of compactness to turn the sequence of regular pairs into an uncountable free sequence of points of X contradicting the countable tightness of X.

Case 2: There is $\alpha_0 < \omega_1$ such that for every $\beta \geq \alpha_0$ there exist finite $K \subseteq \alpha_0$ and $L \subseteq \omega_1 \backslash \beta$ such that $F_{KL}^0 = \varnothing$. Then we can find finite $K_0 \subseteq \alpha_0$ and a strictly increasing sequence $L_\beta(\beta < \omega_1)$ of finite subsets of $\omega_1 \backslash \alpha_0$, such that $F_{K_0 L_\beta}^0 = \varnothing$ for all β. For $\beta < \omega_1$, set

$$\hat{F}_\beta^1 = \bigcap_{\xi \in L_\beta} F_\xi^1 \quad \text{and} \quad \hat{G}_\beta^1 = \bigcap_{\xi \in L_\beta} G_\xi^1.$$

CLAIM 2: $\langle \hat{F}_\beta^1, \hat{G}_\beta^1 \rangle$ $(\beta| < \omega_1)$ is a free sequence of regular pairs of X.

Proof: Suppose $M < N$ are given finite subsets of ω_1. Let

$$K = K_0 \cup \bigcup_{\beta \in M} L_\beta \quad \text{and} \quad L = \bigcup_{\beta \in N} L_\beta.$$

We know that F_{KL} is a nonempty subset of X^2, so we fix an element $y = (y_0, y_1)$ from that set. Note that for every $\beta \in N$, $F_{K_0 L_\beta}^0 = \varnothing$, so in particular y_0 is not in this set. So there must be $\xi_\beta \in L_\beta$ such that $y_0 \in G_{\xi_\beta}^0$. We also know that (y_0, y_1) is an element of $X^2 \backslash G_{\xi_\beta}^0 \times G_{\xi_\beta}^1$, so we must have that $y_1 \in X \backslash G_{\xi_\beta}^1$. This means that y_1 is an element of the set

$$\bigcup_{\xi \in L_\beta} (X \backslash G_\xi^1) = X \backslash \hat{G}_\beta^1$$

for all $\beta \in N$. It follows that y_1 is an element of $\bigcap_{\alpha \in M} \hat{F}_\alpha^1 \cap \bigcap_{\beta \in N}(X \backslash \hat{G}_\beta^1)$, which finishes the proof of Claim 2. $\square$

We have already observed that an uncountable free sequence of regular pairs cannot exist in a countably tight initially ω_1-compact space, so Claim 2 finishes the proof of Theorem 20. $\square$

It is clear that this proof gives the following more general result.

THEOREM 21: Suppose that every open cover of X of size at most $t(X)^+$ has a finite subcover. Then the tightness of every finite power of X is equal to the tightness of X.

REMARK: Note that the proof of Theorem 20 shows that we are always able to pull down an uncountable free sequence $\langle F_\xi, G_\xi \rangle$ $(\xi < \omega_1)$ of regular pairs of some product $X \times Y$ to one of its factors provided that every F_ξ and G_ξ can be written as a finite union of "rectangles," that is, sets of the form $A \times B$ (with both A and B closed in the case of F_ξ and open in the case of G_ξ). The free sequence from the third alternative of Theorem 12 is separated by such a sequence of regular pairs, so we can state a corollary of this, saying that if X contains no uncountable free sequence of regular pairs, then a subspace of X is hereditarily separable iff it is hereditarily Lindelöf (assuming, of course, that $MA_{\aleph_1}$ holds). Unfortunately, this formulation is far less useful than Theorem 12 itself since countably tight initially ω_1-compact spaces are the only class of spaces that we know of to which this result would apply. In this context, note that a normal space X contains no uncountable free sequence of regular pairs iff its maximal compactification βX is countably tight. In fact, the freedom of choice of the elements of the free sequence of regular pairs from the first part of the proof of Theorem 20 also shows, for example, that a completely regular initially ω_1-compact space X is countably tight iff its maximal compactification βX is countably tight. Looking even closer at that construction reveals the following.

LEMMA 3: Suppose X is a regular space such that for some space Y the product $X \times Y$ contains a point (a, b) of tightness $> \theta$, while the tightness of b in Y is $\leq \theta$. Then there is a free θ^+-sequence of regular pairs of X.

Proof: Let $A \subseteq X \times Y$ be such that $(a, b) \in \overline{A}$, but $(a, b) \notin \overline{A_0}$ for every $A_0 \subseteq A$ of size $\leq \theta$. Call $U \subseteq X$ *large* if the set

$$A[U] = \{y \in Y : (x, y) \in A \text{ for some } x \in U\}$$

has b in its closure; otherwise U is called a *small* subset of X.

CLAIM 1: For every family $\mathscr{U}$ of size $\leq \theta$ of large subsets of X there is an open set G containing a such that $U \backslash G$ is large for every $U \in \mathscr{U}$.

Proof: Suppose the claim is false and fix a family $\mathscr{U}$ of large sets such that for every open set G containing a there is $U \in \mathscr{U}$ such that $U \backslash G$ is small. For every $U \in \mathscr{U}$ fix $D_U \subseteq A[U]$ of size $\leq \theta$ accumulating to b. Fix also a map $f_U : D_U \to U$ such that

$$(f_U(y), y) \in A \quad \text{for all} \quad y \in D,$$

and set

$$A_0 = \{(f_U(y), y) : U \in \mathscr{U} \text{ and } y \in D_U\}.$$

Then A_0 is a subset of A of size $\leq \theta$ and we claim that it accumulates to (a, b), contradicting our initial assumption about A and the point (a, b). So, suppose we are given an open rectangle $G \times H$ containing (a, b). Then we can find an U in $\mathscr{U}$ such that $F = U \backslash G$ is a small set. This means that $A[F]$ does not accumulate to b, so we may also assume (shrinking H) that $H \cap A[F] = \varnothing$. On the other hand, we know that $H \cap D_U \neq \varnothing$, so fix a point y in this set and let $x = f_U(y)$. Since y is not in $A[F]$, the point x cannot be in F, so x must be in G. Hence $(x, y) \in A_0 \cap (G \times H)$, so the intersection is nonempty. This finishes the proof of Claim 1.   □

Recursively on $\xi < \theta^+$ construct a sequence $\langle F_\xi, G_\xi \rangle$, $(\xi < \theta^+)$ of regular pairs of X as follows. Having determined $\langle F_\xi, G_\xi \rangle$ $(\xi < \alpha)$, set

$$\mathscr{U}_\alpha = \{F_{KL} : K, L \subseteq \alpha, K < L\}.$$

Then $\mathscr{U}_\alpha$ has size $\leq \theta$ so by Claim 1 there is an open set G_α containing a such that $U \backslash G_\alpha$ is large for every large U in $\mathscr{U}_\alpha$. Choose now a closed set F_α such that $F_\alpha \subseteq G_\alpha$ and such that a is in the interior of F_α.

CLAIM 2: $\langle F_\xi, G_\xi \rangle$ $(\xi < \theta^+)$ is a free sequence of regular pairs of X.

Proof: Fix finite subsets K and L of θ^+ such that $K < L$. By induction on the size of L we shall prove that F_{KL} is in fact a large subset of X. If $L = \varnothing$, the conclusion follows from the fact that

$$a \in \text{int}\left(\bigcap_{\xi \in K} F_\xi\right).$$

So suppose $L \neq \varnothing$ and let $M = L \setminus \{\alpha\}$, where $\alpha = \max(L)$. By the inductive hypothesis $U = F_{KM}$ is a large element of $\mathcal{U}_\alpha$, so by our choice of G_α,

$$F_{KL} = F_{KM} \setminus G_\alpha$$

is also large. This finishes the proof of Claim 2 and also the proof of Lemma 3. □

Note that Lemma 3 gives the following version of Theorem 21, which includes even the most general form of Malykhin's result.

THEOREM 22: Suppose that every open cover of size at most $t(X)^+$ of a regular space X has a finite subcover. Then for every space Y, the product $X \times Y$ has tightness equal to the maximum of $t(X)$ and $t(Y)$.

We should point out, however, that our objective in this section was to present an instance of the combinatorics of pulling down free sequences from products to their factors, a procedure that is often more useful than the productiveness of tightness itself.

REFERENCES

1. BURKE, M. R. & M. MAGIDOR. 1990. Shelah's pcf theory and its applications. Ann. Pure Appl. Logic **50**: 207–254.
2. COMFORT, W. W. 1990. Problems on topological groups and other homogeneous spaces. *In* Open Problems in Topology, J. van Mill and G. M. Reed, Eds. North-Holland. Amsterdam, the Netherlands.
3. KANAMORI, A. & M. MAGIDOR. 1978. The Evolution of Large Cardinal Axioms in Set Theory. *In* Lecture Notes in Mathematics **669**. Springer-Verlag. New York/Berlin.
4. MALYKHIN, V. I. 1987. Nonpreservation of properties of topological groups on taking their square. Siberian Math. J. **28**: 639–645.
5. ROITMAN, J. 1979. Adding a random on a Cohen real: Topological consequences and effect on Martin's axiom. Fundam. Math. **103**: 47–60.
6. ———. 1980. Easy *S* and *L* groups. Proc. Am. Math. Soc. **78**: 424–428.
7. TODORČEVIĆ, S. 1980. Remarks on cellularity in products. Compos. Math. **57**: 357–372.
8. ———. 1989. Partition Problems in Topology. American Mathematical Society, Providence, R.I.
9. ———. 1989. Tightness in products. Interim Report of the Prague Topological Symposium 4: 7–8; Densities of closed sets, *ibid.*, p. 16.
10. RUDIN, M. E. 1985. Lectures on set theoretic topology. CBMS. Regional Conferences Series in Mathematics **23**. American Mathematical Society, Providence, R.I.
11. ———. 1972. A normal hereditarily separable non-Lindelöf space. Ill. J. Math. **16**: 621–626.
12. ———. 1974. A non-normal hereditarily separable space. Ill. J. Math. **18**: 481–483.
13. ———. 1979. The undecidability of the existence of a perfectly normal nonmetrizable manifold. Houston J. Math. **5**: 249–252.
14. RUDIN, M. E. & P. ZENOR. 1970. A perfectly normal nonmetrizable manifold. Houston J. Math. **2**: 129–134.
15. JUHÁSZ, I., K. KUNEN & M. E. RUDIN. 1976. Two more hereditarily separable non-Lindelöf spaces. Can. J. Math. **28**: 998–1005.
16. RUDIN, M. E. 1980. *S* and *L* spaces. *In* Surveys in General Topology, G. M. Reed, Ed.: 432–444. Academic Press. New York.
17. TALL, F. D. 1977. A reduction of the hereditary separable non-Lindelöf problem. *In* Set-Theoretic Topology: 349–351. Academic Press. New York.
18. SZENTMIKLOSSY, Z. 1980. *S*-spaces and *L*-spaces under MA. Colloq. Math. Soc. János Bolyai **23**: 1139–1145. North-Holland. Amsterdam, the Netherlands.
19. ENGELKING, R. 1989. General Topology. Heldermann Verlag. Berlin.

20. ARHANGEL'SKII, A. V. 1988. Some problems and lines of investigations in general topology. CMUC 29(4): 611–629.
21. TODORČEVIĆ, S. 1988. On the Lindelöf Property of Aronszajn Trees. *In* General Topology and its Relation to Analysis and Algebra VI, Proceedings of the Sixth Prague Topology Symposium (1986), Z. Frolik, Ed.: 577–588. Heldermann Verlag. Berlin.
22. KOPPELBERG, S. 1989. General Theory of Boolean Algebras. *In* Handbook on Boolean Algebras, Vol. I, J. D. Monk, Ed. Elsevier. Amsterdam, the Netherlands.
23. VAN DOUWEN, E. 1976. A technique for constructing honest locally compact submetrizable examples. Preprint.
24. WARREN, N. M. 1972. Properties of Stone-Čech compactifications of discrete spaces. Proc. Am. Math. Soc. 33: 599–606.
25. TODORČEVIĆ, S. 1987. Partitioning pairs of countable ordinals. Acta Math. 159: 261–294.
26. MALYKHIN, V. I. 1987. Spaces of continuous functions in simplest generic extensions. Math. Notes 41: 301–304.
27. TODORČEVIĆ, S. 1983. Forcing positive partition relations. Trans. Am. Math. Soc. 280: 703–720.
28. ———. 1985. Directed sets and cofinal types. Trans. Am. Math. Soc. 290: 711–723.
29. ASANOV, M. O. 1983. About the space of continuous functions. Colloq. Math. Soc. János Bolyai 41: 31–34.
30. LEIDERMAN, A. G., V. I. MALYKHIN. 1988. On nonpreservation of Lindelöfness under taking products of the spaces of the form $C_p(X)$. Sib. Math. J. 29(1): 65–72.
31. GALVIN, F. 1977. On Gruenhage's generalization of first countable spaces II. Not. Am. Math. Soc. 24: A-257.
32. TODORČEVIĆ, S. A note on the proper forcing axiom. Contemp. Math. 31: 209–218.
33. FREMLIN, D. H. 1988. Perfect pre-images of ω_1 and the PFA. Topol. Appl. 29: 151–166.
34. BATUROV, D. P. 1987. On subspaces of function space. Vestn. Moskovskogo Univ. Ser. Mat.-Mekh. 42(4): 66–69.
35. PESTOV, V. G. 1986. Compactly generated topological groups. Mat. Z. 40(5): 671–676.
36. GALVIN, F. & A. W. MILLER. 1984. On γ-sets and other singular sets of real numbers. Topol. Appl. 17: 145–155.
37. McCOY, R. A. 1980. K-space function spaces. Int. J. Math Math. Sci. 3: 701–711.
38. GERLITS, J. 1983. Some properties of $C(X)$, II. Topol. Appl. 15: 255–262.
39. McCOY, R. A. & I. NTANTU. 1988. Topological properties of spaces of continuous functions. *In* Lecture Notes in Mathematics 1315. Springer-Verlag. New York/Berlin.
40. GROTHENDIECK, A. 1952. Critères de compacticité dans les espaces fonctionnels généreaux. Am. J. Math. 74: 168–186.
41. ———. 1953. Sur les application linéaires faiblement compactes d'espaces du type $C(K)$. Can. J. Math. 5: 129–173.
42. TODORČEVIĆ, S. 1984. Trees and Linearly Ordered Sets. *In* Handbook of Set-Theoretic Theory, K. Kunen and J. E. Vaughan, Eds.: 235–293. Elsevier. Amsterdam, the Netherlands.
43. CORSON, H. H. 1959. Normality in subsets of product spaces. Am. J. Math. 81: 785–796.
44. SHAKHMATOV, D. B. Compact spaces and their generalization. Preprint.
45. TODORČEVIĆ, S. 1990. Free sequences. Topol. Appl. 35: 235–238.
46. COMFORT, W. W. & S. NEGREPONTIS. 1982. Chain Conditions in Topology. Cambridge Univ. Press. New York.
47. TKAČENKO, M. G. 1984. On topologies of free groups. Czech. Math. J. 34(109): 541–551.
48. SHAKHMATOV. D. B. 1986. Precalibers of σ-compact topological groups. Mat. Z. 39(6): 859–868.
49. KELLEY, J. L. 1959. Measures on Boolean algebras. Pac. J. Math. 9: 1165–1177.
50. HEWITT, E. & K. A. ROSS. 1963. Abstract Harmonic Analysis. Springer-Verlag. New York/Berlin.
51. ARHANGEL'SKII, A. V. 1989. Topological Function Spaces. Moscow Univ. Press. Moscow, Russia.
52. ———. 1978. Structure and classification toplogical spaces and invariants. Russ. Math. Surv. 33(6): 33–96.
53. ———. 1971. On bicompacta hereditarily satisfying Souslin's condition. Tightness and free sequences. Sov. Math. Dokl. 12: 1253–1257.

54. SHAPIROVSKII, B. V. 1974. On spaces with the Souslin and Shanin conditions. Math. Notes **15:** 161–167.
55. ———. 1972. On discrete subspaces of topological spaces; Weight tightness and the Souslin numbers. Sov. Math. Dokl. **13:** 215–219.
56. RANCIN, D. V. 1977. Tightness, sequentiality and closed coverings. Sov. Math. Dokl. **18:** 196–200.
57. BALOGH, Z. 1989. On compact Hausdorff spaces of countable tightness. Proc. Am. Math. Soc. **105:** 755–769.
58. KUNEN, K. 1977. Strong *S* and *L* spaces under MA. *In* Set-Theoretic Topology, G. M. Reed, Ed. Academic Press. New York.
59. DOW, A. 1989. An introduction to application of elementary submodels in topology. Topol. Proc. 17–72.
60. ARHANGEL'SKII, A. V. 1989. On hereditarily Lindelöf spaces of continuous functions. Vestn. Moskovskogo Univ. Ser. 1, Mat. Mekh. **3:** 67–69.
61. VELICKO, N. V. 1981. Weak topology of spaces of continuous functions. Mat. Z. **30**(5): 703–712.
62. ZENOR, P. 1980. Hereditary *m*-separability and the hereditary *m*-Lindelöf property in product spaces and functions spaces. Fundam. Math. **106:** 175–180.

Mary Ellen Rudin's Early Work on Suslin Spaces[a]

STEPHEN WATSON

Department of Mathematics
York University
North York, ONT M3W 1P3
Canada

ABSTRACT. In the early 1950s, Mary Ellen Rudin wrote her first three papers: Concerning Abstract Spaces (1950), Separation in Non-Separable Spaces (1951), and Concerning a Problem of Souslin's (1952). A modern interpretation of these articles is given.

1. INTRODUCTION

In the early 1950s, Mary Ellen Rudin wrote her first three papers: Concerning Abstract Spaces (1950), Separation in Non-Separable Spaces (1951), and Concerning a Problem of Souslin's (1952). This cycle represents one of the greatest accomplishments in set-theoretic topology. However, the mathematics in these papers is of such depth that, even forty years later, they remain impenetrable to all but the most diligent and patient of readers. Thus, few of their fundamental insights have reached the topological community.

The Pixley–Roy topology is the best known idea that was germinated in these early papers; however, a contemporary reading brings to light that it is Rudin's implicit use of Skolem functions (that is, elementary submodels) that is of greater significance and a key to understanding her work. For although her unusual ideas can rarely be simplified, an explicit deployment of elementary submodels and trees can express her arguments in a more natural form.

As elementary submodels become standard, Mary Ellen Rudin's proofs will become less inscrutable. Her ideas will then emerge and their beauty and depth will finally be appreciated. They will thus take their place at the heart of the subject. These three papers, which appeared forty years ago in the *Duke Mathematical Journal,* should be treated like rare preprints, written in an obscure language, full of secrets, burying ideas that have not been further explored, waiting to provide strength to any readers willing to devote themselves to their study.

A few comments on Mary Ellen Rudin's style may be useful. In her proofs you will find many long sequences of definitions such as: "There is a well-ordered sequence . . . , Let M = . . . , There is a well-ordered sequence . . . , Let β_g = . . . , There is a well-ordered sequence . . .". Often, these are just explicit definitions of

Mathematics Subject Classification. Primary 54F15, 54G20, 54A35, 54A25; Secondary 03E05, 03E35.

[a]This work was supported by the National Sciences and Engineering Research Council of Canada.

"witnesses for Skolem functions." We shall explain this. Suppose ϕ is some statement about some x that also mentions certain parameters $\vec{a}$. Whenever it is true that there is some x, depending on $\vec{a}$, that makes $\phi(x, \vec{a})$ true, we can choose such an x. This usually requires the axiom of choice since there are infinitely many (maybe even uncountably many) choices for the parameters $\vec{a}$ and there is no "rule" for picking an x. To choose these x's, there are at least two formal alternatives. The choices can be made in a transfinite induction (which requires the axiom of choice) or we can explicitly well order the set of all x's and then use the rule that says "choose the least x such that $\phi(x, \vec{a})$ is true." A *Skolem function* is just a function that assigns, to each possible parameter $\vec{a}$ in some set, some $x(\vec{a})$ so that $\phi(x(\vec{a}), \vec{a})$.

Another common kind of phrase in her proofs is "Let E be a countable point set which for each n, contains a point in every region α_n." This is a more succinct way of saying: "Well-order the set of points p for which there is an integer n such that $p \in \alpha_n$. Let p_n be the least element of α_n. Let $E = \{p_n : n \in \omega\}$," but again the same thing is accomplished: a Skolem function is explicitly defined.

Yet another phrase to be found has the form "Each of the countable components of $[S - (\overline{E} + \overline{K})]$ is a space satisfying all the conditions placed on S." Here she is saying that, if she wanted to, she could go back and make the same definitions and repeat the same process over and over again. Of course, it is this juncture that makes the explicit use of elementary submodels most useful. The reader may compare our proofs with that in [6].

Nothing in this article will argue that Rudin's proofs admit mathematical simplification; in fact, I hope to argue that even supposed simplifications like the Pixley–Roy space leave something behind. There may be easier ways to obtain certain results that she obtains, but her proofs never become obsolete. I do claim however, that her proofs can admit tremendous psychological simplification. Contemporary language like elementary submodels, maximal almost disjoint families, selections, and trees can help to do this.

But even after taking advantage of the innovations of the last forty years and after translating those of her proofs that are written in the language of R. L. Moore into the language of the European tradition, the proofs are still hard. Some say that there are easier proofs of her theorems and that it is best to wait until these proofs are available; to wait until a royal road to her work has been built. But the excitement lies in building that royal road, in following the turns these arguments take, turns still surprising after forty years. Reading the articles of Mary Ellen Rudin, studying them until there is no mystery takes hours and hours; but those hours are rewarded, the student obtains power to which few have access. They are not hard to read, they are just hard mathematics, that's all.

2. COMPLETENESS AND SEMICOMPLETENESS

Much of this section uses the useful notions of *strong* containment and variants of that notion as well as the notion of a *selection*.

DEFINITION 1: • b *strongly contains* a if $\overline{a} \subset b$;
• A sequence $\{a_n : n \in \omega\}$ *strongly contains* a sequence $\{b_n : n \in \omega\}$ if, for each $n \in \omega$, a_n strongly contains b_n;

- A sequence $\{a_n : n \in \omega\}$ is said to be *strongly decreasing* if $(\forall n \in \omega)\overline{a_{n+1}} \subset a_n$;
- A family $\mathscr{U}$ *strongly refines* a family $\mathscr{V}$ if $(\forall U \in \mathscr{U})(\exists V \in \mathscr{V})\overline{U} \subset V$.

DEFINITION 2: A *selection* from a sequence $\{A_n : n \in \omega\}$ is a sequence $\{a_n : n \in \omega\}$ where $(\forall n \in \omega)a_n \in A_n$. The closure of a selection (or a sequence) $\{a_n : n \in \omega\}$ is the sequence $\{\overline{a_n} : n \in \omega\}$. A sequence (or selection) $\{a_n : n \in \omega\}$ is contained in another sequence (or selection) $\{b_n : n \in \omega\}$ if $(\forall n \in \omega)a_n \subset b_n$.

The idea behind the proof of Lemma 1 is that, if we define, for each element α of an ordinal η, a well-ordered set $(\mathscr{H}_\alpha, \lhd_\alpha)$ and order $Z = \{(\alpha, H) : \alpha \in \eta, H \in \mathscr{H}_\alpha\}$ lexicographically, then Z is also a well-ordered set.

LEMMA 1: Suppose $\{U_\alpha : \alpha \in \eta\}$ is an open cover of a regular space X and that $\mathscr{V}$ is a base for X. There is $\{V_\beta : \beta \in \delta\}$, which is a subfamily of $\mathscr{V}$ and covers X such that each V_β is the least to contain some element of X and, if $x \in X$ and α, β are minimal such that $x \in U_\alpha \cap V_\beta$, then U_α strongly contains V_β.

Proof: Proceed by induction on $\alpha \in \eta$. For each $\alpha \in \eta$, choose a subfamily $\mathscr{H}_\alpha$ from $\mathscr{V}$ that covers U_α and each element of which is strongly contained in U_α. Well order $\mathscr{H}_\alpha$ and assume, without loss of generality, that, for each element H of $\mathscr{H}_\alpha$, there is some $x_H \in H$ such that no earlier element of $\mathscr{H}_\alpha$ contains x_H and no element of any $\mathscr{H}_\gamma$, where $\gamma < \alpha$ contains x_H. Let $\{V_\beta : \beta \in \delta\}$ be a well-ordering of Z. $\square$

COROLLARY 1 ([5]): Suppose $X' \subset X$ are Moore spaces, $\{\mathscr{U}_i : i \in \omega\}$ and $\{\mathscr{U}'_i : i \in \omega\}$ are developments[b] for X and X', respectively.

- Suppose $i \in \omega$ and $\{U_\alpha : \alpha \in \eta\}$ is a subfamily of $\mathscr{U}_i$ that covers X'. There is $\{V'_\beta : \beta \in \delta\}$ that is a subfamily of $\mathscr{U}'_i$, which covers X' such that each V'_β is the least to contain some element of X' and, if $x \in X'$ and α, β are minimal such that $x \in U_\alpha \cap V'_\beta$, then U_α strongly contains V'_β.
- Suppose $i \in \omega$ and $\{U'_\alpha : \alpha \in \eta\}$ is a subfamily of $\mathscr{U}'_i$ that covers X'. There is $\{V_\beta : \beta \in \delta\}$, which is a subfamily of $\mathscr{U}_{i+1}$ that covers X' such that each V_β is the least to contain some element of X' and, if $x \in X'$ and α, β are minimal such that $x \in U'_\alpha \cap V_\beta$, then U'_α strongly contains V_β.[c] Furthermore, if $\{W_\gamma : \gamma \in \xi\}$, a subfamily of $\mathscr{U}_i$ that covers X', is also given, then $\{V_\beta : \beta \in \delta\}$ can be chosen so that, in addition, if γ is minimal for $x \in W_\gamma$, then W_γ strongly contains V_β.

Proof: To get the first part of the corollary, take $\{U_\alpha \cap X' : \alpha \in \eta\}$, work in X', and apply Lemma 1. Use the fact that $\mathscr{U}'_i$ is a base for X'. To get the second part of the corollary, give $\{U'_\alpha \cap W_\gamma : \alpha \in \eta, \gamma \in \xi\}$ a well-ordering $\lhd$ that extends the well-founded product ordering, but apply the proof of Lemma 1 instead. That is, for each $\alpha \in \eta$, $\gamma \in \xi$, choose a subfamily $\mathscr{H}_{\alpha,\gamma}$ from $\mathscr{U}_{i+1}$, which covers $U_\alpha \cap W_\gamma$ such that $(\forall H \in \mathscr{H}_{\alpha,\gamma})U'_\alpha \supset \overline{H \cap X'}$ and $W_\gamma \supset \overline{H}$. Well order $\mathscr{H}_{\alpha,\gamma}$ and assume, without loss of generality, that, for any $H \in \mathscr{H}_{\alpha,\gamma}$, there is $x_H \in H \cap X'$ such that no earlier element of $\mathscr{H}_{\alpha,\gamma}$ and no element of $\mathscr{H}_{\alpha',\gamma'}$ contains x_H where $U'_{\alpha'} \cap W_{\gamma'} \lhd U'_\alpha \cap W_\gamma$. $\square$

---

[b] $\{\mathscr{U}_i : i \in \omega\}$ is said to be a *development* for a topological space X if each $\mathscr{U}_i$ is a family of open sets that cover X so that, for each $i \in \omega$, $\mathscr{U}_{i+1}$ is contained in $\mathscr{U}_i$ and, for each $x \in X$, $\{\cup\{U \in \mathscr{U}_i : x \in U\} : i \in \omega\}$ is a neighborhood base at x.
[c] That is, that contains $\overline{V_\beta \cap X'}$

The idea behind the proof of Lemma 3 is a combinatorial fact about decreasing partial selections from well-ordered sets that is encapsulated in Lemma 2.

LEMMA 2: If $\{\alpha_n : n \in \omega\}$ is a sequence of ordinals and, for each $n \in \omega$, f_n is an element of $\Pi \{\alpha_i : i \in n\}$, and $(\forall i \in n)f_{n+1}(i) \leq f_n(i)$, then there is $f \in \Pi \{\alpha_i : i \in \omega\}$ and an increasing sequence of integers $\{n_i : i \in \omega\}$ such that $(\forall i \in \omega)f \upharpoonright i + 1 \subset f_{n_i}$.

Proof: Let $f(i) = \inf \{f_n(i) : i < n\}$. For each $i \in \omega$, let n_{i+1} be the least integer greater than n_i such that $f \upharpoonright i \subset f_{n_{i+1}}$, which exists since $\Pi \{\alpha_j : j \in i\}$ is well founded. $\square$

LEMMA 3 (Key Lemma [5]): Suppose $X' \subset X$ are Moore spaces and $\mathcal{U}$ and $\mathcal{U}'$ are developments for X and X', respectively. We can then choose subdevelopments[d] $\mathcal{V}$ and $\mathcal{V}'$, respectively, such that

- Each $\mathcal{V}_n$ and each $\mathcal{V}'_n$ cover X';
- Each $\mathcal{V}'_n$ strongly refines $\mathcal{V}_n$;
- For each decreasing selection from $\mathcal{V}'$, there is a strongly decreasing selection from $\mathcal{V}$ that strongly contains an infinite subsequence of it;
- For each decreasing selection from $\mathcal{V}$, there is a strongly decreasing selection from $\mathcal{V}'$ that strongly contains[e] an infinite subsequence of it.

Proof: Construct $\mathcal{V}$ and $\mathcal{V}'$ by applying the claim repeatedly. Start with $\mathcal{V}_0^* = \mathcal{U}_0$. Apply the first half of Corollary 1 to $\{U_\alpha : \alpha \in \eta\} = \mathcal{V}_i^*$ to define $\{V'_\beta : \beta \in \delta\}$ to be $\mathcal{V}_i'^*$. Apply the second half of Corollary 1 to $\{U'_\alpha : \alpha \in \eta\} = \mathcal{V}_i'^*$ and $\{W_\gamma : \gamma \in \xi\} = \mathcal{V}_i^*$ to define $\{V_\beta : \beta \in \delta\}$ to be $\mathcal{V}_{i+1}^*$. Now let $\mathcal{V}_i = \cup \{\mathcal{V}_j^* : j \geq i\}$ and $\mathcal{V}_i' = \cup \{\mathcal{V}_j'^* : j \geq i\}$.

Suppose that we are given a decreasing selection $\{U_i : i \in \omega\}$ from $\mathcal{V}'$ (the case of a decreasing selection from $\mathcal{V}$ is similar). Suppose n is fixed. Find the least element V_n^n of $\mathcal{V}_n$ that strongly contains U_n. Find the least element V_{n-1}^n of $\mathcal{V}_{n-1}$ which strongly contains V_n^n and so forth. This sequence $\{V_i^n : i \leq n\}$ is a finite initial selection[f] from $\mathcal{V}$. Since the U_n are decreasing, these finite selections as n varies are coordinatewise decreasing and so we can apply Lemma 2 to obtain a strongly decreasing selection from $\mathcal{V}$ that strongly contains an infinite subsequence of $\{U_i : i \in \omega\}$. $\square$

Now Rudin obtains a remarkable variety of consequences of Lemma 3. We begin with the definition of a *complete* Moore space.

DEFINITION 3: We say a Moore space X is *complete* if X has a development such that any decreasing sequence of closed sets contained in a selection from the development[g] has nonempty intersection.

PROPOSITION 1 ([5]): X is complete if and only if any decreasing sequence of closed sets contained in the closure of a selection from some development has nonempty intersection.

---

[d] A subdevelopment of a development $\{\mathcal{U}_n : n \in \omega\}$ is a sequence $\{\mathcal{V}_n : n \in \omega\}$ such that $\mathcal{V}_n$ is a subfamily of $\mathcal{U}_n$. Note that the topology induced by any subdevelopment of a development, each of whose elements is a cover, equals the topology induced by the development.

[e] In this condition, the strong containment is when all sets are intersected with the subspace X'.

[f] That is, a selection from $\{\mathcal{V}_i : i \leq n\}$.

[g] Note that we assume, like Moore, that developments are decreasing.

Proof: Apply Lemma 3 for $X = X'$. Assume X is complete. Let u be a decreasing sequence of closed sets and s a selection from $\mathcal{V}$ such that $u \subset \bar{s}$. Thus we can find a selection t from $\mathcal{V}'$ that strongly contains s (applying only the fact that $\mathcal{V}'_n$ strongly refines $\mathcal{V}_n$ in Lemma 3). Now $u \subset \bar{s} \subset t$. Thus, by completeness, $\cap u \neq \varnothing$ as required. □

We continue with the definition of a *semicomplete* or a *Rudin-complete space*.

DEFINITION 4 ([5]): We say X is *semicomplete* if X has a development such that any strongly decreasing selection has nonempty intersection.

Note that for the rest of the section, we commonly say that we have taken a subsequence of a sequence $\{a_n : n \in \omega\}$ and do not mention that we have enumerated this sequence as $\{a_{n_m} : m \in \omega\}$, except that later we suddenly use this sequence of numbers $\{n_m : m \in \omega\}$. We believe that since the actual enumeration matters little, this allows the main ideas to emerge more effectively.

PROPOSITION 2 ([5]): X is semicomplete if and only if X has a development such that the closures of any decreasing selection have nonempty intersection.

Proof: Apply Lemma 3 for $X = X'$. Assume X is semicomplete. Let s be a decreasing selection from $\mathcal{V}$. There is a strongly decreasing selection t from $\mathcal{V}'$ so that t strongly contains an infinite subsequence of s. Now $\cap t \neq \varnothing$ by semicompleteness, so suppose $\cap t = \{x\}$. Suppose x is not in the closure of some $s(n)$. Choose m so that $St(x, \mathcal{V}'_m) \cap s(n) = \varnothing$. Assume, without loss of generality, that $n_m \geq n$. Thus $St(x, \mathcal{V}'_m) \cap s(n_m) = \varnothing$. Now $x \in t(m) \in \mathcal{V}'_m$ so $t(m) \cap s(n_m) = \varnothing$, but $t(m) \supset s(n_m)$, which is impossible. □

PROPOSITION 3 ([5]): Let X' be a subspace of the semicomplete Moore space X. X' is semicomplete if and only if X' is a G_δ subset of X.

Proof: First, we show that any semicomplete subspace is a G_δ subset. Apply Lemma 3 and then let $K_n = \cup \mathcal{V}_n$. We show that $X' = \cap\{K_n : n \in \omega\}$. If x is in a selection u of $\mathcal{V}$, then we can find a decreasing subsequence t of u since u is a neighborhood base at x. We can also find a strongly decreasing selection s in $\mathcal{V}'$ that strongly contains an infinite subsequence w of t. Now s has nonempty intersection, so we can find y with $\cap s = \{y\}$. If $x = y$, then we can deduce $x \in X'$ as required. By Lemma 3, find a strongly decreasing selection v in $\mathcal{V}$ that strongly contains an infinite subsequence r of s. Now $\cap v = \{z\}$. Also $\cap v \supset \cap r = \cap s = \{y\}$ so $y = z$. Find $n \in \omega$ so that $St(y, \mathcal{V}_j) \cap St(x, \mathcal{V}_n) = \varnothing$ where $n = n_{i_j}$ (using the notation $t(i) = u(n_i)$ since t is a subsequence of u and $w(j) = t(i_j)$ since w is a subsequence of t). Now $x \in u(n) = t(i_j) = w(j)$ so $w(j) \cap St(y, \mathcal{V}_j) = \varnothing$ but $s(j) \supset w(j) \cap X' \neq \varnothing$ and $y \in s(j)$ so $s(j) \cap w(j) = \varnothing$, which is impossible.

Second, we show that semicomplete is inherited by any G_δ subset X' of X. Suppose $X' = \cap\{V_n : n \in \omega\}$. We can apply Lemma 3 and ensure, in the application of Corollary 1, that $\mathcal{V}_n$ is a subcollection of $\{W : W$ is open in $X, W \subset V_n\}$. So suppose we take a decreasing selection s in $\mathcal{V}'$. Find a strongly decreasing selection t in $\mathcal{V}$ that strongly contains an infinite subsequence of s. Thus $\cap t = \{x\}$. Furthermore, $x \in \cap\{V_n : n \in \omega\} = X'$. If x is not in the closure of some $s(n)$, then apply the argument in the proof of Proposition 2, working in X instead of X'. □

COROLLARY 2: Any semicomplete Moore space that is not complete is not completable.

Proof: Suppose X' is a semicomplete subspace of a complete Moore space X. Now X is semicomplete, so by Proposition 3, X' must be a G_δ subset of X. Now complete is also inherited by G_δ subsets and so X' is complete, which is impossible. $\square$

COROLLARY 3 ([5]): For metric spaces, complete is equivalent to semicomplete.

Proof: Any completely metrizable space is complete as a Moore space. Suppose X is a semicomplete but not complete (as a Moore space) metric space. Take Y to be the metric completion of X. Then Y is complete as a Moore space and shows that X is completable, which contradicts Corollary 2. $\square$

PROPOSITION 4 (Moore [1]): No countable chain condition (ccc) nonseparable space X' can be embedded in a semicomplete Moore space X.

Proof: Suppose otherwise. Apply Lemma 3 to get developments $\mathscr{V}$ and $\mathscr{V}'$. Inductively on $n \in \omega$, define a tree T of height ω whose nth level is a disjoint maximal subfamily of $\mathscr{V}'_n$ and whose branches are strongly decreasing sequences. Let V be an open set with $V \cap X' \neq \varnothing$ and choose an open set U such that $\overline{U} \subset V$ and $U \cap X' \neq \varnothing$. We can define a branch b of T each of whose elements intersect U. There is a strongly decreasing selection c in $\mathscr{V}$ that strongly contains an infinite subsequence of b. By semicompleteness, we know $\cap c \neq \varnothing$, so suppose $\{x\} = \cap c$. Now if $x \notin \overline{U}$, then find m such that $St(x, \mathscr{V}_m) \cap U = \varnothing$. Now $x \in c(m) \supset b(n_m)$ and $c(m) \in \mathscr{V}_m$, which is impossible. So $x \in \overline{U} \subset V$. Choose n so that $St(x, \mathscr{V}_n) \subset V$. We have $b(n_m) \subset c(n) \subset V$, and so the elements of T are a π-base for X'. Since X' is ccc, we know that T is countable, and so X' has a countable π-base. $\square$

3. MAD FAMILIES

EXAMPLE 1 ([5]): There is a Moore space that cannot be embedded in a semicomplete Moore space.

Proof: Take an dense-in-itself subset $\mathbb{Q}$ of the reals such that every element of $\mathbb{Q}$ is a right endpoint of an open interval disjoint from $\mathbb{Q}$. Define $P = \mathbb{R} - \mathbb{Q}$. Take a maximal almost disjoint family $\mathscr{F}$ of functions from ω to $\mathbb{P}$. Take $X = ((\mathbb{Q} \cup (\mathbb{P} \times (0, 1))) \times \omega) \cup \mathscr{F}$ and let $(\mathbb{Q} \cup (\mathbb{P} \times (0, 1))) \times \omega$ be open and have the Euclidean topology where $\mathbb{Q}$ is viewed as a subspace of the x-axis except that each $\{p\} \times (0, 1)$ is open. Let a neighborhood of $f \in \mathscr{F}$ be $\{f\} \cup \cup\{f(n) \times (0, 1) \times \{n\} : n > m\}$.

Suppose X is embedded in a semicomplete Moore space Y and that $\mathscr{U}$ is a development for Y that witnesses this. Construct an increasing sequence $\{q_i^n : i \in \omega\}$, a decreasing sequence $\{r_i^n : i \in \omega\}$, and a strongly decreasing selection $\{U_i^n : i \in \omega\}$ from $\mathscr{U}$ such that each $q_i^n \in \mathbb{Q}$ and, for each $i \in \omega$, $q_i^n < r_i^n$, $X \cap ([q_i^n, r_i^n] \times [0, \epsilon) \times \{n\})$ is contained in U_i^n for some $\epsilon > 0$. Let $p^n = \lim \{q_i^n : i \in \omega\}$. Now $p^n \in \mathbb{P}$. Let $\{x^n\} = Y \cap \cap\{U_i^n : i \in \omega\}$. Assume without loss of generality that $\{p^n : n \in \omega\} = f \in \mathscr{F}$. Now each U_i^n contains $\{p^n\} \times (0, \epsilon_i) \times \{n\}$ for some $\epsilon_i > 0$ and so $\{(p^n, 1/j, n) : j \in \omega\}$ converges to x^n. We know that $\{q_i^n : i, n \in \omega\}$ does not have f in its closure in X

and thus also not in Y. However, $\{(q_i^n, n) : i \in \omega\}$ does converge to x^n. Thus $\{x^n : n \in \omega\}$ does not have f in its closure. Choose an open U such that $f \in U$ and $\overline{U} \cap \{x_n : n \in \omega\} = \varnothing$. Now U contains all but finitely many $\{p^n\} \times (0, 1) \times \{n\} = \{f(n)\} \times (0, 1) \times \{n\}$, and thus all but finitely many x_n. $\square$

EXAMPLE 2 ([5]): There is a semicomplete Moore space that is not complete.

Proof: Take a dense-in-itself subset $\mathbb{Q}$ of the reals such that every element of $\mathbb{Q}$ is a right endpoint of an open interval disjoint from $\mathbb{Q}$. Define $\mathbb{P} = \mathbb{R} - \mathbb{Q}$. For each $q \in \mathbb{Q}$, take a maximal almost disjoint family $\mathscr{F}_q$ of sequences from $\mathbb{P}$ that approach q. Take $X = (\mathbb{P} \times (0, 1)) \cup \bigcup \{\mathscr{F}_x : x \in \mathbb{Q}\}$, let $\mathbb{P} \times (0, 1)$ be open and have the Euclidean topology except that each $\{p\} \times (0, 1)$ is open. Given $f \in \mathscr{F}_x$ and a cofinite subset f' of f, let $\{f\} \cup (\text{rng} f' \times (0, 1/n))$ be a neighborhood of f. That X is a semicomplete Moore space is straightforward. Suppose that $\mathscr{U}$ is a development for X that witnesses that X is complete. Construct an increasing sequence $\{q_i : i \in \omega\}$ and a decreasing sequence $\{r_i : i \in \omega\}$ such that, for each $i \in \omega$, $q_i \in \mathbb{Q}$, $q_i < r_i$, and for each $p \in \mathbb{P} \cap (q_i, r_i)$, there is $U \in \mathscr{U}_i$ and $\epsilon(p, i) > 0$ such that $\{p\} \times (0, \epsilon(p, i)) \subset U$. Note that this is possible since each element of $\mathscr{F}_{q_i}$ is contained in some element of $\mathscr{U}_i$ and $\mathscr{F}_x$ is maximal. Let $x = \lim \{q_i : i \in \omega\}$. Note that $x \in \mathbb{P}$. Now $\{\{x\} \times (0, \epsilon(x, i)) : i \in \omega\}$ refines a selection from $\mathscr{U}$ and has a nontrivial refinement that is a decreasing sequence of closed sets. Thus if $\mathscr{U}$ witnessed completeness, this sequence must have a point in its intersection, but since $\epsilon(x, i) \to 0$, there is no such point. $\square$

4. THE PIXLEY–ROY SPACES

In 1941, Očan [2] defined a topology σ on the family $\mathscr{F}(X)$ of all closed subsets of a space (X, τ). If $F \in \mathscr{F}(X)$ and U is an open subset of X, then $\{F' \in \mathscr{F}(X) : F \subset F' \subset U\}$ is open. Očan showed that if (X, τ) is T_1, then $(\mathscr{F}(X), \sigma)$ is a zero-dimensional Hausdorff space. He also showed that if (X, τ) is a compact second-countable space, then$(\mathscr{F}(X), \sigma)$ has the countable chain condition.

In 1969, Pixley and Roy [3] defined the Pixley–Roy space over the reals. This space is just the subspace $\mathrm{PR}(X)$ of $(\mathscr{F}(X), \sigma)$ consisting of the nonempty finite subsets of X. They restricted themselves to the case where (X, τ) is the real line. Of course, Očan's results imply that if (X, τ) is T_1, then $\mathrm{PR}(X)$ is a zero-dimensional Hausdorff space. A simple argument shows that if (X, τ) has a countable network, then $\mathrm{PR}(X)$ has the countable chain condition.

In 1950 in theorem 1 of [5], Mary Ellen Rudin constructed a ccc nonseparable Moore space M. Now M is a subspace of $\mathrm{PR}(\mathbb{R})$. Rather than taking the reals, Rudin takes an almost increasing (mod finite) sequence of irrationals $I = \{f_\alpha : \alpha \in \omega_1\} \subset \omega^\omega$. Rather than taking all nonempty finite subsets of I, she takes those finite subsets that are *nice*. Roughly, *nice* is defined inductively by declaring the union of two nice "isomorphic" sets to be nice. Any uncountable subset of $\mathrm{PR}(\mathbb{R})$ is not separable and so M is not separable. If (X, τ) is first countable (and T_1), then $\mathrm{PR}(X)$ is a Moore space. However, the countable chain condition is not usually inherited by subspaces. In fact, whether M has the countable chain condition depends on what it means for a finite set to be nice.

A dramatic way for a space to have the countable chain condition is for any uncountable family of basic open sets to have an uncountable subfamily each two of which intersect, but each three of which have empty intersection. Since Rudin defines nice by declaring the union of two nice "isomorphic" sets to be nice, she gets precisely this property. Of course, the Pixley–Roy space over the reals does not have this property.

Her use of an uncountable almost increasing family $\mathscr{F}$ of functions from ω to ω as a combinatorial structure each uncountable subfamily of which is likely to possess "bad" finite behavior is an early ancestor of Todorčević's theorem 1.1 in [8] (and many other similar results of Todorčević) that shows that countably directed unbounded subsets of *unbounded* almost increasing families of functions from ω to ω have "all possible" finite behaviors in a specific sense.

In this section, we give, under $b = \omega_1$, a description of a ccc nonseparable Moore space X that is not homeomorphic to M. However, X will be very similar to M. The only differences are first, we redefine her *breakdowns* in a similar, slightly different fashion reminiscent of Δ-system arguments in forcing, and second, we simplify the notation by suppressing the first coordinate of each breakdown. While X is not identical to M, X does possess the same kind of control over intersection of open sets that M possesses, but that the Pixley–Roy space does not possess. We do not know of any advantage of M over the Pixley–Roy space in the absence of $b = \omega_1$, except that, in ZFC, we can use an unbounded family of size b and obtain a space analogous to M with properties analogous to those below. Of course M has the ccc in ZFC, but that relies on the weaker definition of breakdown given there.

DEFINITION 5: Let $\mathscr{F} = \{f_\alpha : \alpha \in \omega_1\}$ be an almost increasing sequence of functions from ω to ω.

DEFINITION 6: Suppose $n \in \omega, n \geq 1$, and A_0, A_1 are nonempty elements of $[\mathscr{F}]^n$. We say that A_0 and A_1 are *isomorphic* if there is $m \in \omega$ and listings $A_i = \{f_j^i : j \in n\}$ and $E \subset n$ so that

- $(\forall j \in n) f_j^0 \upharpoonright m = f_j^1 \upharpoonright m$;
- $(\forall j \in E) f_j^0 = f_j^1$;
- $j < j' \vee (j = j' \notin E \wedge i < i') \Rightarrow (\forall k \geq m) f_j^i(k) < f_{j'}^{i'}(k)$.

DEFINITION 7: We say that a finite nonempty subset A of $\mathscr{F}$ is nice if either $|A| = 1$ or else there are nice isomorphic finite nonempty subsets A_0, A_1 such that $A = A_0 \cup A_1$.

Note that the definition of nice is inductive and that each nice set has size 2^n for some n.

DEFINITION 8: Let $X = \{A \in PR(\mathscr{F}, \tau) : A$ is nice$\}$, where τ is the separable metrizable subspace topology on $\mathscr{F} \subset \omega^\omega$.

We need a lemma of Todorčević ([8, lemma 0.7]):

LEMMA 4: If A is an almost increasing unbounded subset of ω^ω, $k \in \omega$, and $F \subset A^k$ satisfies $(\forall a \in A)(\exists f \in F)(\forall i \in k) a <^* f_i$, then there are $f, g \in F$ such that $(\forall i \in k) f_i \leq g_i$.

PROPOSITION 5 $(b = \omega_1) : X$ is a zero-dimensional nonseparable Moore space with the countable chain condition. Furthermore, any uncountable family of nonempty open sets has an uncountable open refinement each three elements of which have empty intersection.

Proof: X inherits the property of being a zero-dimensional Moore space. Since X is uncountable, X fails to be separable.

Suppose that $\{U_\alpha : \alpha \in \omega_1\}$ is an uncountable family of nonempty open sets. We can assume that there is $n \in \omega$ and an open subset $U \subset \mathbb{R}$ each U_α has the form $\{F' \in X : F_\alpha \subset F' \subset U_\alpha\}$ where each F_α has size n and each $U_\alpha = U$. We can assume $\{F_\alpha : \alpha \in \omega\}$ is a Δ system with root Δ. Now any three of these basic open sets have empty intersections. To show that these basic open sets are not disjoint, we find two isomorphic F_α. Suppose $(\forall \alpha \in \omega_1)F_\alpha = \{f_i^\alpha : i \in n\}$. Find $m \in \omega$ and $E \subset n$ and assume that

- $(\forall \alpha \in \omega)\{f_i^\alpha : i \in n\} = \Delta$
- $(\forall \alpha \in \omega_1)(\forall i < i')(\forall j \geq m)f_i^\alpha(j) > f_{i'}^\alpha(j)$
- $(\forall i \in n)(\forall \alpha, \alpha' \in \omega_1)(\forall j \in m)f_i^\alpha(j) = f_i^{\alpha'}(j)$.

Apply Lemma 4 to $\{\{f_i^\alpha : i \in n - E\} : \alpha \in \omega_1\}$ to obtain $\alpha, \alpha' \in \omega_1$. Now F_α and $F_{\alpha'}$ are isomorphic. $\square$

5. A LOCALLY CONNECTED HYPERSPACE

In 1951, Rudin gave a construction of a locally connected ccc nonseparable Moore space [6]. This construction is based on an equivalent description of the Pixley–Roy space in terms of the characteristic functions. Let $PR(X)$ be the set of all characteristic functions of finite nonempty subsets of a space (X, τ). If $f : X \to 2$ and $f \in PR(X)$ and $\eta : f^{-1}(1) \to \tau$ such that $x \in \eta(x)$ and $rng(\eta)$ is a pairwise disjoint family, then $\{f' \in PR(X) : f'^{-1}(1) \subset \cup rng\eta\}$ is a basic open set in the Pixley–Roy topology. The locally connected variant of this construction is to take, not all characteristic functions of finite nonempty subsets, but instead all $[0, 1]$-valued functions with finite nonempty support (that is, satisfying $0 < |f^{-1}(0, 1)]| < \omega$).

DEFINITION 9: Suppose X is a topological space. Let $LC(X) = \{f \in [0, 1]^X : 0 < |f^{-1}(0, 1]| < \omega$. If $f \in LC(X)$ and $\eta : f^{-1}(0, 1] \to \tau$ such that $(\forall x, x' \in f^{-1}(0, 1])$ $x \in \eta(x)$, $\eta(x) \cap \eta(x') = \varnothing$, then we call η a choice for f. If η is a choice for f and $\epsilon > 0$, then $U(\eta, \epsilon, f) = \{g \in LC(X) : (\forall x \in X)g(x) < \epsilon$ or $(\exists x' \in f^{-1}(0, 1])((x \in \eta(x'),$ $x \neq x', g(x) < f(x') + \epsilon) \vee (x' = x, | g(x) - f(x) | < \epsilon))$. The space $LC(X)$ is topologized by declaring each $U(\eta, \epsilon, f)$ to be open.

PROPOSITION 6: If (X, τ) is T_1, then $LC(X)$ is a locally (pathwise-) connected hereditarily metacompact completely regular space.

Proof: Suppose $f \in LC(X)$ and $\epsilon_0 > \epsilon_1 > 0$ and η is a choice for f. Let $G_i = U(\eta, \epsilon_i, f)$ for each $i \in 2$. We claim that $G_1 \subset G_0$. To see this, note that if g does not lie in G_0, then there is some $x \in X$ such that $g(x)$ does not lie in some interval of the form $[0, \epsilon_0)$ or $(r - \epsilon_0, r + \epsilon_0)$ or $[0, r + \epsilon_0)$. Since $g(x) \neq 0$ in each of these cases, we can find a neighborhood B of g so that if $h \in B$, then $h(x)$ does not lie in the corresponding interval of the form $[0, \epsilon_1)$ or $(r - \epsilon_1, r + \epsilon_1)$ or $[0, r + \epsilon_1)$. A little work shows that there is, in fact, a continuous real-valued function that maps $LC(X) - G_0$ to 1 and G_1 to 0.

Next we show that $LC(X)$ is locally pathwise connected. Suppose that $\pi_x :$ $LC(X) \rightarrow [0, 1]$ is the projection mapping. Suppose $\phi : [0, 1] \rightarrow LC(X)$ and $x \in X$ are so that $\pi_x \circ \phi$ is continuous and $(\forall y \neq x)\pi_y \circ \phi$ is constant. Then ϕ is a path in $LC(X)$. Since each basic open neighborhood G of $LC(X)$ is coordinatewise convex, we deduce that, by changing the nonzero coordinates one at a time, we can find a path from any point in G to any other point in G. $\square$

PROPOSITION 7: If X is first countable and T_1, then $LC(X)$ is a Moore space.

Proof: Since X is first countable, let $\{V(j, x) : j \in \omega\}$ list a neighborhood base at each $x \in X$. Let $\mathcal{U}_{\epsilon, n, m} = \{U(\eta, \epsilon, f) : |f^{-1}(0, 1]| = |f^{-1}[\epsilon, 1]| \geq n, (\forall x \in f^{-1}(0, 1])(\exists j \geq m)\eta(x) = V(j, x)\}$. Suppose $g, h \in U(\eta, \epsilon, f) \in \mathcal{U}_{\epsilon, n, m}$ and $|g^{-1}(0, 1]| = n$. Then, for each $x \in f^{-1}(0, 1], f(x) \geq \epsilon$, and so $g(x), h(x) > 0$. Thus $g^{-1}(0, 1] = f^{-1}(0, 1] \subset h^{-1}(0, 1]$. Thus $h \in U(\eta, 2\epsilon, g)$. Thus $\{\mathcal{U}_{\epsilon, n, m} : \epsilon > 0, \epsilon \in \mathbb{Q}, n, m \in \omega\}$ is a countable sequence of open families so that, for each $f \in LC(X), \{St(f, \mathcal{U}_{\epsilon, n, m}) : \epsilon > 0, \epsilon \in \mathbb{Q}, n \in \omega\}$ is a neighborhood base at f.

Let $G_n = \cup \mathcal{U}_{\epsilon, n, m}$. Note that $G_n = \{f \in LC(X) : |f^{-1}(0, 1]| \geq n\}$. We show that $LC(X) - G$ is a G_δ-set. Let $H_k = \cup\{U(\eta, 1/k, f) : f \in LC(X) - G, \eta \subset \{V(i, x) : i \geq k, x \in f^{-1}(0, 1]\}, f^{-1}(0, 1] = f^{-1}(1/k, 1]\}$. Suppose $g \in G$ and $g \in H_k$. Then $g \in U(\eta, 1/k, f)$ where $g^{-1}(0, 1] \supset f^{-1}(0, 1]$ and $g^{-1}(0, 1] \subset \cup\eta$. There are therefore only finitely many possible f and so, by choosing k large enough, we can arrange that no $\eta \subset \{V(i, x) : i \geq k, x \in f^{-1}(0, 1]\}$ can cover $g^{-1}(0, 1]$ since $g^{-1}(0, 1]$ has size at least n, whereas each $f^{-1}(0, 1]$ has size less than n. So let $LC(X) - G_n = \cap\{G_n^p : p \in \omega\}$. Now $\{\mathcal{U}_{\epsilon, n, m} \cup \{G_n^p\} : p, n, m \in \omega, \epsilon \in \mathbb{Q}, \epsilon > 0\}$ is a development. $\square$

PROPOSITION 8: If (X, τ) is uncountable, then $LC(X)$ is not separable.

Proof: If there were a countable dense set $\{f_n : n \in \omega\}$, then let $Y = \cup\{f_n^{-1}(0, 1] : n \in \omega\}$. Since Y is countable, we have $x \in X - Y$. Now take some $f \in LC(X)$ so that $f(x) > 0$ and take a neighborhood of f that contains only elements g of $LC(X)$ so that $g(x) > 0$. This neighborhood misses the supposedly dense set. $\square$

PROPOSITION 9: If (X, τ) has a countable network, then $LC(X)$ has the countable chain condition (and the topology is even σ-centred).

Proof: Suppose $\mathcal{N}$ is a countable network that we assume to be closed under finite unions. Suppose that $\{U(\eta_\alpha, \epsilon_\alpha, f_\alpha) : \alpha \in \omega_1\}$ is a family of open sets. We can assume that each ϵ_α is rational and thus fixed at ϵ. We can also assume $|f_\alpha^{-1}(0, 1]| = n_0$ is fixed. We can also find, by a pigeonhole argument, without loss of generality, a subset of $\mathcal{N}$ of size n_0 so that, for each $\alpha \in \omega_1$ and $x \in X$ such that $f_\alpha(x) > 0$, there is $N \in \mathcal{N}$ such that $x \in N \subset \eta_\alpha(x)$. We can also assume that, whenever $x \in N \cap f_\alpha^{-1}(0, 1]$ and $x' \in N \cap f_{\alpha'}^{-1}(0, 1]$, we have $|f_\alpha(x) - f_{\alpha'}(x')| < \epsilon$. Now these supposedly disjoint neighborhoods are centred, which is a contradiction. $\square$

6. CUT SETS

In 1939, Burton Jones showed that a ccc nonseparable Moore space cannot be ordered. There is now a short proof of this fact. Ordered spaces are collectionwise normal. Collectionwise normal Moore spaces are metrizable (proved by Bing only in 1952) and metrizable ccc spaces are separable. The main concern of Separation in Non-Separable Spaces *and* Concerning a Problem of Souslin's can be summarized as "How linear can a ccc nonseparable Moore space be?" The idea behind Mary Ellen Rudin's solution of this question was to restrict herself to the class of locally connected Moore spaces and to measure linearity by calculating the size and position of cut-sets in those spaces. In Separation in Non-Separable Spaces, she showed that existence of a countable or even separable cut-set arbitrarily close to the first of any two points is impossible. She also showed, in that article, that the existence of a finite cut-set between any two points is impossible. She showed in Concerning a Problem of Souslin's that the existence of a separable (or countable) cut-set between any two points is equiconsistent with Suslin's hypothesis. We give a modern exposition of all of Mary Ellen Rudin's arguments on these matters except for her construction, from a Suslin line, of a locally connected Moore space in which each two points can be separated by a countable set.

We begin with some facts that will be needed in the proof of Theorems 1 and 2.

LEMMA 5: Every Moore space has a σ-discrete dense subset.

Proof: Let $\{ \mathcal{U}_n : n \in \omega \}$ be a development for X. Construct a sequence of maximal families $\{(U_\alpha^n, x_\alpha^n) : \alpha \in \gamma_n\}$ such that each $U_\alpha^n \in \mathcal{U}_n, x_\alpha^n \in U_\alpha^n, x_\alpha^n$ is not an element of any earlier U_β^n, x_α^n is not in the closure of all earlier x_β^n.

The subset $\{x_\alpha^n : \alpha \in \gamma_n , n \in \omega\}$ is dense. To see this, suppose $St(x, \mathcal{U}_n)$ misses all the x_α^m. The reason x is not a x_α^n must be that x is an element of some U_β^n. Now $x_\beta^n \in U_\beta^n$ as well so $x_\beta^n \in St(x, \mathcal{U}_n)$, which is impossible.

However, for each n, $\{x_\alpha^n : \alpha \in \gamma_n\}$ is discrete. $\square$

DEFINITION 10: If X is a topological space and A, B, and C are disjoint subsets of X, then we say that A separates B and C if $X - A$ with the subspace topology contains a clopen subset D such that $D \supset B$ and $D \cap C = \emptyset$.

In all uses of this definition, A is closed and B and C are either open sets or points.

PROPOSITION 10 (The Key Idea): Suppose that X is a ccc space and $V \subset X$ is an open set. There is a countable $M \subset X - V$ such that, for any countable $K \subset X$ and any nonempty connected open set U with $K \cap (U \cup V) = \emptyset$ and $U \cap M = \emptyset$, if $\overline{K}$ separates U from V, then $\overline{M}$ separates U from V.

Proof: Suppose otherwise. Take an increasing sequence $\{\mathcal{M}_\alpha : \alpha \in \omega_1\}$ of countable elementary submodels containing X and V with $\mathcal{M}_\alpha \in \mathcal{M}_{\alpha+1}$. Applying $M = (\mathcal{M}_\alpha \cap X) - V$, we can find $U_\alpha, K_\alpha, \in \mathcal{M}_{\alpha+1}$ such that the hypothesis is satisfied. Since K_α is countable, $K_\alpha \subset \mathcal{M}_{\alpha+1}$. Thus $\overline{(\mathcal{M}_{\alpha+1} \cap X) - (U_\alpha \cup V)} \supset \overline{K_\alpha}$ separates U_α from V.

We claim that $\{U_\alpha : \alpha \in \omega_1\}$ is a disjoint family of nonempty open sets. Suppose that $\beta < \alpha$. Now $\overline{M_\alpha} \supset \overline{(\mathcal{M}_{\beta+1} \cap X) - (U_\beta \cup V)}$ separates $U_\beta - \overline{M_\alpha}$ from V and yet does not separate U_α from V. Since $\overline{M_\alpha}$ is disjoint from both U_α and V, and U_α is connected, we deduce that $\overline{M_\alpha}$ separates no point of U_α from V and $U_\beta \cap U_\alpha = \emptyset$. $\square$

COROLLARY 4: If X is locally connected and ccc and, for any $x, y \in X$ and any open neighborhood U of x, some separable subset of U separates x and y, and V is a nonempty connected open subset of X, then there is a separable closed set $E = E(V) \subset X - V$ such that V is dense in its component in $X - E$.

Proof: Let M be as in Proposition 10. Suppose $x \notin \overline{M} \cup \overline{V}$. Choose $y \in V$. Find a separable subset $\overline{K}$ (where K is countable) of $X - \overline{V}$, which separates x and y. Choose a connected open set U containing x, which is disjoint from K, V, and $\overline{M}$. Proposition 10 says that $\overline{M}$ separates U from V and so x is not in the same component as in $X - \overline{M}$. Thus, since x was arbitrary, V is dense in its component in $X - \overline{M}$. $\square$

DEFINITION 11: K is said to be a minimal finite cut-set if K is a finite set that separates some two points that are separated by no proper subset of K.

PROPOSITION 11: Any point in a minimal finite cut-set lies in the closure of at least two components of its complement.

COROLLARY 5: For any connected open set V in a locally connected ccc space X, there is a countable $M = M_V$ with $M \cap V = \varnothing$ such that, if K is a minimal finite cut-set in X and $K \cap V = \varnothing$, then $(\forall x \in K - M)$, x is separated from V by $\overline{M}$.

Proof: Let M be as in Proposition 10. Suppose that K is a minimal finite cut-set in X disjoint from V and that $x \in K - \overline{M}$ is not separated from V by $\overline{M}$. By Proposition 11, x is in the closure of at least two components of $X - K$. Let C be one of these components that is disjoint from V. Choose any open connected $U \subset C$ that lies in the same component as x in $X - \overline{M}$. Proposition 10 applies to show that $\overline{M}$ separates U from V. We deduce that $\overline{M}$ separates x from V as well. $\square$

THEOREM 1 ([6]): In any locally connected ccc nonseparable Moore space X, there are x, y, and an open neighborhood U of x such that no separable subset of U separates x and y.

Proof: By Lemma 5, there is a discrete B that is not in the closure of any countable set. Suppose otherwise so that we can apply Corollary 4. Define a maximal sequence of disjoint connected open sets $\{U_\alpha : \alpha \in \gamma\}$ such that the closure of each U_α contains exactly one element of B and no U_α intersects either U_β or $E(U_\beta)$ whenever $\beta < \alpha$. Of course γ is a countable ordinal. Thus we get that any element b of B must be either in the closure of the union of the $E(U_\alpha)$'s or else in the closure of the union of the U_α. Since the $E(U_\alpha)$'s are all separable, there is $b \in B$, which is in the closure of the U_α and yet not in the closure of any single U_α. So some U_α intersects a connected open neighborhood of b that misses all the $E(U_\alpha)$'s. Yet removing $E(U_\alpha)$ puts the component of U_α, and therefore b, inside $\overline{U_\alpha}$, which is a contradiction. $\square$

THEOREM 2 ([6]): In any locally connected ccc nonseparable Moore space X, some two points cannot be separated by a finite set.

Proof: Let $\{\mathscr{U}_n : n \in \omega\}$ be a development of X with each $\mathscr{U}_n$ consisting of connected open sets and $(\forall n \in \omega) \mathscr{U}_{n+1} \subset \mathscr{U}_n$. Take a continuous increasing sequence $\{\mathscr{M}_n : n \leq \omega\}$ of countable elementary submodels that contain X and $\{\mathscr{U}_n : n \in \omega\}$ such that $\mathscr{M}_n \in \mathscr{M}_{n+1}$. Since the union of all the minimal finite cut-sets are dense in X, we shall take a particular minimal finite cut-set F and show that $F \subset \overline{X \cap \mathscr{M}_\omega}$. Suppose $x \in F - \overline{X \cap \mathscr{M}_\omega}$.

CLAIM: If $p \in \omega$ and W is an open subset of X with $W \in \mathscr{M}_p$, and K is a minimal finite cutset in W, then $K \cap \overline{X \cap \mathscr{M}_p}$ is nonempty.

Proof of Claim: Suppose otherwise. For each $n \in \omega$, take any maximal disjoint family $\mathscr{D}_n \in \mathscr{M}_p$ of elements of $\mathscr{U}_n$, which are subsets of W and disjoint from K. Each family must be countable by the countable chain condition, and thus $\mathscr{D}_n \subset \mathscr{M}_p$ as well. Applying Corollary 5, if $K \cap \overline{\cup\{M_V : V \in \mathscr{D}_n\}} = \varnothing$ and $K \cap \cup \mathscr{D}_n = \varnothing$, then every $z \in K$ is separated in W from each $V \in \mathscr{D}_n$ by M_V. Now z has a connected neighborhood that misses M_V when $V \in \mathscr{D}_n$, but by maximality must intersect some $V \in \mathscr{D}_n$, which is a contradiction.

Thus either K intersects $\overline{\cup\{M_V : V \in \mathscr{D}_n, n \in \omega\}}$ or else K intersects each $\cup \mathscr{D}_n$. In the latter case a fixed $z \in K$ belongs to infinitely many $\cup \mathscr{D}_n$, which means that there are $x_n \in \mathscr{M}_p \cup \cap \mathscr{D}_n$ for infinitely many n such that $x_n \in St(z, \mathscr{U}_n)$, and so z is in the closure of $X \cap \mathscr{M}_p$. In the former case, some element of K lies in $\overline{\cup\{M_V : V \in \mathscr{D}_n, n \in \omega\}} \subset \overline{X \cap \mathscr{M}_p}$. □

Proof of Theorem 2 Continued: Construct a sequence of open sets $\{U_n : n \in \omega\}$ and minimal finite cut-sets F_n in U_n with $X = U_0$, $F_0 = F$ and $U_{n+1} = X - \overline{X \cap \mathscr{M}_n}$. Note that $U_n \in \mathscr{M}_n$. Given U_n and F_n, declare F_{n+1} to be a minimal finite cut-set containing x and contained in $F_n \cap U_{n+1}$. Of course, the intersection of a minimal finite cut-set with an open set may not be a minimal finite cut-set with respect to that open set; however, by applying Proposition 11, we can express it as the union of minimal finite cut-sets with respect to that open set. The claim says $F_n - F_{n+1} \supset F_n - U_{n+1} = F_n \cap \overline{X \cap \mathscr{M}_{n+1}} \neq \varnothing$, which yields an infinite decreasing sequence of finite sets, which is impossible. □

The basic idea in the proof of Theorem 3 is here already. Construct an object. To show that it is complex, suppose it is simple, construct a witness and then see how the object treats the witness, and show that the object has to be complex to handle the witness.

THEOREM 3 ([7]): If there is a locally connected ccc nonseparable space such that any two points can be separated by a separable set, then there is a Suslin space.

Proof: Suppose X is such a space. We construct a Suslin tree by induction. The nodes of the tree are open connected subsets of X. The root of the tree is X. Given a vertex U of the tree, we subtract a separable closed subset that disconnects, and make every component of the resulting open set a successor of U. At limit stages, take the interior of the branches.

Since antichains are disjoint families of open sets, the tree has no uncountable antichains. We must show that it is an uncountable tree. If it were countable, then removing a countable family of separable sets leaves some connected interior in X by nonseparability, so follow this interior up the tree. □

7. THE IMPOSSIBLE

EXAMPLE 3 ([7]): If there is a Suslin line, then there is a locally connected ccc nonseparable Moore space in which each two points can be separated by a countable set.

I am unable to understand this construction and I have been unable to locate anyone who has studied it in any detail. The definition of the space itself takes six pages! I have the impression that she has defined a kind of machine[h] that takes a first-countable locally connected space and turns it into a "nearly linear" Moore space while preserving local connectedness and chain conditions. This particular space seems to be the application of this machine to a Suslin continuum. Thus a first-countable locally connected ccc nonseparable ordered continuum is turned into a locally connected ccc nonseparable "nearly ordered" (in the sense that any two points can be separated by a countable set) Moore space.

PROBLEM 1: Understand Rudin's 1952 Suslin space.

PROBLEM 2: Find a locally connected preserving machine that converts first-countable ordered spaces into "nearly ordered" Moore spaces while preserving chain-condition-type properties.

8. BAD FUNCTIONS

The idea of the next example is to construct a "bad" function from the reals to the reals whose support is the irrationals. Then partition f into continuum many partial functions in a "bad" way. Finally, unfold the graph of f like a hedgehog.

PROPOSITION 12: There is a function f from the irrationals to the reals that can be partitioned into partial functions $\{f_\alpha : \alpha \in 2^\omega\}$ such that, for any $\alpha \in 2^\omega$ and any nonempty open $U \subset \mathbb{R}^2$ that is bounded above, there is $p \in \mathbb{P}$ such that $(p, f_\alpha(p)) \in \partial U$.

Proof: Let $\{(\alpha_\gamma, U_\gamma) : \gamma \in 2^\omega\}$ list all pairs (α, U) where $\alpha \in 2^\omega$ and $U \subset \mathbb{R}^2$ is nonempty open and bounded above. We construct $\{f_\alpha : \alpha \in 2^\omega\}$, partial functions with disjoint domain, by putting irrationals into the domains of these partial functions by transfinite induction, and we choose distinct irrationals $\{p_\alpha : \alpha \in 2^\omega\}$ along the way. At stage γ, find $p_\gamma \notin \{p_\alpha : \alpha \in \gamma\}$ so that $U_\gamma \cap (\{p_\gamma\} \times \mathbb{R}) \neq \varnothing$, then find r with $(p_\gamma, r) \in \partial U_\gamma$ (since U_γ is bounded, its boundary is nonempty), and put $f_{\alpha_\gamma}(p_\gamma) = r$. $\square$

EXAMPLE 4: There are, on the reals, a connected metrizable topology τ, a separable metrizable topology ν and an order topology σ and a disjoint family of sets $\{U_\alpha : \alpha \in 2^\omega\}$ such that each U_α is τ-open and ν-dense and $\tau \supset \nu \supset \sigma$.

Proof: Let σ be the Euclidean topology. Choose a function $f \in \mathbb{R}^\mathbb{P}$ and a partition of f into partial functions $\{f_\alpha : \alpha \in 2^\omega\}$ by Proposition 12. Let U_α be the domain of f_α and let $U = \cup\{U_\alpha : \alpha \in 2^\omega\}$. To construct ν, declare U to be open and have the topology inherited from its graph as a subspace of the plane. If V is σ-open and $n \in \omega$, then declare $(V \cap \mathbb{Q}) \cap \{p \in V \cap \mathbb{P} : (\exists \alpha \in 2^\omega) f_\alpha(p) > n\}$ to be open. Note that ν is homeomorphic to a subspace of the plane in which each U_α is dense. To construct τ, declare each U_α to be open as well. Note that τ is homeomorphic to a subspace of the product of the reals with hedgehog with 2^ω spines.

We claim that τ is connected. Suppose U is a nontrivial τ-clopen set. Let $N = U \cap \mathbb{Q}$. First, suppose $q \in N$ and $\alpha \in 2^\omega$. For some interval I and q, there is $n \in \omega$ with $\{x \in I : (\exists \alpha \in 2^\omega) f_\alpha(x) > n\} \subset U$. Thus $\{f_\alpha(x) : \alpha \in 2^\omega, x \in I - U\}$ is bounded above.

---

[h] In the sense of Reed [4].

Now U contains I, since otherwise there must be $p \in \mathbb{P}$ and $r \in \mathbb{R}$ such that $(p, r) \in \partial(I - U)$ and $f_\alpha(p) = r$, which is a contradiction.

Second, suppose $f_\alpha(p) = r$, $(p, r) \notin U$ and $p \in \overline{N}^\sigma$. Choose $q_n \to p$ such that each $q_n \in N$. Since $q_n \in U$, choose I_n, which contains q_n so that U contains I_n. Also choose bounded open rectangles $X_n \times Y_n$ such that $X_n \to p$, $\overline{X_n} \subset I_n$, and $Y_n \to r$. Since the graph of each f_α is dense in the plane, we can find $(p_n, r_n) \in f_\alpha \cap (X_n \times Y_n)$. Thus $(p_n, r_n) \to (p, r)$. Since each $p_n \in U$, we get $(p, r) \in \overline{U}^\tau$ and that is impossible.

We deduce that U contains $\overline{N}^\sigma$. Now, similarly, U^c contains $\overline{\mathbb{Q} - N^\sigma}$. We deduce that $\overline{N}^\sigma$ and $\overline{\mathbb{Q} - N^\sigma}$ partition $\mathbb{R}$ and no such partition exists. □

The next construction shows that for mere connectedness, even a single point suffices to separate.

EXAMPLE 5 ([6]): There is a connected ccc nonseparable Moore space Y such that for any two points y_0, y_1 and any open set U about y_0, there is $y \in U$, which disconnects y_0 and y_1.

Proof: Let ϱ be any ccc nonseparable Moore topology on 2^ω (see [3]). Apply Example 4 to obtain $\tau \supset \nu \supset \sigma$. Define a topology $\rho \supset \nu$ by declaring, for each $W \in \varrho$, $\cup\{U_\alpha : \alpha \in W\}$ to be open. Note that $\tau \supset \rho$ and so ρ is connected. Since each U_α is ν-dense, U is ρ-dense (and ρ-open) and U, as a subspace of $(\mathbb{R}, \rho)$ is homeomorphic to a dense subspace of the product of ϱ and ν. Since ϱ is ccc nonseparable and ν has a countable base, we deduce that ρ is ccc nonseparable. For any two points y_0, y_1 and any ρ-open set U about y_0, there is $y \in U$, which disconnects y_0 and y_1 since $\rho \supset \sigma$. □

REFERENCES

1. MOORE, R. L. 1942. Concerning separability. Proc. Natl. Acad. Sci. U.S.A. **28:** 56–58.
2. OCHAN, J. S. 1941. Spaces of subsets of a topological space. C. R. Acad. Sci. URSS **32**(2): 107–109.
3. PIXLEY, C. & P. ROY. 1969. Uncompletable Moore spaces. Proc. Auburn Topol. Conf. 65–69.
4. REED, G. M. 1971. Concerning normality, metrizability and the Souslin property in subspaces of Moore spaces. Gen. Topol. Appl. **1:** 223–246.
5. RUDIN, M. E. (ESTILL). 1950. Concerning abstract spaces. Duke Math J. **17**(4): 317–327.
6. ———. 1951. Separation in non-separable spaces. Duke Math J. **18:** 623–629.
7. ———. 1952. Concerning a Problem of Souslin's. Duke Math J. **19:** 629–639.
8. TODORČEVIĆ, S. Partition problems in topology. *In* Contemporary Mathematics. American Mathematical Society. Providence, R.I.

Index of Contributors